Handbook of Quantum Field Theory

Handbook of Quantum Field Theory

Edited by **Richard Burrows**

NY RESEARCH
PRESS
New York

Published by NY Research Press,
23 West, 55th Street, Suite 816,
New York, NY 10019, USA
www.nyresearchpress.com

Handbook of Quantum Field Theory
Edited by Richard Burrows

International Standard Book Number: 978-1-63238-276-4 (Hardback)

Printed in the United States of America.

Contents

Preface

The book provides state-of-the-art information describing the Quantum Field Theory. Quantum Field Theory is now identified as an effective tool not only in Particle Physics but also in Nuclear Physics, Condensed Matter Physics, Solid State Physics and even in Mathematics. This book is a compilation of some present applications of Quantum Field Theory to these mentioned fields of modern physics and mathematics, in order to give a more profound understanding of known facts and unsolved problems.

After months of intensive research and writing, this book is the end result of all who devoted their time and efforts in the initiation and progress of this book. It will surely be a source of reference in enhancing the required knowledge of the new developments in the area. During the course of developing this book, certain measures such as accuracy, authenticity and research focused analytical studies were given preference in order to produce a comprehensive book in the area of study.

This book would not have been possible without the efforts of the authors and the publisher. I extend my sincere thanks to them. Secondly, I express my gratitude to my family and well-wishers. And most importantly, I thank my students for constantly expressing their willingness and curiosity in enhancing their knowledge in the field, which encourages me to take up further research projects for the advancement of the area.

Editor

Part 1

Quantization

Quantizing with a Higher Time Derivative

Sergei V. Ketov[1,2], Genta Michiaki[1] and Tsukasa Yumibayashi[1]
[1]*Department of Physics, Graduate School of Science, Tokyo Metropolitan University, Hachioji-shi, Tokyo*
[2]*Institute for the Physics and Mathematics of the Universe, The University of Tokyo, Kashiwa-shi, Chiba*
Japan

1. Introduction

It is commonplace in *Quantum Field Theory* (QFT) that a QFT with higher (time) derivatives is believed to be doomed from the point of view of physics, because of ghosts or states of negative norm, and thus it should be dismissed. The standard reference is the very old result (known in the literature as the Ostrogradski theorem (1)) claiming a linear instability in any Hamiltonian system associated with the Lagrangian having the higher (ie. more than one) time derivative that cannot be eliminated by partial integration.

The key point of the Ostrogradski method (1) is a canonical quantization of the clasically equivalent theory without higher derivatives via considering the higher derivatives of the initial coordinates as the *independent* variables.

The interest in the higher-derivative QFT was recently revived due to some novel developments in the gravitational theory, related to the so-called $f(R)$-gravity theories – see eg., ref. (2) for a review. The $f(R)$ gravity theories are defined by replacing the scalar curvature R in the Einstein action by a function $f(R)$. The $f(R)$ gravity theories give the self-consistent non-trivial alternative to the standard Λ-CDM Model of Cosmology, by providing the geometrical phenomenological description of inflation in the early universe and Dark Energy in the present universe. Despite of the apparent presence of the higher derivatives, a classical $f(R)$ gravity theory can be free of ghosts and tachyons. A supersymmetric extension of $f(R)$ gravity was recently constructed in superspace (3).

Already the simplest model of $(R + R^2)$ gravity (4) is known as the viable model of chaotic inflation, because it is consistent with the recent WMAP measurements of the Cosmic Macrowave Background (CMB) radiation (5). Its supersymmetric extension was recently constructed in refs. (6; 7).

On the one side, any quadratically generated (with respect to the curvature) quantum theory of gravity has ghosts in its perturbative quantum propagator (8). However, on the other side, any $f(R)$ gravity theory is known to be *classically* equivalent to the scalar-tensor gravity (ie. to the usual quintessence) (9–11), while the stability conditions in the $f(R)$ gravity ensure the ghost-and-tachyon-freedom of the classically equivalent quintessence theory (12; 13). It now appears that in some cases the presence of the higher derivatives may be harmless (14). It also gives rise to the non-trivial natural question of how to make sense out of the quantized $f(R)$ gravity?

The $f(R)$ gravity theories are just the particular case of the higher-derivative quantum gravity theories which have been investigated in the past. They were found to be renormalizable (15) and asymptotically free (16). A generic higher-derivative gravity suffers, however, from the presence of ghosts and states of negative norm which apparently spoil those QFT from physical applications. However, the issue of ghosts and their physical interpretation deserves a more detailed study. The complexity of the higher-derivative gravity is the formidable technical obstacle for that. It is, therefore, of interest to consider simpler QFT as the toy-models.

Similar features (like renormalizability and asymptotic freedom) exhibit the quantum Non-Linear Sigma-Models with higher derivatives, which have striking similarities to the higher-derivative quantum gravity (17–19). However, even those QFT are too complicated because of their high degree of non-linearity.

Perhaps, the simplest toy-model is given by the *Pais-Uhlenbeck* (PU) quantum oscillator in Quantum Mechanics (20). As was demonstrated by Hawking and Hertog (21), it may be possible to give physical meaning to the Euclidean path integral of the PU oscillator, as the set of consistent rules for calculation of observables, even when "living with ghosts". The basic idea of ref. (21) is to abandom unitarity, while never producing and observing negative norm states.

The idea of Hawking and Hertog found further support in refs. (22; 23) where the physical propagator of the PU oscillator was calculated by using the van Vleck-Pauli approach (the saddle point method for the Euclidean path integral) and Forman's theorem (24). In this Chapter we systematically review the classical and quantum theory of the PU oscillator from the first principles, along the lines of refs. (14; 21–23).

2. Ostrogradski method with higher derivatives

Consider a one-dimensional mechanical system with the action

$$S[q] = \int dt\; L(q, Dq, \cdots, D^n q) \tag{2.1}$$

in terms of the Lagrange function L of $q(t)$ and its time derivatives, where $n \geq 2$ and $D = \frac{d}{dt}$. The Euler-Lagrange equation reads

$$\sum_{i=0}^{n} (-D)^i \frac{\partial L}{\partial (D^i q)} = 0 \tag{2.2}$$

The Ostrogradski method (1) gives the Hamiltonian formulation of the higher derivative Lagrange formulation by introducing more independent variables.

The independent generalized coordinates Q_i are defined by

$$Q_i = D^{i-1} q \qquad (i = 1, \cdots, n) \tag{2.3}$$

The generalized momentum P_n is defined by

$$\left.\frac{\partial L}{\partial(D^n q)}\right|_{\substack{D^{i-1}q=Q_i \\ D^n q=A}} = P_n \tag{2.4}$$

There are $n+1$ independent variables $\{Q_1, \cdots, Q_n, P_n\}$ that are in correspondence to the $n+1$ variables $\{D^0 q, \cdots, D^n q\}$ of the higher derivative action (2.1).

By solving eq.(2.4) with respect to $A = D^n q$ (assuming that it is possible), one gets

$$D^n q = A(Q_1, \cdots, Q_n, P_n) \tag{2.5}$$

Therefore, the Lagrange dynamics can be represented in terms of the $n+1$ independent variables $\{Q_1, \cdots, Q_n, P_n\}$ as

$$L = L\big(Q_1, \cdots, Q_n, A(Q_1, \cdots, Q_n, P_n)\big) \tag{2.6}$$

A Legendre transformation is used to pass from the Lagrange formulation to the Hamiltonian one. With the generalized coordinates $\{Q_1, \cdots, Q_n\}$ and the generalized momentum P_n as the independent variables, the total differential of the Lagrangian is given by

$$\begin{aligned} dL &= \sum_{j=1}^{n} \left.\frac{\partial L}{\partial(D^{j-1}q)}\right|_{\substack{D^{i-1}q=Q_i \\ D^n q=A}} dQ_j + P_n dA \\ &= \frac{\partial L}{\partial q} dQ_1 + \sum_{j=2}^{n} \frac{\partial L}{\partial(D^{j-1}q)} dQ_j + P_n dA \\ &= D \sum_{j=1}^{n} (-D)^{j-1} \frac{\partial L}{\partial(D^j q)} dQ_1 + \sum_{j=1}^{n-1} \frac{\partial L}{\partial(D^j q)} dQ_{j+1} + P_n dA \end{aligned} \tag{2.7}$$

where we have used eqs. (2.2) and (2.4), and

$$dA = \sum_{j=1}^{n} \frac{\partial A}{\partial Q_j} dQ_j + \frac{\partial A}{\partial P_n} dP_n \tag{2.8}$$

Let us now define the $n-1$ generalized momenta as

$$P_i = \sum_{j=i}^{n} (-D)^{j-i} \frac{\partial L}{\partial(D^j q)} \qquad (i = 1, \cdots, n-1) \tag{2.9}$$

They satisfy the relations

$$\frac{\partial L}{\partial(D^i q)} = P_i + DP_{i+1} \tag{2.10}$$

Therefore, eq. (2.7) can be rewritten to the form

$$d\left[\sum_{i=1}^{n-1} P_i(DQ_i) + P_n A - L\right] = -\sum_{i=1}^{n} (DP_i) dQ_i + \sum_{i=1}^{n} (DQ_i) dP_i \tag{2.11}$$

Equation (2.11) gives rise to the Hamiltonian in the form

$$H = \sum_{j=1}^{n-1} P_j(DQ_j) + P_n A - L \tag{2.12}$$

The Hamilton equations of motion are given by

$$DQ_i = \frac{\partial H}{\partial P_i} \qquad \text{and} \qquad DP_i = -\frac{\partial H}{\partial Q_i} \tag{2.13}$$

3. PU oscillator

The PU oscillator (20) is an extension of the harmonic oscillator with the higher time derivatives, and is the particular case of the higher-derivative theory introduced in Sec. 2. The special features of the PU opscilator are
(i) the equation of motion is *linear*:

$$F(D)q = 0 \tag{3.1}$$

where F is a linear differential operator;
(ii) the F is *polynomial* (with respect to D) with *constant* coefficients:

$$F(D) = \sum_{i=0}^{n} a_i D^i \tag{3.2}$$

where $a_0, \cdots, a_n$ are the real constants;
(iii) there is the time reversal invariance with respect to $t \to -t$. Hence, the polynomial F has only even powers of the time derivative D.

The Lagrangian of the one-dimensional PU oscillator reads

$$L(q, Dq, \cdots, D^n q) = -\sum_{i=0}^{n} \frac{a_i}{2}(D^i q)^2 \qquad (a_0 \neq 0, a_n \neq 0) \tag{3.3}$$

where a_i $(i = 0, \cdots, n)$ are real constants. The Euler-Lagrange equation of motion is given by

$$\begin{aligned} 0 &= \sum_{i=0}^{n} (-D)^i \left[-a_i D^i q \right] \\ &= -a_0 \left[\sum_{i=0}^{n} (-1)^i \frac{a_i}{a_0} D^{2i} \right] q \end{aligned} \tag{3.4}$$

Accordingly, the differential operator $F(D)$ reads

$$F(D) = \sum_{i=0}^{n} (-1)^i \frac{a_i}{a_0} D^{2i} \tag{3.5}$$

The equation of motion can be rewritten to the form

$$F(D)q = 0 \tag{3.6}$$

The PU Lagrangian takes the form (up to a boundary term)

$$\bar{L} = -\frac{a_0}{2} q F(D) q \tag{3.7}$$

The differential operator $F(D)$ can be brought to the factorized form

$$F(D) = \prod_{i=1}^{n} \left(1 + \frac{D^2}{\omega_i^2}\right) \tag{3.8}$$

where the constants ω_i $(i = 1, \cdots, n)$ are the solutions (roots) of the equation $F(i\omega) = 0$. Let us introduce n new operators

$$G_i(D) = \prod_{\substack{j=1 \\ j\neq i}}^{n} \left(1 + \frac{D^2}{\omega_j^2}\right) \qquad (i = 1, \cdots, n) \tag{3.9}$$

and define the n generalized coordinates as

$$Q_i = G_i(D) q \qquad (i = 1, \cdots, n) \tag{3.10}$$

Those generalized coordinates Q_j are called *harmonic* coordinates. By using the harmonic coordinates, the PU Euler-Lagrange eq. (3.6) can be rewitten to the n equations

$$\left[1 + \frac{D^2}{\omega_i^2}\right] Q_i = 0 \tag{3.11}$$

It means that the PU oscillator can be interpreted as n harmonic oscillators. Accordingly, the PU Lagrangian (3.7) can be rewritten to the form

$$\bar{L} = -\frac{a_0}{2} \sum_{i=1}^{n} \eta_i Q_i \left(1 + \frac{D^2}{\omega_i^2}\right) Q_i \tag{3.12}$$

where the n constants η_i have been introduced as

$$\eta_i = \left(\omega_i^2 \left. \frac{dF}{d(D^2)} \right|_{D^2 = -\omega_i^2} \right)^{-1} \tag{3.13}$$

To prove eq. (3.13), we first notice that it amounts to

$$\sum_{i=1}^{n} \eta_i G_i(D) = 1 \tag{3.14}$$

By the definiton of $G(D)$ in eq.(3.9) we have

$$\begin{aligned} G_i(D^2 = -\omega_j^2) &= \prod_{\substack{k=1 \\ k\neq i}}^{n} \left(1 - \frac{\omega_j^2}{\omega_k^2}\right) \\ &= \delta_{ij} \prod_{\substack{k=1 \\ k\neq j}}^{n} \left(1 - \frac{\omega_j^2}{\omega_k^2}\right) \end{aligned} \tag{3.15}$$

so that

$$\sum_{i=1}^{n} \eta_i G_i(D) = 1 \tag{3.16}$$

$$\sum_{i=1}^{n} \eta_i G_i(D^2 = -\omega_j^2) = \eta_j \prod_{\substack{k=1 \\ k\neq j}}^{n} \left(1 - \frac{\omega_j^2}{\omega_k^2}\right) = 1 \tag{3.17}$$

indeed. Therefore, the constants η_i are given by

$$\eta_i = \left[\prod_{\substack{k=1 \\ k\neq i}}^{n} \left(1 - \frac{\omega_i^2}{\omega_k^2}\right)\right]^{-1} \tag{3.18}$$

Next, we prove that

$$\omega_i^2 \frac{dF}{dD^2}\bigg|_{D^2=-\omega_i^2} = \prod_{\substack{k=1 \\ k\neq i}}^{n} \left(1 - \frac{\omega_i^2}{\omega_k^2}\right) \tag{3.19}$$

By the use of eq.(3.8) we find

$$\begin{aligned} \frac{dF}{dD^2} &= \frac{d}{dD^2} \prod_{j=1}^{n} (1 + \frac{D^2}{\omega_j^2}) \\ &= \sum_{k=1}^{n} \frac{1}{\omega_k^2} \prod_{\substack{j=1 \\ j\neq k}}^{n} (1 + \frac{D^2}{\omega_j^2}) \\ &= \sum_{j=1}^{n} \frac{1}{\omega_j^2} G_j(D) \end{aligned} \tag{3.20}$$

so that

$$\begin{aligned} \frac{dF}{dD^2}\bigg|_{D^2=-\omega_i^2} &= \sum_{j=1}^{n} \frac{1}{\omega_j^2} G_j(D^2 = -\omega_i^2) \\ &= \sum_{j=1}^{n} \frac{1}{\omega_j^2} \delta_{ji} \prod_{\substack{k=1 \\ k\neq i}}^{n} (1 - \frac{\omega_i^2}{\omega_k^2}) \\ &= \frac{1}{\omega_i^2} \prod_{\substack{k=1 \\ k\neq i}}^{n} (1 - \frac{\omega_i^2}{\omega_k^2}) \end{aligned} \tag{3.21}$$

Equation (3.19) is now confirmed and, hence, via eq. (3.18) also eq. (3.13) follows.

In terms of the harmonic coordinates (3.10), the Lagrangian $\bar{L}$,

$$\begin{aligned} \bar{L} &= -\frac{a_0}{2} q F(D) q \\ &= -\frac{a_0}{2} \sum_{i=1}^{n} \eta_i Q_i (1 + \frac{D^2}{\omega_i^2}) Q_i \end{aligned} \tag{3.22}$$

with the constants η_i given by eq. (3.13), can be rewritten to the form

$$\tilde{L} = \frac{a_0}{2}\sum_{i=1}^{n}\eta_i\left(\frac{1}{\omega_i^2}(DQ_i)^2 - Q_i^2\right) \tag{3.23}$$

up to a boundary term.

The Lagrangian (3.23) is just a sum of the Lagrangians of n harmonic oscillators. Hence, similarly to a free system of n particles, we can change the Lagrangian formulation into the Hamiltonian formulation. We define the generalized momenta P_i by taking the harmonic coordinates Q_i and the velocities DQ_i as the Lagrange variables,

$$\begin{aligned} P_i &= \frac{\partial \tilde{L}}{\partial (DQ_i)} \\ &= \frac{a_0\eta_i}{\omega_i^2} DQ_i \qquad (i = 1, \cdots, n) \end{aligned} \tag{3.24}$$

The system of n free particles does not have higher derivatives, so its Hamiltonian is

$$H = \sum_{i=1}^{n} P_i(DQ_i) - L \tag{3.25}$$

Equations (3.23) and (3.24) imply

$$H = \sum_{i=1}^{n}\left(\frac{\omega_i^2}{2a_0\eta_i}P_i^2 + \frac{a_0\eta_i}{2}Q_i^2\right) \tag{3.26}$$

By rescaling the harmonic coordinates and the generalized momenta as

$$Q_i \to \tilde{Q}_i = \frac{\sqrt{a_0|\eta_i|}}{\omega_i}Q_i \qquad \text{and} \qquad P_i \to \tilde{P}_i = \frac{\omega_i\sqrt{|\eta_i|}}{\eta_i\sqrt{a_0}}P_i \tag{3.27}$$

we get the final Hamiltonian

$$H = \frac{1}{2}\sum_{i=1}^{n}\frac{\eta_i}{|\eta_i|}(\tilde{P}_i^2 + \omega_i^2\tilde{Q}_i^2) \tag{3.28}$$

The presence of both positive and negative values of the constants η_i in the Hamiltonian implies both positive and negative values of energy. The constants η_i are given by eq. (3.18). If ω_i satisfy $i < j \Rightarrow \omega_i < \omega_j$, the constants η_i are positive for the odd number i, and are negative for the even number i. Therefore, the Hamiltonian is

$$H = \frac{1}{2}\sum_{i=1}^{n}(-1)^{i-1}\left(\tilde{P}_i^2 + \omega_i^2\tilde{Q}_i^2\right) \tag{3.29}$$

This Hamiltonian can be interpreted as that of n harmonic oscillators, with the positive and negative energy levels appearing alternatively. Because of that reason, the PU oscillator has an instability (for any interaction). It is related to a possible ghost state of negative norm in PU quantum theory (see Sec. 6). In what follows we consider the simplest case of PU oscillator with $n = 2$ only.

4. PU oscillator for $n = 2$: explicit results

Let us consider the Lagrangian

$$L = \frac{1}{2}\left(\frac{dq}{dt}\right)^2 - V(q) - \frac{\alpha^2}{2}\left(\frac{d^2q}{dt^2}\right)^2 \qquad (\text{where } \alpha \neq 0) \tag{4.1}$$

with a scalar potential $V(q)$. In the case of the PU oscillator, the potential $V(q)$ is a quadratic function of q. Since the (mass) dimension of time is -1 (in the natural units $\hbar = c = 1$), the dimension of the Lagrangian L is 1, the dimension of q is $-1/2$, and that of the constant α is -1.

Let the trajectory q be a sum of the classical trajectory q_{cl} and the displacement $\tilde{q}$, ie. $q = q_{cl} + \tilde{q}$, where the classical trajectory q_{cl} is a solution to the *equation of motion* (EOM) with the boundary conditions (21)

$$\mathscr{A}: \quad q(0) = q_0, \quad q(T) = q_T, \quad \dot{q}(0) = \dot{q}_0, \quad \dot{q}(T) = \dot{q}_T \tag{4.2}$$

where the dots above stand for the time derivatives.

With the boundary conditions (4.2), the boundary condition of $\tilde{q}$ is

$$\tilde{\mathscr{A}}: \quad \tilde{q}(0) = 0, \quad \tilde{q}(T) = 0, \quad \dot{\tilde{q}}(0) = 0, \quad \dot{\tilde{q}}(T) = 0 \tag{4.3}$$

The action of $q_{cl} + \tilde{q}$ is given by

$$S[q_{cl} + \tilde{q}] = S[q_{cl}] + \int_0^T dt \left(\frac{1}{2}\dot{\tilde{q}}^2 - V(q_{cl} + \tilde{q}) + V(q_{cl}) + \tilde{q}V'(q_{cl}) - \frac{\alpha^2}{2}\ddot{\tilde{q}}^2\right) \tag{4.4}$$

where we have introduced the notation

$$V'(q_{cl}) = \left.\frac{dV}{dq}\right|_{q=q_{cl}} \tag{4.5}$$

In eq.(4.4) the term $V(q_{cl} + \tilde{q}) - V(q_{cl}) - \tilde{q}V'(q_{cl})$ represents the *gap* between the full action $S[q]$ and the classical action $S[q_{cl}]$, which generically depends on both the classical trajectory q_{cl} and the displacement $\tilde{q}$. After expanding the scalar potential V in Taylor series,

$$V(q_{cl} + \tilde{q}) = V(q_{cl}) + \tilde{q}V'(q_{cl}) + \frac{1}{2!}\tilde{q}^2 V''(q_{cl}) + \cdots \tag{4.6}$$

we find that, when the second derivative V'' is constant, the gap $V(q_{cl} + \tilde{q}) - V(q_{cl}) - \tilde{q}V'(q_{cl})$ does *not* depend on the classical trajectory q_{cl}. It is the case when the potential V is a quadratic function of q, like the PU oscillator.

In the path integral quantization (sec. 7), the gap between the full action and the classical action is a quantum effect. When the potential is a quadratic function (like that of the PU oscillator), that quantum effect does depend on $\tilde{q}$, but does not depend on the classical trajectory. In what follows, we only consider a quadratic function for the scalar potential in the form

$$V(q) = \frac{m^2}{2}q^2 \tag{4.7}$$

ie. the scalar potential of a harmonic oscillator with the mass $m > 0$, The Lagrangian is given by

$$L_{PU} = \frac{1}{2}\dot{q}^2 - \frac{m^2}{2}q^2 - \frac{\alpha^2}{2}\ddot{q}^2 \tag{4.8}$$

The parameter α measures a contribution of the second derivative to the harmonic oscillator. Therefore, we can expect the classical trajectory to behave just like that of the harmonic oscillator when α is small.

The Euler-Lagrange EOM of the Lagrangian (4.8) are given by eq.(2.2),

$$\begin{aligned} 0 &= \sum_{i=0}^{2}(-D)^i \frac{\partial L}{\partial(D^i q)} \\ &= -m^2 q - \ddot{q} - \alpha^2 \ddddot{q} \end{aligned} \tag{4.9}$$

or, equivalently,

$$\left(m^2 + D^2 + \alpha^2 D^4\right) q = 0 \tag{4.10}$$

It is not difficult to find clasical solutions to the EOM in eq. (4.10). When searching for the classical trajectory in the oscillatory form $q_{cl} = \exp(i\lambda t)$, the EOM reads

$$\left(m^2 - \lambda^2 + \alpha^2\lambda^4\right) e^{i\lambda t} = 0 \tag{4.11}$$

and, therefore, we have

$$\lambda^2 = \frac{1 \pm \sqrt{1 - 4\alpha^2 m^2}}{2\alpha^2} \tag{4.12}$$

When λ is real, the Lagrangian $L_(PU)$ is an extension of the harmonic oscillator indeed. Hence, we need the condition

$$0 < \alpha m < \frac{1}{2} \tag{4.13}$$

It means that the Lagrangina L_{PU} has the oscillating solution which is similar to the trajectory of the harmonic oscillator. A general solution reads

$$q(t) = A_+ \cos(\lambda_+ t) + B_+ \sin(\lambda_+ t) + A_- \cos(\lambda_- t) + B_- \sin(\lambda_- t) \tag{4.14}$$

where A_+, B_+, A_-, B_- are the integration constants, and

$$\lambda_\pm = \sqrt{\frac{1 \mp \sqrt{1 - 4\alpha^2 m^2}}{2\alpha^2}} \tag{4.15}$$

The values of the constants (A_+, B_+, A_-, B_-) are determined by the boundary conditions.

The Hamiltonian formulation for the Lagrangian (4.8) can be obtained by the Ostrogradski method. The generalized coodinates and momenta are given in Sec. 2, ie.

$$\begin{aligned} Q_1 &= q \qquad \text{and} \qquad P_1 = \frac{\partial L}{\partial \dot{q}} - D\frac{\partial L}{\partial \ddot{q}} \\ Q_2 &= \dot{q} \qquad \text{and} \qquad P_2 = \frac{\partial L}{\partial \ddot{q}} \end{aligned} \tag{4.16}$$

which imply

$$P_1 = \dot{q} + \alpha^2 \dddot{q}$$
$$P_2 = -\alpha^2 \ddot{q} \tag{4.17}$$

The Hamiltonian is given by eq.(2.12). ie.

$$\begin{aligned} H &= P_1(DQ_1) + P_2 A - L \\ &= P_1 Q_2 - \frac{1}{2\alpha^2} P_2^2 - \frac{1}{2} Q_2^2 + \frac{m^2}{2} Q_1^2 \end{aligned} \tag{4.18}$$

or, equivalently,

$$H = \alpha^2 \dot{q} \dddot{q} - \frac{\alpha^2}{2} \ddot{q}^2 + \frac{1}{2} \dot{q}^2 + \frac{m^2}{2} q^2 \tag{4.19}$$

Since the Hamiltonian does not evolve with time, we can find the energy by substituting $q(t)$ of eq. (4.14) at $t = 0$ into eq. (4.19), as well as $q, \dot{q}, \ddot{q}$ and $\dddot{q}$ at $t = 0$, ie.

$$\begin{aligned} q(0) &= A_+ + A_- \\ \dot{q}(0) &= B_+ \lambda_+ + B_- \lambda_- \\ \ddot{q}(0) &= -A_+ \lambda_+^2 - A_- \lambda_-^2 \\ \dddot{q}(0) &= -B_+ \lambda_+^3 - B_- \lambda_-^3 \end{aligned} \tag{4.20}$$

It is now straightforward to calculate the Hamiltonian (4.19). We find

$$\begin{aligned} H &= \alpha^2 \dot{q}(0) \dddot{q}(0) - \frac{\alpha^2}{2} \ddot{q}(0)^2 + \frac{1}{2} \dot{q}(0)^2 + \frac{m^2}{2} q(0)^2 \\ &= \frac{1}{2} \lambda_+^2 \sqrt{1 - 4\alpha^2 m^2} (A_+^2 + B_+^2) - \frac{1}{2} \lambda_-^2 \sqrt{1 - 4\alpha^2 m^2} (A_-^2 + B_-^2) \end{aligned} \tag{4.21}$$

To get the Hamiltonian formulation in the harmonic coordinates, we begin with the EOM in the form (4.10), whose differential operator $F(D)$ is defined by

$$F(D) = 1 + \frac{D^2}{m^2} + \frac{\alpha^2 D^4}{m^2} \tag{4.22}$$

It can be factorized as

$$F(D) = \left(1 + \frac{D^2}{\lambda_+^2}\right)\left(1 + \frac{D^2}{\lambda_-^2}\right) \tag{4.23}$$

where $\lambda_\pm$ are given by eq. (4.15). Therefore, the harmonic coodinates are given by

$$Q_+ = \left(1 + \frac{D^2}{\lambda_-^2}\right) q \qquad \text{and} \qquad Q_- = \left(1 + \frac{D^2}{\lambda_+^2}\right) q \tag{4.24}$$

The constants η_i of eq. (3.13) can be computed as follows. We have

$$\frac{dF}{dD^2} = \frac{1}{m^2} + \frac{2\alpha^2 D^2}{m^2} \tag{4.25}$$

so that

$$\begin{aligned}\eta_\pm &= \left(\lambda_\pm^2 \left.\frac{dF}{dD^2}\right|_{D^2=-\lambda_\pm^2}\right)^{-1} \\ &= \left(\frac{\lambda_\pm^2}{m^2}(1-2\alpha^2\lambda_\pm^2)\right)^{-1} \\ &= \left(\pm\frac{\lambda_\pm^2}{m^2}\sqrt{1-4\alpha^2m^2}\right)^{-1} \\ &= \pm\frac{m^2}{\lambda_\pm^2\sqrt{1-4\alpha^2m^2}} \end{aligned} \tag{4.26}$$

Therefore, the generalized momenta in eq. (3.24) are

$$\begin{aligned}P_\pm &= \frac{m^2\eta_\pm}{\lambda_\pm^2}DQ_\pm \\ &= \pm\frac{m^4}{\lambda_\pm^4\sqrt{1-4\alpha^2m^2}}DQ_\pm\end{aligned} \tag{4.27}$$

and the Hamiltonian is given by

$$\begin{aligned}H &= \sum_{j=\pm}\left(\frac{\lambda_j^2}{2m^2\eta_j}P_j^2+\frac{m^2\eta_j}{2}Q_j^2\right) \\ &= \sum_{j=\pm} j\frac{m^4}{2\lambda_j^4\sqrt{1-4\alpha^2m^2}}\left((DQ_j)^2+\lambda_jQ_j^2\right)\end{aligned} \tag{4.28}$$

where we have substituted the classical solution (4.14).

The harmonic coodinates (4.24) read

$$Q_+ = A_+\left(1-\frac{\lambda_+^2}{\lambda_-^2}\right)\cos(\lambda_+t)+B_+\left(1-\frac{\lambda_+^2}{\lambda_-^2}\right)\sin(\lambda_+t) \tag{4.29}$$

$$Q_- = A_-\left(1-\frac{\lambda_-^2}{\lambda_+^2}\right)\cos(\lambda_-t)+B_-\left(1-\frac{\lambda_-^2}{\lambda_+^2}\right)\sin(\lambda_-t) \tag{4.30}$$

where

$$\begin{aligned}1-\frac{\lambda_\pm^2}{\lambda_\mp^2} &= \lambda_\pm^2\left(\frac{1}{\lambda_\pm^2}-\frac{1}{\lambda_\mp^2}\right) \\ &= \pm\frac{\lambda_\pm^2}{m^2}\sqrt{1-4\alpha^2m^2}\end{aligned} \tag{4.31}$$

Hence, we find

$$Q_\pm = \pm\frac{\lambda_\pm^2}{m^2}\sqrt{1-4\alpha^2m^2}\Big(A_\pm\cos(\lambda_\pm t)+B_\pm\sin(\lambda_\pm t)\Big) \tag{4.32}$$

Substituting them into the Hamiltonian (4.28), we get

$$H = \frac{1}{2}\lambda_+^2 \sqrt{1 - 4\alpha^2 m^2}(A_+^2 + B_+^2) - \frac{1}{2}\lambda_-^2 \sqrt{1 - 4\alpha^2 m^2}(A_-^2 + B_-^2) \tag{4.33}$$

Equations (4.22) and (4.33) are *the same.* Therefore, we conclude that the Hamiltonian formulation by the Ostrogradski method is consistent with the Hamiltonian formulation in the harmonic coordinates, as they should.

The integration constants (A_+, B_+) correspond to the harmonic oscillator with positive energy, while the integration constants (A_-, B_-) correspond to the harmonic oscillator with negative energy.

5. Boundary conditions and spectrum

Going back to the Lagrangian (4.8), let us consider its action over a finite time period T,

$$S[q] = \int_0^T dt\, L_{PU} \tag{5.1}$$

with the trajectory q being a sum of the classical trajectory q_{cl} and the displacement $\tilde{q}$, $q = q_{cl} + \tilde{q}$. In quantum theory, the displacement $\tilde{q}$ is a quantum coordinate. The action can be rewritten as

$$S[q] = S[q_{cl}] + S[\tilde{q}] - \int_0^T dt \left(\ddot{q}_{cl} + m^2 q_{cl} + \alpha^2 \ddddot{q}_{cl} \right) \tilde{q} + \left[\dot{q}_{cl}\tilde{q} - \alpha^2 \ddot{q}_{cl}\dot{\tilde{q}} + \alpha^2 \dddot{q}_{cl}\tilde{q} \right]_0^T \tag{5.2}$$

Here the first term is the action of the classical trajectory q_{cl}, and the second term is the action of the quantum part $\tilde{q}$. The integrand of the third term vanishes because the classical trajectory is a solution of the (Euler-Lagrange) EOM. The fourth term depends on the boundary. However, if the boundary condition on $\tilde{q}$ is given by

$$\tilde{\mathscr{A}}: \quad \tilde{q}(0) = 0, \quad \tilde{q}(T) = 0, \quad \dot{\tilde{q}}(0) = 0, \quad \dot{\tilde{q}}(T) = 0 \tag{5.3}$$

the fourth term in eq. (5.2) also vanishes. That boundary condition is the same as that of

$$\mathscr{A}: \quad q(0) = q_0, \quad q(T) = q_T, \quad \dot{q}(0) = \dot{q}_0, \quad \dot{q}(T) = \dot{q}_T \tag{5.4}$$

which was proposed in ref. (21). The quantum action now takes the form

$$\begin{aligned} S[\tilde{q}] &= \int_0^T dt \left(\frac{1}{2}\dot{\tilde{q}}^2 - \frac{m^2}{2}\tilde{q}^2 - \frac{\alpha^2}{2}\ddot{\tilde{q}}^2 \right) \\ &= -\frac{1}{2}\int_0^T dt\, \tilde{q}\left(D^2 + m^2 + \alpha^2 D^4 \right)\tilde{q} + \frac{1}{2}\left[\tilde{q}\dot{\tilde{q}} - \alpha^2 \dot{\tilde{q}}\ddot{\tilde{q}} + \alpha^2 \tilde{q}\dddot{\tilde{q}} \right]_0^T \end{aligned} \tag{5.5}$$

where the (last) boundary term vanishes due to the boundary condition (5.3).

The boundary term in eq. (5.5) also vanishes by another boundary condition,

$$\tilde{\mathscr{A}}': \quad \tilde{q}(0) = 0, \quad \tilde{q}(T) = 0, \quad \ddot{\tilde{q}}(0) = 0, \quad \ddot{\tilde{q}}(T) = 0 \tag{5.6}$$

As a result, the action (5.5) takes the Gaussian form, which is quite appropriate for a path integral quantization with the Gaussian functional

$$-\frac{1}{2}\int_0^T dt\,\tilde{q}\left(D^2+m^2+\alpha^2D^4\right)\tilde{q} \tag{5.7}$$

Let us now compute the *spectrum* of the operator $D^2+m^2+\alpha^2D^4$. For this purpose, we need to find the solutions u_k to the eigenvalue equation

$$(D^2+m^2+\alpha^2D^4)u_k(t)=ku_k(t) \tag{5.8}$$

with the eigenvalues k. A general solution is

$$\begin{aligned} u_k(t) &= A_1\cos(\omega_+t)+A_2\sin(\omega_+t)+A_3\cos(\omega_-t)+A_4\sin(\omega_-t)\\ \omega_\pm &= \sqrt{\frac{1\mp\sqrt{1-4\alpha^2(m^2-k)}}{2\alpha^2}} \end{aligned} \tag{5.9}$$

where A_1, A_2, A_3, A_4 is the constants of integration. The function $\tilde{q}$ can be expanded in terms of u_k,

$$\tilde{q}=\int dk\;u_k(t) \tag{5.10}$$

The spectrum of k is now determined by appying the physical boundary conditions (5.3) or (5.6) to u_k in the form of eq. (5.9). Applying the boundary condition (5.3) at $t=0$ yields

$$\tilde{q}(0)=A_1+A_3=0, \qquad \dot{\tilde{q}}(0)=A_2\omega_++A_4\omega_-=0 \tag{5.11}$$

The boundary condition (5.3) at $t=T$ then takes the form

$$\begin{aligned} \tilde{q}(T) &= A_1\cos(\omega_+T)+A_2\sin(\omega_+T)-A_1\cos(\omega_-T)\\ &\quad -A_2\frac{\omega_+}{\omega_-}\sin(\omega_-T)=0\\ \dot{\tilde{q}}(T) &= -A_1\omega_+\sin(\omega_+T)+A_2\omega_+\cos(\omega_+T)\\ &\quad +A_1\omega_-\sin(\omega_-T)-A_2\omega_+\cos(\omega_-T)=0 \end{aligned}$$

In particular, the determinant of the matrix on the left side of this equation,

$$\begin{aligned} &\det\begin{pmatrix} \omega_-\Big[\cos(\omega_+T)-\cos(\omega_-T)\Big] & \omega_-\sin(\omega_+T)-\omega_+\sin(\omega_-T)\\ -\omega_+\sin(\omega_+T)+\omega_-\sin(\omega_+T) & \omega_+\Big[\cos(\omega_+T)-\cos(\omega_-T)\Big] \end{pmatrix}\\ &=\omega_+\omega_-\Big[\cos(\omega_+T)-\cos(\omega_-T)\Big]^2\\ &\quad+\omega_+\omega_-\Big[\sin^2(\omega_+T)+\sin^2(\omega_-T)\Big]-(\omega_+^2+\omega_-^2)\sin(\omega_+T)\sin(\omega_-T)\\ &=2\omega_+\omega_-\Big[1-\cos(\omega_+T)\cos(\omega_-T)\Big]-(\omega_+^2+\omega_-^2)\sin(\omega_+T)\sin(\omega_-T) \end{aligned} \tag{5.12}$$

must vanish. We find

$$
\begin{aligned}
2\omega_+\omega_-\Big[1-\cos(\omega_+T)\cos(\omega_-T)\Big] &= (\omega_+^2+\omega_-^2)\sin(\omega_+T)\sin(\omega_-T)\\
\frac{2\sqrt{m^2-k}}{\alpha}\Big[1-\cos(\omega_+T)\cos(\omega_-T)\Big] &= \frac{1}{\alpha^2}\sin(\omega_+T)\sin(\omega_-T)\\
1-\cos(\omega_+T)\cos(\omega_-T) &= \frac{1}{\alpha\sqrt{m^2-k}}\sin(\omega_+T)\sin(\omega_-T)
\end{aligned}
\tag{5.13}
$$

where $\omega_\pm(k)$ ar given by eq. (5.9). Apparently, there is no simple solution here.

When employing the boundary conditions (5.6) with eq. (5.9) on u_k, the boundary condition in $t = 0$ yields

$$\tilde{q}(0) = A_1 + A_3 = 0, \qquad \ddot{\tilde{q}}(0) = -A_1\omega_+^2 - A_3\omega_-^2 = 0 \tag{5.14}$$

so that we find $A_1 = A_3 = 0$ when $\omega_+ \neq \omega_-$. Now the boundary condition at $t = T$ reads

$$
\begin{aligned}
\tilde{q}(T) &= A_2\sin(\omega_+T) + A_4\sin(\omega_-T) = 0\\
\ddot{\tilde{q}}(T) &= -A_2\omega_+^2\sin(\omega_+T) - A_4\omega_-^2\sin(\omega_-T) = 0
\end{aligned}
\tag{5.15}
$$

To get a nontrivial solution, the corresponding determinant must vanish, which yields the condition

$$(\omega_+^2 - \omega_-^2)\sin(\omega_+T)\sin(\omega_-T) = 0 \tag{5.16}$$

Since $\omega_+ \neq \omega_-$, we find

$$\sin(\omega_+T) = 0 \qquad \text{or} \qquad \sin(\omega_-T) = 0 \tag{5.17}$$

It means

$$\omega_+ = \frac{n\pi}{T} \qquad \text{or} \qquad \omega_- = \frac{n\pi}{T} \qquad (\text{where } n \text{ is an integer}) \tag{5.18}$$

and $\omega_\pm$ are the solutions to the equation

$$x^2 + m^2 + \alpha^2x^4 = k \tag{5.19}$$

Therefore, the spectrum of k with the boundary condition $\tilde{\mathscr{A}}'$ has the simple form

$$k = \left(\frac{n\pi}{T}\right)^2 + m^2 + \alpha^2\left(\frac{n\pi}{T}\right)^4 \tag{5.20}$$

6. Canonical quantization and instabilities

In this section we recall about istabilities and ghosts in the quantum PU oscillator (14). The most straightforward way is based on identifying the energy rasing and lowering operators (14). The classical solution (4.14) can be rewritten to the form

$$
\begin{aligned}
q(t) &= \frac{1}{2}(A_+ - iB_+)e^{i\lambda_+t} + \frac{1}{2}(A_+ + iB_+)e^{-i\lambda_+t}\\
&+\frac{1}{2}(A_- - iB_-)e^{i\lambda_-t} + \frac{1}{2}(A_- + iB_-)e^{-i\lambda_+t}
\end{aligned}
\tag{6.1}
$$

Since the λ_- modes have negative energy, the lowering operator must be proportional to the $(A_- - iB_-)$ amplitude. Similarly, since the λ_+ modes have negative energy, the raising operator must be proportional to the $(A_+ + iB_+)$ amplitude, ie.

$$\begin{aligned}\alpha_\pm &\sim A_\pm \pm iB_\pm \\ &\sim \frac{\lambda_\pm}{2}(1 \pm \sqrt{1-4\alpha^2 m^2})Q_1 \pm iP_1 \mp \frac{i}{2}(1 \mp \sqrt{1-4\alpha^2 m^2}) - \lambda_\pm P_2\end{aligned} \tag{6.2}$$

where we have used

$$A_\pm = \frac{\ddot{q}_0 + \lambda_\mp^2 q_0}{\lambda_\mp^2 - \lambda_\pm^2} \tag{6.3}$$

and

$$B_\pm = \frac{\dddot{q}_0 + \lambda_\mp^2 \dot{q}_0}{\lambda_\pm(\lambda_\mp^2 - \lambda_\pm^2)} \tag{6.4}$$

as well as [1]

$$Q_1 = q_0 \tag{6.5}$$

$$Q_2 = \dot{q}_0 \tag{6.6}$$

$$P_1 = \dot{q}_0 + \alpha^2 \dddot{q}_0 \tag{6.7}$$

$$P_2 = -\alpha^2 \ddot{q}_0 \tag{6.8}$$

It is now straightforward to derive the commutation relations,

$$[\alpha_\pm, \alpha_\pm^\dagger] = 1 \tag{6.9}$$

The next step depends upon physical interpretation (14).

(I) The 'empty' (or 'ground') state may be defined by the condition

$$\alpha_+ \left|\bar{\Omega}\right\rangle = \alpha_-^\dagger \left|\bar{\Omega}\right\rangle = 0 \tag{6.10}$$

Then the 'empty' state wave function $\bar{\Omega}(Q_1, Q_2)$ (in the Q-representation, with $P = -i\partial/\partial Q$) reads

$$\bar{\Omega}(Q_1, Q_2) = N \exp\left[-\frac{\sqrt{1-4\alpha^2 m^2}}{2(\lambda_- - \lambda_+)}(\lambda_+\lambda_- Q_1^2 - Q_2^2) - im\alpha Q_1 Q_2\right] \tag{6.11}$$

and is *infinite or not normalizable*, because the size of the wave function gets bigger with the increase of Q_2, so that the integral over the whole space diverges.

In addition, when the eigenstate $|\bar{N}_+, \bar{N}_-\rangle$ with the eigenvalues $\bar{N} = (\bar{N}_+, \bar{N}_-)$ is defined by

$$|\bar{N}_+, \bar{N}_-\rangle = \frac{a_+^\dagger}{\sqrt{N_+!}} \frac{a_-}{\sqrt{N_-!}} |\bar{\Omega}\rangle \tag{6.12}$$

[1] The canonical variables were calculated at the initial time value because the operators in Schrodinger picture do not depend upon time.

the norm of the $(0,1)$ state is given by

$$\begin{aligned} <0,\bar{1}|0,\bar{1}> &= \langle\bar{\Omega}|\,\alpha_-^\dagger\alpha_-\,|\bar{\Omega}\rangle \\ &= \langle\bar{\Omega}|\,(-1+\alpha_-\alpha_-^\dagger)\,|\bar{\Omega}\rangle \\ &= -<\bar{\Omega}|\bar{\Omega}> \\ &= -1 \end{aligned} \tag{6.13}$$

which is a ghost. The non-normalizable quantum 'states' are physically unacceptable, so the interpretation (I) should be dismissed (14).

(II) It is, however, possible to treat all particles (with positive or negative energy) as the truly ones by defining the 'empty' state Ω differently, namely, as

$$\alpha_\pm\,|\Omega\rangle = 0 \tag{6.14}$$

In this interpretation the negative energy can arbitrarily decrease and the Hamiltian is unbounded from below. The 'empty' state solution $\Omega(Q_1, Q_2)$ in the Q representation is now given by

$$\Omega(Q_1,Q_2) = N\exp\left[-\frac{\sqrt{1-4\alpha^2m^2}}{2(\lambda_-+\lambda_+)}(\lambda_+\lambda_-Q_1^2+Q_2^2)+im\alpha Q_1Q_2\right] \tag{6.15}$$

and is apparently *finite or normalizable*, because the first term in the exponential is negative.

The eigenstate $|\bar{N}_+,\bar{N}_-\rangle$ of the eigenvalues $\bar{N}=(\bar{N}_+,\bar{N}_-)$ is now given by

$$|N_+,N_-\rangle = \frac{a_+^\dagger}{\sqrt{N_+!}}\frac{a_-^\dagger}{\sqrt{N_-!}}\,|\Omega\rangle \tag{6.16}$$

while the norm of the $(0,1)$ state is

$$\begin{aligned} <0,1|0,1> &= \langle\Omega|\,\alpha_-\alpha_-^\dagger\,|\Omega\rangle \\ &= \langle\Omega|\,(1-\alpha_-^\dagger\alpha_-)\,|\Omega\rangle \\ &= <\Omega|\Omega> \\ &= 1 \end{aligned} \tag{6.17}$$

ie. it is not a ghost.

In the correct physical interpretation (II) the correspondence principle between the classical and quantum states is preserved, but the system has indefinite energy. When interactions are switched on, mixing the negative and positive energy states would lead to instabilities in the classical theory, and the exponentially growing and decaying states in quantum theory (25; 26). Excluding the negative energy states would lead to the loss of unitarity (21).

7. Path integral quantization and Forman theorem

The idea of ref. (21) is to define the quantum theory of the PU oscillator as the *Euclidean* path integral and then Wick rotate it back to Minkowski case. It makes sense since the Euclidean action of the PU oscillator — see eq. (8.3) below — is positively definite. It can also make the difference to the canonical quantization and the Ostrogradski method (Sec. 2) when one integrates over the path *only*, but not over its derivatives.

Let us first recall some basic facts about a path integral in QFT, according to the standard textbooks in Quantum Field Theory – see, for example, ref. (27).

The definition of the probability amplitude for a one-dimensional quantum particle by Feynman path integral is given by

$$Z(q_b, t_b; q_a, t_a) = \int_{q_a}^{q_b} \mathscr{D}q \exp\left[\frac{i}{\hbar}\int_{t_a}^{t_b} dtL\right] \tag{7.1}$$

where the integration goes over all paths $q(t)$ between q_a and q_b. After Wick rotation

$$t \to t = -i\tau \tag{7.2}$$

the path integral takes the form[2]

$$Z(q_b, t_b; q_a, t_a) = \int_{q_a}^{q_b} \mathscr{D}q \exp\left[-\frac{1}{\hbar}\int_{\tau_a}^{\tau_b} d\tau L_E\right] \tag{7.3}$$

It is called the Euclidean path integral. In the case of the PU oscillator the Euclidean path integral is *Gaussian*. Let us recall some basic properties of the Gaussian integrals.

The simplest Gaussian integral reads

$$\int_{-\infty}^{\infty} dxe^{-ax^2} = \sqrt{\frac{\pi}{a}} \qquad a > 0 \tag{7.4}$$

It can be easily extended to a quadratic form in the exponential as

$$\int_{-\infty}^{\infty} dxe^{-ax^2-bx} = \sqrt{\frac{\pi}{a}}\exp\left(\frac{b^2}{4a}\right) \tag{7.5}$$

It can also be easily extended to the case of several variables with the diagonal quadratic form as

$$\int_{-\infty}^{\infty} [d^n x] \exp\left(-\sum_{i=1}^{n} a_i x_i^2\right) = \frac{1}{\prod_{i=1}^{n} a_i^{\frac{1}{2}}} \tag{7.6}$$

where we have introduced the normalized measure $[dx] = dx/\sqrt{\pi}$.

By diagonalizing a generic (non-degenerate) quadratic form, one can prove a general finite-dimensional formula,

$$\begin{aligned} \int_{\infty}^{\infty} [d^n x] \exp\left(-x^t Ax - b^t x\right) &= \frac{1}{\prod_{i=1}^{n} \lambda_i} \exp\left(\frac{1}{4} b^t A^{-1} b\right) \\ &= \frac{1}{det A} \exp\left(\frac{1}{4} b^t A^{-1} b\right) \end{aligned} \tag{7.7}$$

Finally, when formally sending the number of integrations to infinity, one gets the Gaussian path integral,

$$\begin{aligned} &\int_{q_a}^{q_b} \mathscr{D}q \exp\left[-\int_{t_a}^{t_b} dt(q(t)F(D)q(t) + q(t)J(t))\right] \\ &= \frac{1}{\sqrt{DetF(D)}} \exp\left[-\frac{1}{4}\int_{t_a}^{t_b} dtJ(t)F^{-1}(D)J(t)\right] \end{aligned} \tag{7.8}$$

[2] The sign factor in the Wick rotation is chosen to make the path integral converging.

where $DetF(D)$ is now the functional determinant.

A generic functional determinant diverges since it is defined as the product of all the eigenvalues in the spectrum of a differential operator. Therefore, one needs a regularization. It is most convenient to use the zeta function regularization in our case — see, for example, ref. (28) for a comprehensive account. The Riemann zeta function is defined by

$$\zeta(s) = \sum_{n=1}^{\infty} \frac{1}{n^s} \tag{7.9}$$

in the convergence area of the series. It is then expanded for $\mathrm{Re}(s) > 1$ by analytic continuation. It is often useful to employ an integral representation of the zeta function in the form

$$\zeta(s) = \frac{1}{\Gamma(s)} \int_0^{\infty} dt\, t^{s-1} \sum_{n=1}^{\infty} e^{-nt} \tag{7.10}$$

where the (Euler) gamma function has been introduced,

$$\Gamma(s) = \int_0^{\infty} dt\, t^{s-1} e^{-t} \tag{7.11}$$

Equation (7.10) allows one to define the zeta function for an elliptic operator L as

$$\zeta(s|L) = \frac{1}{\Gamma(s)} \int_0^{\infty} dt\, t^{s-1} \mathrm{tr} e^{-tL} \tag{7.12}$$

where $\mathrm{tr} e^{-tL}$ is given by

$$\mathrm{tr} e^{-tL} = \mathrm{tr} \begin{pmatrix} e^{-\lambda_1 t} & & \\ & e^{-\lambda_2 t} & \\ & & \ddots \end{pmatrix} = \sum_{n=1}^{\infty} e^{-\lambda_n t} \tag{7.13}$$

in terms of the positive eigenvalues λ_n of L. One easily finds

$$\zeta(s|L) = \sum_{n=1}^{\infty} \frac{1}{\lambda_n^s} = \sum_{n=0}^{\infty} e^{-s \ln \lambda_n} \tag{7.14}$$

Differentiating both sides of this equation with respect to s at $s = 0$, one finds

$$\begin{aligned} \left.\frac{d\zeta(s|L)}{ds}\right|_{s=0} &= -\sum_{n=1}^{\infty} \ln \lambda_n \\ &= -\ln \prod_{n=1}^{\infty} \lambda_n \\ &= -\ln DetL \end{aligned} \tag{7.15}$$

so that the functional determinant of an elliptic operator L is given by

$$DetL = e^{-\zeta'|_{s=0}} \tag{7.16}$$

The zeta function regularization of the right hand side of this equation is

$$\begin{aligned}\ln Det\frac{L(\epsilon)}{\mu^2} &= -\frac{1}{\epsilon}\zeta\left(\epsilon\left|\frac{L}{\mu^2}\right.\right) = -\frac{1}{\Gamma(\epsilon+1)}\int_0^\infty dt\, t^{\epsilon-1}\mathrm{tr}e^{-t\frac{L}{\mu^2}} \\ &= -\frac{1}{\epsilon}\sum_{n=1}^{\infty}\frac{1}{\left(\frac{\lambda_n}{\mu^2}\right)^{\epsilon}} = -\frac{\mu^{2\epsilon}}{\epsilon}\sum_{n=1}^{\infty}\frac{1}{\lambda_n^{\epsilon}} = -\frac{\mu^{2\epsilon}}{\epsilon}\zeta(\epsilon|L) \\ &= -\frac{1}{\epsilon}(1+\epsilon\ln\mu^2)(\zeta(0|L)+\epsilon\zeta'(0|L))+O(\epsilon^2) \\ &= -\frac{1}{\epsilon}\zeta(0|L)-\zeta'(0|L)-\ln\mu^2\zeta(0|L)+O(\epsilon^2) \end{aligned} \tag{7.17}$$

where we have introduced the regularization parameter ϵ and the dimension parameter μ.

The zeta-function renormalization amounts to deleting the first term in eq. (7.17), since it UV-diverges in the limit $\epsilon \to 0$, as well as the third term since it IR-diverges in the limit $\mu \to 0$.

To put equation (7.17) into a more explicit form, without resorting to the spectrum of the differential operator, it is convenient to use Forman's theorem (24): [3]

Let $K_{\mathscr{A}}$ and $\bar{K}_{\bar{\mathscr{A}}}$ are the differential operators defined by

$$\begin{cases} K = P_0(\tau)\frac{d^n}{d\tau^n} + O(\frac{d^{n-1}}{d\tau^{n-1}}) \\ \bar{K} = P_0(\tau)\frac{d^n}{d\tau^n} + O(\frac{d^{n-1}}{d\tau^{n-1}}) \end{cases} \tag{7.18}$$

over the domain $[0, T]$. Consider a linear differential equation

$$Kh(\tau) = 0 \tag{7.19}$$

with a boundary condition

$$\mathscr{A} : M\begin{pmatrix} h(0) \\ h^{(1)}(0) \\ \vdots \\ h^{(n-1)}(0)\end{pmatrix} + N\begin{pmatrix} h(0) \\ h^{(1)}(0) \\ \vdots \\ h^{(n-1)}(0)\end{pmatrix} = 0 \tag{7.20}$$

and take the boundary condition $\bar{\mathscr{A}}$ to be *smoothly* connected to $\mathscr{A}$. The time evolution operator $Y_K(\tau)$ is introduced as

$$\begin{pmatrix} h(\tau) \\ \vdots \\ h^{(n-1)}(\tau)\end{pmatrix} = Y_K(\tau)\begin{pmatrix} h(0) \\ \vdots \\ h^{(n-1)}(0)\end{pmatrix} \tag{7.21}$$

so that the boundary condition can be written to

$$(M + NY_K(T))\begin{pmatrix} h(0) \\ \vdots \\ h^{(n-1)}(0)\end{pmatrix} = 0 \tag{7.22}$$

[3] Forman theorem is an extension of the Gel'fand-Yaglom theorem.

The Forman theorem is given by the statement:

$$\frac{DetK_{\mathscr{A}}}{Det\bar{K}_{\mathscr{A}}} = \frac{det\,(M + NY_K(T))}{det\,(\bar{M} + \bar{N}Y_{\bar{K}}(T))} \tag{7.23}$$

This theorem is effective for finding the functional determinant of the operator K with unknown spectrum by connecting it to the one with a simple spectrum via changing the boundary conditions.

8. Path integral of PU oscillator

The Euclidean path integral of the PU oscillator over a domain $[0, T]$ was calculated in refs. (21–23). Here we confirm the results of ref. (22) by our calculation.

The path integral of PU oscillator with the action

$$S_{PU} = \int_0^T dt \left(\frac{1}{2}\dot{q}(t)^2 - \frac{m^2}{2} q(t)^2 - \frac{\alpha^2}{2}\ddot{q}(t)^2 \right) \tag{8.1}$$

after the Wick rotation ($t \to it$) takes the form

$$Z(q_T, T; q_0, 0) = \int_{q_0}^{q_T} \mathscr{D}q \exp\left(-S_E\right) \tag{8.2}$$

where the Euclidean PU action is given by

$$S_E = \int_0^T dt \left(\frac{1}{2}\dot{q}(t)^2 + \frac{m^2}{2} q(t)^2 + \frac{\alpha^2}{2}\ddot{q}(t)^2 \right) \tag{8.3}$$

This S_E is *positively definite*, so that the Euclidean path integral is well defined.

Since our discussion of the classical theory (Sec. 4), the integral trajectory is a sum of a classical trajectory q_{cl} and quantum fluctuations $\hat{q}$, $q = q_{cl} + \hat{q}$. Accordingly, the action can be also written down as a sum,

$$S_E[q] = S_{cl} + S[\hat{q}] \tag{8.4}$$

and the path integral of the PU oscillator takes the form

$$Z(q_T, T; q_0, 0) = e^{-S_{cl}} \int_0^0 \mathscr{D}\hat{q} \exp\left(-S[\hat{q}]\right) \tag{8.5}$$

where the quantum action $S[\hat{q}]$ is given by

$$S[\hat{q}] = \frac{1}{2}\int_0^T dt\,\hat{q} \left(\alpha^2 \frac{d^4}{dt^4} - \frac{d^2}{dt^2} + m^2 \right) \hat{q} \tag{8.6}$$

after integration by parts.

Let us denote the differential operator $\alpha^2 \frac{d^4}{dt^4} - \frac{d^2}{dt^2} + m^2$ with the boundary condition $\mathscr{A}$ as $K_{\mathscr{A}}$. Then the path integral can be written down in the form

$$Z(q_T, T; q_0, 0) = e^{-S_{cl}} \int_0^0 \mathscr{D}\hat{q} \exp\left(-\frac{1}{2}\int_0^T dt\,\hat{q}K_{\mathscr{A}}\hat{q} \right) \tag{8.7}$$

The path integral of the PU oscillator is Gaussian and, therefore, can be computed along the lines of Sec. 7 as

$$Z(q_T,T;q_0,0) = \frac{N}{\sqrt{DetK_{\mathscr{A}}}} \exp\left(-S_{cl}\right) \tag{8.8}$$

where N is the normalization constant. The classical part S_{cl} was found in ref. (21), and it is quite involved. The functional determinant is the key part of a quantum propagator of PU oscillator, which is of primary physical interest. It can be computed by the use of Forman theorem (Sec. 7).

First, one calculates the time evolution operator Y_K. It is given by

$$Y_K(t) = \begin{pmatrix} u_1(t) & u_2(t) & u_3(t) & u_4(t) \\ \dot{u}_1(t) & \dot{u}_2(t) & \dot{u}_3(t) & \dot{u}_4(t) \\ \ddot{u}_1(t) & \ddot{u}_2(t) & \ddot{u}_3(t) & \ddot{u}_4(t) \\ \dddot{u}_1(t) & \dddot{u}_2(t) & \dddot{u}_3(t) & \dddot{u}_4(t) \end{pmatrix} \tag{8.9}$$

where

$$Ku_i(t) = 0\,(i = 1,\ldots,4) \tag{8.10}$$

and the inital condition is

$$Y_k(0) = 1 \tag{8.11}$$

The operator $K_{\bar{\mathscr{A}}}$ is equal to $K_{\mathscr{A}}$, so they have $Y_K(t)$ is common.

By solving the equation $Ku_i = 0$ for u_i with

$$K = \alpha^2 \frac{d^4}{dt^4} - \frac{d^2}{dt^2} + m^2 \tag{8.12}$$

one gets its general solution in the form

$$u_i(t) = A_i \sinh(\lambda_+ t) + B_i \cosh(\lambda_+ t) + C_i \sinh(\lambda_- t) + D_i \cosh(\lambda_- t) \tag{8.13}$$

The boundary condition $Y_K(0) = 1$ amounts to the relations

$$B_i + D_i = \delta_{1i} \tag{8.14}$$

$$\lambda_+ A_i + \lambda_- C_i = \delta_{2i} \tag{8.15}$$

$$\lambda_+^2 B_i + \lambda_-^2 D_i = \delta_{3i} \tag{8.16}$$

$$\lambda_+^3 A_i + \lambda_-^3 C_i = \delta_{4i} \tag{8.17}$$

Therefore, the solutions are

$$u_1 = \frac{\lambda_-^2}{\lambda_-^2 - \lambda_+^2} \cosh(\lambda_+ t) + \frac{\lambda_+^2}{\lambda_+^2 - \lambda_-^2} \cosh(\lambda_- t) \tag{8.18}$$

$$u_2 = \frac{\lambda_-^2}{\lambda_+(\lambda_-^2 - \lambda_+^2)} \sinh(\lambda_+ t) + \frac{\lambda_+^2}{\lambda_2(\lambda_+^2 - \lambda_-^2)} \sinh(\lambda_- t) \tag{8.19}$$

$$u_3 = -\frac{1}{\lambda_-^2 - \lambda_+^2} \cosh(\lambda_+ t) - \frac{1}{\lambda_+^2 - \lambda_-^2} \cosh(\lambda_- t) \tag{8.20}$$

$$u_4 = -\frac{1}{\lambda_+(\lambda_-^2 - \lambda_+^2)} \sinh(\lambda_+ t) + \frac{1}{\lambda_2(\lambda_+^2 - \lambda_-^2)} \sinh(\lambda_- t) \tag{8.21}$$

Next, one writes down the boundary conditions $\mathscr{A}$ and $\bar{\mathscr{A}}$ in terms of the matrices M and N appearing in the Forman theorem. The boundary condition $\mathscr{A}$ is

$$\mathscr{A} \,:\, \hat{q}(0) = 0\,,\, \hat{q}(T) = 0\,,\, \dot{\hat{q}}(0) = 0\,,\, \dot{\hat{q}}(T) = 0 \tag{8.22}$$

so that its matrices M and N are given by

$$M = \begin{pmatrix} 1\,0\,0\,0 \\ 0\,1\,0\,0 \\ 0\,0\,0\,0 \\ 0\,0\,0\,0 \end{pmatrix} \tag{8.23}$$

and

$$N = \begin{pmatrix} 0\,0\,0\,0 \\ 0\,0\,0\,0 \\ 1\,0\,0\,0 \\ 0\,1\,0\,0 \end{pmatrix} \tag{8.24}$$

In the same way, the boundary condition $\bar{\mathscr{A}}$ is

$$\bar{\mathscr{A}} \,:\, \hat{q}(0) = 0\,,\, \hat{q}(T) = 0\,,\, \ddot{\hat{q}}(0) = 0\,,\, \ddot{\hat{q}}(T) = 0 \tag{8.25}$$

so that its matrices M and N are given by

$$\bar{M} = \begin{pmatrix} 1\,0 & 0 & 0 \\ 0\,0 & -1 & 0 \\ 0\,0 & 0 & 0 \\ 0\,0 & 0 & 0 \end{pmatrix} \tag{8.26}$$

and

$$\bar{N} = \begin{pmatrix} 0\,0\,0\,0 \\ 0\,0\,0\,0 \\ 1\,0\,0\,0 \\ 0\,0\,1\,0 \end{pmatrix} \tag{8.27}$$

Having found M, N and Y_K, as well as $\bar{M}$, $\bar{N}$ and $Y_{\bar{K}}$, we calculate

$$\begin{aligned} & det(M + NY_K(T)) \\ & = \frac{\alpha^3}{m}\left[\frac{1}{1+2m\alpha}\sinh^2\left(\frac{\sqrt{1+2m\alpha}}{2\alpha}T\right) - \frac{1}{1-2m\alpha}\sinh^2\left(\frac{\sqrt{1-2m\alpha}}{2\alpha}T\right)\right] \end{aligned} \tag{8.28}$$

and

$$\begin{aligned} & det(\bar{M} + \bar{N}Y_{\bar{K}}(T)) \\ & = \frac{\alpha}{m}\left[\sinh^2\left(\frac{\sqrt{1+2m\alpha}}{2\alpha}T\right) - \sinh^2\left(\frac{\sqrt{1-2m\alpha}}{2\alpha}T\right)\right] \end{aligned} \tag{8.29}$$

A calculation of $DetK_{\bar{\mathscr{A}}}$ goes along the standard lines (21–23),

$$DetK_{\bar{\mathscr{A}}} = \prod_{n=1}^{\infty} k_n = \prod_{n=1}^{\infty}\left(\alpha^2\left(\frac{n\pi}{T}\right)^4 + \left(\frac{n\pi}{T}\right)^2 + m^2\right) \tag{8.30}$$

$$= \frac{\alpha}{mT^2}\left[\sinh^2\left(\frac{\sqrt{1+2m\alpha}}{2\alpha}T\right) - \sinh^2\left(\frac{\sqrt{1-2m\alpha}}{2\alpha}T\right)\right] \tag{8.31}$$

By using the Forman formula, one gets the final answer:

$$DetK_{\mathscr{A}} = \frac{det\left(M + NY_K(T)\right)}{det\left(\bar{M} + \bar{N}Y_{\bar{K}}(T)\right)} DetK_{\bar{\mathscr{A}}}$$

$$= \frac{\alpha^3}{mT^2}\left[(1+2\alpha m)^{-1}\sinh^2\left(\frac{\sqrt{1+2m\alpha}}{2\alpha}T\right)\right.$$

$$\left. -(1-2\alpha m)^{-1}\sinh^2\left(\frac{\sqrt{1-2m\alpha}}{2\alpha}T\right)\right] \tag{8.32}$$

in full agreement with ref. (22) in its last (v2) version. In the large T limit one finds

$$DetK_{\mathscr{A}} \approx \frac{\alpha}{m}\left[(1+2\alpha m)^{-1}\frac{\exp\left(\sqrt{1+2m\alpha}\frac{T}{\alpha}\right)}{\left(\frac{2T}{\alpha}\right)^2} - (1-2\alpha m)^{-1}\frac{\exp\left(\sqrt{1-2m\alpha}\frac{T}{\alpha}\right)}{\left(\frac{2T}{\alpha}\right)^2}\right], \tag{8.33}$$

and in the small T limit one gets

$$DetK_{\mathscr{A}} \approx \frac{T^2}{12} + \mathcal{O}(T^4) \tag{8.34}$$

The ground state probability amplitude (or the Euclidean quantum propagator) of PU oscillator, is given by

$$< q_T, \dot{q}_T; \tau = T | q_0, \dot{q}_0; \tau = 0 >= \sqrt{\frac{2\pi}{DetK_{\mathscr{A}}}} \exp\left(-S_E[q_{cl}]\right), \tag{8.35}$$

The classical Euclidean action $S_E[q_{cl}]$ was calculated in Appendix of ref. (21). It is finite for large $T \gg 1$ and behaves like $\frac{1}{2T}$ for small $T \ll 1$. Hence, the transition amplitude (or the quantum Euclidean propagator) is exponentially suppressed both for small and large T, ie. the transition amplitude is normalizable and the Euclidean path integral is well defined indeed.

9. Conclusion

The procedure of calculating Euclidean transition probabilities (for observables) in the quantum PU theory was outlined in ref. (21). The probabilities in the Minkowski space can be obtained by analytic continuation. It is, therefore, possible to make physical sense out of the quantum PU theory.

In classical PU theory with interactions, even at a very small value of the parameter $\alpha > 0$, one gets runaway production of states with negative and positive energy. However, as was suggested in ref. (21), the Euclidean formulation of the quantum theory implicitly imposes certain restrictions that can remove classical instabilities. The price of removing the instabilities is given by an apparent violation of unitarity (21). Indeed, integrating over the basic trajectory, and not over its derivatives in the Euclidean path integral formulation of the quantum PU oscillator given above is not in line with the canonical quantization and the Ostrogradski method. By doing it, one looses some information and, hence, one loses unitarity. As was argued in ref. (21), one can, nevertheless, never produce a negative norm state or get a negative probability, so that the departure from unitarity may be very small at the low energies (say, in the present universe), but important at the very high energies (say, in the early universe). Of course, it is debateable whether the 'price' of loosing unitarity is too high or not.

Apparently, the $f(R)$ gravity theories are special in the sense that for each of them there exist the classically equivalent scalar-tensor field theory without higher derivatives, under the physical stability conditions. Still, as the quantum field theories, they may be very different. It may be possible to quantise $f(R)$ gravity without loosing unitarity. Figuring out details is still a challenge.

10. References

[1] M. Ostrogradski, Mem. Ac. St. Petersbourg VI 4 (1850) 385
[2] A. De Felice and S. Tsujikawa, Living Rev. Rel. 13 (2010) 3
[3] S. James Gates, Jr., and S.V. Ketov, Phys. Lett. B674 (2009) 59
[4] A.A. Starobinsky, Phys. Lett. B91 (1980) 99
[5] E. Komatsu et al., Astrophys. J. Suppl. 192 (2011) 18
[6] S.V. Ketov and A.A. Starobinsky, Phys. Rev. D83 (2011) 063512
[7] S.V. Ketov and N. Watanabe, Phys. Lett. B705 (2011) 410
[8] S.W. Hawking, in "Quantum Field Theory and Quantum Statistics: Essays in Honer of the 60th Birthday of E.S. Fradkin", A. Batalin et al. (Eds.), Hilger, Bristol, UK, 1987
[9] B. Whitt, Phys. Lett. B145 (1984) 176
[10] J.D. Barrow and S. Cotsakis, Phys. Lett. B214 (1988) 515
[11] K.-I. Maeda, Phys. Rev. D39 (1989) 3159
[12] S. Kaneda, S.V. Ketov and N. Watanabe, Mod. Phys. Lett. A25 (2010) 2753
[13] S.V. Ketov and N. Watanabe, JCAP 1103 (2011) 011
[14] R. P. Woodard, Lect. Notes Phys. 720 (2007) 403
[15] K.S. Stelle, Phys. Rev. D16 (1977) 953
[16] E.S. Fradkin and A.A. Tseytlin, Nucl. Phys. B201 (1982) 469
[17] I.L. Buchbinder and S.V. Ketov, Teor. Mat. Fiz. 77 (1988) 42, Fortschr. Phys. 39 (1991) 1
[18] R. Percacci and O. Zanusso, Phys. Rev. D81 (2010) 065012
[19] S.V. Ketov, *Quantum Non-linear Sigma-models*, Springer-Verlag, 2000
[20] A. Pais and G.E. Uhlenbeck, Phys. Rev. 79 (1950) 145
[21] S.W. Hawking and T. Hertog, Phys. Rev. D 65 (2002) 103515
[22] R. Di Criscienzo and S. Zerbini, J. Math. Phys. 50 (2009) 103517; arXiv:0907.4265 v2 [hep-th]
[23] K. Andrzejewski, J. Gonera and P. Maslanka, Prog. Theor.Phys. 125 (2011) 247
[24] R. Forman, Invent. Math. 88 (1987) 447
[25] E. Tomboulis, Phys. Lett. B70 (1977) 361
[26] V.V. Nesterenko, Phys. Rev. D75 (2007) 087703
[27] L.H. Ryder, *Quantum Field Theory*, Cambridge University Press, 1985
[28] E. Elizalde, *Ten Physical Applications of Spectral Zeta Functions*, Springer-Verlag, 1995.

2

The Method of Unitary Clothing Transformations in Quantum Field Theory: Applications in the Theory of Nuclear Forces and Reactions

A. V. Shebeko
Institute for Theoretical Physics
National Research Center "Kharkov Institute of Physics & Technology"
Ukraine

1. Introduction

In what follows we will show how one can realize the notion of "clothed " particles (Greenberg & Schweber, 1958) for field theoretical treatments based upon the so-called instant form of relativistic dynamics formulated by Dirac (Dirac, 1949). In the context, let us recall that the notion points out a transparent way for including the so-called cloud or persistent effects in a system of interacting fields (to be definite, mesons and nucleons). A constructive step (see surveys (Shebeko & Shirokov, 2000; 2001) and refs. therein) is to express the total field Hamiltonian H and other operators of great physical meaning, e.g., the Lorentz boost generators and current density operators, which depend initially on the creation and destruction operators for the "bare" particles, through a set of their "clothed" counterparts. It is achieved via unitary clothing transformations (UCTs) (see article (Korda et al., 2007)) in the Hilbert space $\mathcal{H}$ of meson-nucleon states and we stress, as before, that each of such transformations remains the Hamiltonian unchanged unlike other unitary transformation methods (Glöckle & Müller, 1981; Kobayashi, 1997; Okubo, 1954; Stefanovich, 2001))[1] for Hamiltonian-based models. In the course of the clothing procedure a large amount of virtual processes associated in our case with the meson absorption/emission, the $N\bar{N}$-pair annihilation/production and other cloud effects turns out to be accumulated in the creation (destruction) operators for the clothed particles. The latter, being the quasiparticles of the method of UCTs, must have the properties (charges, masses, etc.) of physical (observable) particles. Such a bootstrap reflects the most significant distinction between the concepts of clothed and bare particles.

At the same time, after Dirac, any relativistic quantum theory may be so defined that the generator of time translations (Hamiltonian), the generators of space translations (linear momentum), space rotations (angular momentum) and Lorentz transformations (boost operator) satisfy the well-known commutations. Basic ideas, put forward by Dirac with his "front", "instant" and "point" forms of the relativistic dynamics, have been realized in many relativistic quantum mechanical models. In this context, the survey (Keister & Polyzou, 1991), being a remarkable introduction to a subfield called the relativistic Hamiltonian dynamics,

[1] Some specific features of these methods are discussed in (Shebeko & Shirokov, 2001) and (Korda et al., 2007)

represents various aspects and achievements of relativistic direct interaction theories. Among the vast literature on this subject we would like to note an exhaustive exposition in lectures (Bakker, 2001; Heinzl, 2001) of the appealing features of the relativistic Hamiltonian dynamics with an emphasis on "light-cone quantization". Following a pioneering work (Foldy, 1961), the term "direct" is related to a system with a fixed number of interacting particles, where interactions are rather direct than mediated through a field. In the approach it is customary to consider such interactions expressed in terms of the particle coordinates, momenta and spins. Along the guideline the so-called separable interactions and relativistic center-of-mass variables for composite systems were built up by assuming that the generators of the Poincaré group (Π) can be represented as expansions on powers of $1/c^2$ or, more exactly, $(v/c)^2$, where v is a typical nuclear velocity (cf. the (p/m) expansion, introduced in (Friar, 1975), where m is the nucleon mass and p is a typical nucleon momentum). Afterwards, similar expansions were rederived and reexamined (with new physical inputs) in the framework of a field-theoretic approach (Glöckle & Müller, 1981). There, starting from a model Lagrangian for "scalar nucleons" interacting with a scalar meson field (like the Wentzel model (Wentzel, 1949)) the authors showed (to our knowledge first) how the Hamiltonian and the boost generator (these noncommuting operators), determined in a standard manner (Schwinger, 1962), can be blockdiagonalized by one and the same unitary transformation after Okubo (Okubo, 1954). The corresponding blocks derived in leading order in the coupling constant act in the subspace with a fixed nucleon number (the nucleon "sector" of the full space $\mathcal{H}$). In general, the work (Glöckle & Müller, 1981) and its continuation (Krüger & Glöckle, 1999) exemplify applications of local relativistic quantum field theory (RQFT), where the generators of interest, being compatible with the basic commutation rules for fields, are constructed within the Lagrangian formalism using the Nöther theorem and its consequences. Although the available covariant perturbation theory and functional-integral methods are very successful when describing various relativistic and quantum effects in the world of elementary particles, the Hamilton method can be helpful too. As known, it is the case, where one has to study properties of strongly interacting particles, e.g., as in nuclear physics with its problems of bound states for meson-nucleon systems. Of course, any Hamiltonian formulation of field theory, not being manifestly covariant, cannot be *ab initio* accepted as equivalent to the way after Feynman, Schwinger and Tomonaga. However, in order to overcome the obstacle starting from a field Hamiltonian H one can consider it as one of the ten infinitesimal operators (generators) of space-time translations and pure Lorentz transformations that act in a proper Hilbert space. Taken together they compose a basis of the Lie-Poincaré algebra (see below) to ensure relativistic invariance (RI) in the Dirac sense, being referred to the RI as a whole.

The purpose of the present exposition is twofold. First, we consider an algebraic method (Shebeko & Frolov , 2011) to meet the Poincaré commutators for a wide class of field theoretic models (local and nonlocal ones taking into account their invariance with respect to space translations). In particular, this recursive method is appropriate for models with derivative couplings and spins ≥ 1 , typical of the meson theory of nuclear forces, where only some part of the interaction density in the Dirac picture has the property to be a Lorentz scalar. The antiparticle degrees of freedom are included together with such an important issue as mass renormalization vs relativistic invariance in the Dirac sense. Second, special attention is paid to finding analytic expressions for the generators in the clothed-particle representation, in which the so-called bad terms are simultaneously removed from the Hamiltonian and the boosts. Moreover, the mass renormalization terms introduced in the Hamiltonian at the outset turn out to be related to certain covariant integrals that are convergent in the field models with proper cutoff factors. After constructing interactions between the clothed particles and addressing an equivalence theorem for evaluation of the S-matrix we derive the approximate eigenvalue equations for the simplest bound and scattering states. The latter can be found in a nonperturbative way using the methods elaborated in the theory of nuclear structure and reactions that is demonstrated by a few examples.

However, before to apply the UCT method (in particular, beyond the Lagrangian formalism with its local interaction densities) we would like to mention the two algebraic procedures to solve the basic commutator equations of Π (see Sec. 2). One of them, proposed in (Shebeko & Frolov , 2011), has some touching points with the other developed in (Kita, 1966; 1968) and essentially repeated many years after by Chandler (Chandler, 2003). In paper (Kita, 1968) the author considers three kinds of neutral spinless bosons and nonlocal interaction between them in a relativistic version of the Lee model with a cutoff in momentum space. A similar model for two spinless particles has been utilized in (Chandler, 2003) with a Yukawa-type interaction that belongs to the realm of the so-called models with persistent vacuum (see, for instance, (Eckmann, 1970)). Certain resemblance between our and those explorations is that we prefer to proceed within a corpuscular picture (see Chapter IV in monograph (Weinberg, 1995)), where each of the ten generators of the Poincaré group Π (and not only they) may be expressed as a sum of products of particle creation and annihilation operators $a^\dagger(n)$ and $a(n)$ ($n = 1,2,...$), e.g., bosons and/or fermions. Some mathematical aspects of the corpuscular notion were formulated many years ago in (Friedrichs, 1953) (Chapter III). As in (Weinberg, 1995), a label n is associated with all the necessary quantum numbers for a single particle: its momentum $\mathbf{p}_n$[2], spin z-component (helicity for massless particles) μ_n, and species ξ_n. The operators $a^\dagger(n)$ and $a(n)$ satisfy the standard commutation relations such as Eqs. (4.2.5)-(4.2.7) in (Weinberg, 1995).

In the framework of such a picture the Hamiltonian of a system of interacting mesons and nucleons can be written as

$$H = \sum_{C=0}^{\infty} \sum_{A=0}^{\infty} H_{CA}, \tag{1}$$

$$H_{CA} = \not{\sum} H_{CA}(1',2',...,n'_C;1,2,...,n_A) a^\dagger(1') a^\dagger(2') ... a^\dagger(n'_C) a(n_A) ... a(2) a(1), \tag{2}$$

where the capital $C(A)$ denotes the particle-creation (annihilation) number for the operator substructure H_{CA}. Sometimes we say that the latter belongs to the class $[C.A]$. Operation $\not{\sum}$ implies all necessary summations over discrete indices and covariant integrations over continuous spectra.

Further, it is proved (Weinberg, 1995) that the S-matrix meets the so-called cluster decomposition principle, if the coefficient functions H_{CA} embody a single three-dimensional momentum-conservation delta function, viz.,

$$\begin{aligned} H_{CA}(1',2',...,C;1,2,...,A) &= \delta(\mathbf{p}'_1 + \mathbf{p}'_2 + ... + \mathbf{p}'_C - \mathbf{p}_1 - \mathbf{p}_2 - ... - \mathbf{p}_A) \\ &\times h_{CA}(p'_1\mu'_1\xi'_1, p'_2\mu'_2\xi'_2, ..., p'_C\mu'_C\xi'_C; p_1\mu_1\xi_1, p_2\mu_2\xi_2, ..., p_A\mu_A\xi_A), \end{aligned} \tag{3}$$

[2] Or the 4-momentum $p_n = (p_n^0, \mathbf{p}_n)$ on the mass shell $p_n^2 = p_n^{02} - \mathbf{p}_n^2 = m_n^2$ with the particle mass m_n

where the c-number coefficients h_{CA} do not contain delta function.

Following the guideline "to free ourselves from any dependence on pre-existing field theories "(cit. from (Weinberg, 1995) on p.175), the three boost operators $\mathbf{N} = (N^1, N^2, N^3)$ can be written as

$$\mathbf{N} = \sum_{C=0}^{\infty} \sum_{A=0}^{\infty} \mathbf{N}_{CA}, \tag{4}$$

$$\mathbf{N}_{CA} = \not{\sum} \mathbf{N}_{CA}(1', 2', ..., n'_C; 1, 2, ..., n_A) a^\dagger(1') a^\dagger(2') ... a^\dagger(n'_C) a(n_A) ... a(2) a(1). \tag{5}$$

In turn, the operator H, being divided into the no-interaction part H_F and the interaction H_I, owing to its translational invariance allows H_I to be written as

$$H_I = \int H_I(\mathbf{x}) d\mathbf{x}. \tag{6}$$

Our consideration is focused upon various field models (local and nonlocal) in which the interaction density $H_I(\mathbf{x})$ consists of scalar $H_{sc}(\mathbf{x})$ and nonscalar $H_{nsc}(\mathbf{x})$ contributions,

$$H_I(\mathbf{x}) = H_{sc}(\mathbf{x}) + H_{nsc}(\mathbf{x}), \tag{7}$$

where the property to be a scalar means

$$U_F(\Lambda) H_{sc}(x) U_F^{-1} = H_{sc}(\Lambda x), \quad \forall x = (t, \mathbf{x}) \tag{8}$$

for all Lorentz transformations Λ. Henceforth, for any operator $O(\mathbf{x})$ in the Schrödinger (S) picture it is introduced its counterpart $O(x) = \exp(iH_F t) O(\mathbf{x}) \exp(-iH_F t)$ in the Dirac (D) picture.

2. Basic equations in relativistic theory with particle creation and annihilation

When seeking links between the coefficients in the r.h.s. of Eqs. (2) and (5) one considers the fundamental relations of the Lie-Poincaré algebra, which can be divided into the three kinds for:
no-interaction generators

$$[P_i, P_j] = 0, \quad [J_i, J_j] = i\varepsilon_{ijk} J_k, \quad [J_i, P_j] = i\varepsilon_{ijk} P_k, \tag{9}$$

relations linear in H and $\mathbf{N}$

$$[\mathbf{P}, H] = 0, \quad [\mathbf{J}, H] = 0, \quad [J_i, N_j] = i\varepsilon_{ijk} N_k, \quad [P_i, N_j] = i\delta_{ij} H, \tag{10}$$

and ones nonlinear in H and $\mathbf{N}$

$$[H, \mathbf{N}] = i\mathbf{P}, \quad [N_i, N_j] = -i\varepsilon_{ijk} J_k, \tag{11}$$

$$(i, j, k = 1, 2, 3),$$

where $\mathbf{P} = (P^1, P^2, P^3)$ and $\mathbf{J} = (J^1, J^2, J^3)$ are the linear momentum and angular momentum operators, respectively. In this context, let us remind that in the instant form of relativistic dynamics after Dirac (Dirac, 1949) only the Hamiltonian and the boost operators carry interactions with conventional partitions $H = H_F + H_I$ and $\mathbf{N} = \mathbf{N}_F + \mathbf{N}_I$, while $\mathbf{P} = \mathbf{P}_F$ and $\mathbf{J} = \mathbf{J}_F$. In short notations, we distinguish the set $G_F = \{H_F, \mathbf{P}_F, \mathbf{J}_F, \mathbf{N}_F\}$ for free particles and the set $G = \{H, \mathbf{P}_F, \mathbf{J}_F, \mathbf{N}\}$ for interacting particles.

In turn, every operator H_{CA} can be represented as $H_{CA} = \int H_{CA}(\mathbf{x})d\mathbf{x}$, if one uses the formula

$$\delta(\mathbf{p} - \mathbf{p}') = \frac{1}{(2\pi)^3} \int e^{i(\mathbf{p}-\mathbf{p}')\mathbf{x}} d\mathbf{x}.$$

Thus, we come to the form $H = \int H(\mathbf{x})d\mathbf{x}$ well known from local field models with the density

$$H(\mathbf{x}) = \sum_{C=0}^{\infty} \sum_{A=0}^{\infty} H_{CA}(\mathbf{x}). \tag{12}$$

For instance, in case with $C = A = 2$, where $H_{22}(1',2';1,2) = \delta(\mathbf{p}_1' + \mathbf{p}_2' - \mathbf{p}_1 - \mathbf{p}_2)h(1',2';1,2)$, we have

$$H_{22}(\mathbf{x}) = \frac{1}{(2\pi)^3} \sum\!\!\!\!\!\!\int \exp[-i(\mathbf{p}_1' + \mathbf{p}_2' - \mathbf{p}_1 - \mathbf{p}_2)\mathbf{x}]h(1',2';1,2)a^\dagger(1')\, a^\dagger(2')\, a(2)\, a(1). \tag{13}$$

Further, we will employ the transformation properties of the creation and annihilation operators with respect to Π. For example, in case of a massive particle with the mass m and spin j one considers that

$$U_F(\Lambda, b)a^\dagger(p,\mu)U_F^{-1}(\Lambda, b) = e^{i\Lambda p b} D^{(j)}_{\mu'\mu}(W(\Lambda, p))a^\dagger(\Lambda p, \mu'), \tag{14}$$

$$\forall \Lambda \in L_+ \text{ and arbitrary spacetime shifts } \mathrm{b} = (\mathrm{b}^0, \mathbf{b})$$

with D-function whose argument is the Wigner rotation $W(\Lambda, p)$, L_+ the homogeneous (proper) orthochronous Lorentz group. The correspondence $(\Lambda, b) \rightarrow U_F(\Lambda, b)$ between elements $(\Lambda, b) \in \Pi$ and unitary transformations $U_F(\Lambda, b)$ realizes an irreducible representation of Π on the Hilbert space $\mathcal{H}$ (to be definite) of meson-nucleon states. In this context, it is convenient to employ the operators $a(p,\mu) = a(\mathbf{p},\mu)\sqrt{p_0}$ that meet the covariant commutation relations

$$\begin{aligned} [a(p',\mu'), a^\dagger(p,\mu)]_\pm &= p_0 \delta(\mathbf{p} - \mathbf{p}')\delta_{\mu'\mu}, \\ [a(p',\mu'), a(p,\mu)]_\pm &= [a^\dagger(p',\mu'), a^\dagger(p,\mu)]_\pm = 0. \end{aligned} \tag{15}$$

Here $p_0 = \sqrt{\mathbf{p}^2 + m^2}$ is the fourth component of the 4-momentum $p = (p_0, \mathbf{p})$.

3. A possible way for constructing generators of the Poincaré group

Let us recall that within the Lagrangian formalism the 4-vector $P^\mu = (H, \mathbf{P})$ for any local field model, where requirements of relativistic symmetry are *manifestly* provided at the beginning, is determined by the Nöther integrals

$$P^\nu = \int \mathcal{T}^{0\nu}(\mathbf{x})d\mathbf{x} \quad (\nu = 0,1,2,3), \tag{16}$$

where $\mathcal{T}^{0\nu}(\mathbf{x})$ are the components of the energy-momentum tensor density $\mathcal{T}^{\mu\nu}(x)$ at $t=0$.

Other Nöther integrals are expressed through the angular-momentum tensor density

$$\mathcal{M}^{\beta[\mu\nu]}(x) = x^{\mu}\mathcal{T}^{\beta\nu}(x) - x^{\nu}\mathcal{T}^{\beta\mu}(x) + \Sigma^{\beta[\mu\nu]}(x), \tag{17}$$

that contains, in general, so-called polarization part $\Sigma^{\beta[\mu\nu]}$[3] associated with spin degrees of freedom. Namely, the six independent integrals

$$M^{\mu\nu} = \int \mathcal{M}^{0[\mu\nu]}(x)d\mathbf{x}\bigg|_{t=0} \tag{18}$$

are considered as the generators of space rotations

$$J^{i} = \varepsilon_{ikl}M^{kl} \quad (i,k,l=1,2,3) \tag{19}$$

and the boosts

$$N^{k} \equiv M^{0k} = -\int x^{k}\mathcal{T}^{00}(\mathbf{x})d\mathbf{x} + \int \Sigma^{0[0k]}(\mathbf{x})d\mathbf{x},\ (k=1,2,3). \tag{20}$$

The reminder is not accidental as far as we strive to go out beyond the traditional QFT with local Lagrangian densities via special regularization of interactions in a total initial Hamiltonian.

3.1 The Belinfante ansatz. Application to interacting pion and nucleon fields

Regarding an illustration of these general relations let us write, the Lagrangian density

$$\mathcal{L}_{SCH}(x) = \frac{1}{2}\bar{\psi}_H(x)(i\gamma^{\mu}\overrightarrow{\partial}_{\mu} - m_0)\psi_H(x) + \frac{1}{2}\bar{\psi}_H(x)(-i\gamma^{\mu}\overleftarrow{\partial}_{\mu} - m_0)\psi_H(x)$$

$$+\frac{1}{2}[\partial_{\mu}\varphi_H(x)\partial^{\mu}\varphi_H(x) - \mu_0^2\varphi_H^2(x)] - ig_0\bar{\psi}_H(x)\gamma_5\psi_H(x)\varphi_H(x), \tag{21}$$

for interacting pion ϕ and nucleon ψ fields with the PS coupling (see, e.g.,(Schweber, 1961)). Then, one has (omitting argument x): i) energy-momentum tensor density

$$T^{\mu\nu}_{SCH} = \frac{\partial\mathcal{L}_{SCH}}{\partial\bar{\psi}_{H\,\mu}}\bar{\psi}^{\nu}_{H} + \frac{\partial\mathcal{L}_{SCH}}{\partial\psi_{H\,\mu}}\psi^{\nu}_{H} + \frac{\partial\mathcal{L}_{SCH}}{\partial\varphi_{H\,\mu}}\varphi^{\nu}_{H} - g^{\mu\nu}\mathcal{L}_{SCH}$$

$$\equiv T^{\mu\nu}_{N} + T^{\mu\nu}_{\pi} + T^{\mu\nu}_{I}, \tag{22}$$

where

$$T^{\mu\nu}_{N} = \frac{i}{2}\bar{\psi}_H\gamma^{\mu}\partial^{\nu}\psi_H - \frac{i}{2}\gamma^{\mu}\psi_H\partial^{\nu}\bar{\psi}_H - g^{\mu\nu}\mathcal{L}_N, \tag{23}$$

$$T^{\mu\nu}_{\pi} = \partial^{\mu}\varphi_H\partial^{\nu}\varphi_H - g^{\mu\nu}\mathcal{L}_{\pi}, \tag{24}$$

$$T^{\mu\nu}_{I} = ig_0 g^{\mu\nu}\bar{\psi}_H\gamma_5\psi_H\varphi_H, \tag{25}$$

[3] Henceforth, the symbol $[\alpha,\beta]$ for any labels α and β means the property $f^{[\beta,\alpha]} = -f^{[\alpha,\beta]}$ for its carrier f.

and ii) polarization contribution

$$\Sigma_{SCH}^{\beta[\mu\nu]} = \frac{1}{2} i\bar{\psi}_H \{\gamma^\beta \Sigma^{\mu\nu} + \Sigma^{\mu\nu}\gamma^\beta\}\psi_H, \tag{26}$$

where

$$\Sigma^{\mu\nu} = \frac{i}{4}[\gamma^\mu, \gamma^\nu].$$

In formulae (21)-(25) unlike operators $O(x)$ in the D picture, we have operators

$$O_H(x) = e^{iHt} O(x) e^{-iHt},$$

in the Heisenberg picture. We prefer to employ the definitions:

$$\{\gamma^\mu, \gamma^\nu\} = 2g^{\mu\nu}, \gamma_\mu^\dagger = \gamma_0 \gamma_\mu \gamma_0, \{\gamma_\mu, \gamma_5\} = 0, \gamma_5^\dagger = \gamma_0 \gamma_5 \gamma_0 = -\gamma_5.$$

The corresponding Hamiltonian density is given by

$$H_{SCH}(\mathbf{x}) = \mathcal{T}_{SCH}^{00}(\mathbf{x}) = H_{ferm}^0(\mathbf{x}) + H_\pi^0(\mathbf{x}) + V_{ps}^0(\mathbf{x}), \tag{27}$$

where

$$H_{ferm}^0(\mathbf{x}) = \frac{1}{2}\bar{\psi}(\mathbf{x})[-i\overrightarrow{\gamma}\,\overrightarrow{\partial} + m_0]\psi(\mathbf{x}) + \frac{1}{2}\bar{\psi}(\mathbf{x})[+i\overleftarrow{\gamma}\,\overleftarrow{\partial} + m_0]\psi(\mathbf{x}), \tag{28}$$

$$H_\pi^0(\mathbf{x}) = \frac{1}{2}\left[\pi^2(\mathbf{x}) + \nabla\varphi(\mathbf{x})\nabla\varphi(\mathbf{x}) + \mu_0^2\varphi^2(\mathbf{x})\right], \tag{29}$$

$$V_{ps}^0(\mathbf{x}) = ig_0\bar{\psi}(\mathbf{x})\gamma_5\psi(\mathbf{x})\varphi(\mathbf{x}), \tag{30}$$

where, as usually, $\pi(\mathbf{x})$ denotes the canonical conjugate variable for the pion field. One should note that the second integral in the r.h.s. of Eq. (20) does not contribute to the model boost since operator (26) with $\beta = \mu = 0$ and $\nu = k$ is identically equal zero. Thus we arrive to the relation

$$\mathbf{N}_{SCH} = -\int \mathbf{x}\mathcal{T}_{SCH}^{00}(\mathbf{x})d\mathbf{x} = -\int \mathbf{x}H_{SCH}(\mathbf{x})d\mathbf{x}, \tag{31}$$

that exemplifies the so-called Belinfante ansatz:

$$\mathbf{N} = -\int \mathbf{x}H(\mathbf{x})d\mathbf{x}, \tag{32}$$

which, as it has first been shown in (Belinfante, 1940), holds for any local field model with a symmetrized density $\mathcal{T}^{\mu\nu}(x) = \mathcal{T}^{\nu\mu}(x)$. Such a representation helps (Shebeko & Shirokov, 2001) to get simultaneously a sparse structure for the Hamiltonian and the generators of Lorentz boosts in the CPR [4]. We shall come back to this point later.

Further, the Hamiltonian density can be represented as

$$H_{SCH}(\mathbf{x}) = H_F(\mathbf{x}) + H_I(\mathbf{x}) \tag{33}$$

[4] The relation (32) also has turned out to be useful when formulating a local analog of the Siegert theorem in the covariant description of electromagnetic interactions with nuclei (Shebeko, 1990).

with the free part

$$H_F(\mathbf{x}) = H_\pi(\mathbf{x}) + H_{ferm}(\mathbf{x}) \tag{34}$$

and the interaction density

$$H_I(\mathbf{x}) = V_{ps}(\mathbf{x}) + H_{ren}(\mathbf{x}),\ V_{ps}(\mathbf{x}) = ig\bar{\psi}(\mathbf{x})\gamma_5\psi(\mathbf{x})\varphi(\mathbf{x}), \tag{35}$$

where we have introduced the mass and vertex counterterms:

$$H_{ren}(\mathbf{x}) = M_{ren}^{mes}(\mathbf{x}) + M_{ren}^{ferm}(\mathbf{x}) + H_{ren}^{int}(\mathbf{x}), \tag{36}$$

$$M_{ren}^{mes}(\mathbf{x}) = \frac{1}{2}(\mu_0^2 - \mu_\pi^2)\varphi^2(\mathbf{x}),$$

$$M_{ren}^{ferm}(\mathbf{x}) = (m_0 - m)\bar{\psi}(\mathbf{x})\psi(\mathbf{x})$$

and

$$H_{ren}^{int}(\mathbf{x}) = i(g_0 - g)\bar{\psi}(\mathbf{x})\gamma_5\psi(\mathbf{x})\varphi(\mathbf{x}).$$

One should note that the densities in Eqs. (34)-(35) are obtained from Eqs. (28)-(29) replacing the bare values m_0, μ_0 and g_0, respectively, by the "physical" values m, μ_π and g. Such a transition can be done via the mass-changing Bogoliubov-type transformations (details in (Korda et al., 2007)). In particular, the fields involved can be expressed through the set $\alpha = a^\dagger(a), b^\dagger(b), d^\dagger(d)$ of the creation (destruction) operators for the bare pions and nucleons with the physical masses,

$$\varphi(\mathbf{x}) = (2\pi)^{-3/2}\int(2\omega_\mathbf{k})^{-1/2}[a(\mathbf{k}) + a^\dagger(-\mathbf{k})]exp(i\mathbf{kx})d\mathbf{k}, \tag{37}$$

$$\pi(\mathbf{x}) = -i(2\pi)^{-3/2}\int(\omega_\mathbf{k}/2)^{1/2}[a(\mathbf{k}) - a^\dagger(-\mathbf{k})]exp(i\mathbf{kx})d\mathbf{k}, \tag{38}$$

$$\psi(\mathbf{x}) = (2\pi)^{-3/2}\int(m/E\mathbf{p})^{1/2}\sum_\mu[u(\mathbf{p}\mu)b(\mathbf{p}\mu)$$

$$+ v(-\mathbf{p}\mu)d^\dagger(-\mathbf{p}\mu)]exp(i\mathbf{px})d\mathbf{p}. \tag{39}$$

Substituting (33) into (31), we find

$$\mathbf{N} = \mathbf{N}_F + \mathbf{N}_I$$

with

$$\mathbf{N}_F = \mathbf{N}_{ferm} + \mathbf{N}_\pi = -\int\mathbf{x}H_{ferm}(\mathbf{x})d\mathbf{x} - \int\mathbf{x}H_\pi(\mathbf{x})d\mathbf{x}$$

and

$$\mathbf{N}_I = -\int\mathbf{x}H_I(\mathbf{x})d\mathbf{x}.$$

Now, taking into account the transformation properties of the fermion field $\psi(x)$ and the pion field $\varphi(x)$ with respect to Π, it is readily seen that in the D picture density (33) is a scalar, i.e.,

$$U_F(\Lambda, b)H_{SCH}(x)U_F^{-1}(\Lambda, b) = H_{SCH}(\Lambda x + b), \tag{40}$$

so

$$U_F(\Lambda, b)H_I(x)U_F^{-1}(\Lambda, b) = H_I(\Lambda x + b). \tag{41}$$

It is well known (see, e.g., Sect. 5.1 in (Weinberg, 1995)) that for a large class of theories the property (41) with the corresponding interaction densities $H_I(x)$, being supplemented by the condition

$$[H_I(x'), H_I(x)] = 0 \;\; for \;\; (x'-x)^2 \le 0, \tag{42}$$

plays a crucial role for covariant calculations of the S-matrix.

3.2 An algebraic approach within the Hamiltonian formalism

After these preliminaries, let us consider field models with the decomposition

$$H_I = H_{sc} + H_{nsc} \equiv \int H_{sc}(\mathbf{x})d\mathbf{x} + \int H_{nsc}(\mathbf{x})d\mathbf{x}. \tag{43}$$

It means that only the density in the first integral has the property (41), i.e.,

$$U_F(\Lambda, b)H_{sc}(x)U_F^{-1}(\Lambda, b) = H_{sc}(\Lambda x + b). \tag{44}$$

It is the case, where the pseudoscalar (π and η), vector (ρ and ω) and scalar (δ and σ) meson (boson) fields interact with the $1/2$ spin (N and $\bar{N}$) fermion ones via the Yukawa–type couplings $V = \sum_b V_b = V_s + V_{ps} + V_v$ in

$$H_I = V + \text{mass and vertex counterterms} \tag{45}$$

with

$$V_s = g_s \int d\vec{x}\, \bar{\psi}(\vec{x})\psi(\vec{x})\varphi_s(\vec{x}), \tag{46}$$

$$V_{ps} = ig_{ps} \int d\vec{x}\, \bar{\psi}(\vec{x})\gamma_5\psi(\vec{x})\varphi_{ps}(\vec{x}) \tag{47}$$

and

$$V_v = \int d\vec{x} \left\{ g_v\bar{\psi}(\vec{x})\gamma_\mu\psi(\vec{x})\varphi_v^\mu(\vec{x}) + \frac{f_v}{4m}\bar{\psi}(\vec{x})\sigma_{\mu\nu}\psi(\vec{x})\varphi_v^{\mu\nu}(\vec{x}) \right\}$$
$$+ \int d\vec{x} \left\{ \frac{g_v^2}{2m_v^2}\bar{\psi}(\vec{x})\gamma_0\psi(\vec{x})\bar{\psi}(\vec{x})\gamma_0\psi(\vec{x}) + \frac{f_v^2}{4m^2}\bar{\psi}(\vec{x})\sigma_{0i}\psi(\vec{x})\bar{\psi}(\vec{x})\sigma_{0i}\psi(\vec{x}) \right\}, \tag{48}$$

where $\varphi_v^{\mu\nu}(\vec{x}) = \partial^\mu\varphi_v^\nu(\vec{x}) - \partial^\nu\varphi_v^\mu(\vec{x})$ is the tensor of the vector fields involved (details in (Dubovyk & Shebeko, 2010)).

In the context we would like to remind that in "...theories with derivative couplings or spins $j \ge 1$, it is not enough to take Hamiltonian as the integral over space of a scalar interaction density; we also need to add non-scalar terms to the interaction density to compensate non-covariant terms in the propagators" (quoted from Chapter VII in (Weinberg, 1995)).

Then, taking into account that the first relation (11) is equivalent to the equality

$$[\mathbf{N}_F, H_I] = [H, \mathbf{N}_I], \tag{49}$$

we will evaluate its l.h.s.. In this connection, let us regard the operator

$$H_{sc}(t) = \int H_{sc}(x)d\mathbf{x} \tag{50}$$

and its similarity transformation

$$e^{i\vec{\beta}\mathbf{N}_F} H_{sc}(t) e^{-i\vec{\beta}\mathbf{N}_F} = \int H_{sc}(L(\vec{\beta})x) d\mathbf{x}, \tag{51}$$

where $L(\vec{\beta})$ is any Lorentz boost with the parameters $\vec{\beta} = (\beta^1, \beta^2, \beta^3)$.

From (51) it follows that

$$i e^{i\beta^1 N_F^1} [N_F^1, H_{sc}(t)] e^{-i\beta^1 N_F^1} = \frac{\partial}{\partial \beta^1} \int H_{sc}(L(\beta^1)x) d\mathbf{x}, \tag{52}$$

whence, for instance,

$$\begin{aligned} i[N_F^1, H_{sc}(t)] &= \lim_{\beta^1 \to 0} \frac{\partial}{\partial \beta^1} \int H_{sc}(t - \beta^1 x^1, x^1 - \beta^1 t, x^2, x^3) d\mathbf{x} \\ &= -\int (t \frac{\partial}{\partial x^1} H_{sc}(x) + x^1 \frac{\partial}{\partial t} H_{sc}(x)) d\mathbf{x}, \end{aligned} \tag{53}$$

since for the infinitesimal boost

$$L(\vec{\beta})x = (t - \vec{\beta}\mathbf{x}, \mathbf{x} - \vec{\beta}t).$$

In turn, from (53) we get

$$[N_F^1, H_{sc}] = i \lim_{t \to 0} \int (-it[P^1, H_{sc}(x)] + ix^1[H_F, H_{sc}(x)]) d\mathbf{x}$$

so

$$[\mathbf{N}_F, H_{sc}] = -\int \mathbf{x}[H_F, H_{sc}(\mathbf{x})] d\mathbf{x}. \tag{54}$$

By using Eq. (54) equality (49) can be written as

$$-\int \mathbf{x}[H_F, H_{sc}(\mathbf{x})] d\mathbf{x} = [H_F, \mathbf{N}_I] + [H_I, \mathbf{N}_I] + [H_{nsc}, \mathbf{N}_F]. \tag{55}$$

Evidently, this equation is fulfilled if we put

$$\mathbf{N}_I = \mathbf{N}_B \equiv -\int \mathbf{x} H_{sc}(\mathbf{x}) d\mathbf{x} \tag{56}$$

and

$$[H_{sc}, \mathbf{N}_I] = -\int \mathbf{x} d\mathbf{x} \int d\mathbf{x}' [H_{sc}(\mathbf{x}'), H_{sc}(\mathbf{x})] = [\mathbf{N}_F + \mathbf{N}_I, H_{nsc}] \tag{57}$$

or

$$\begin{aligned} &\int d\mathbf{x} \int d\mathbf{x}' (\mathbf{x}' - \mathbf{x}) [H_{sc}(\mathbf{x}'), H_{sc}(\mathbf{x})] \\ &= \int \mathbf{x} d\mathbf{x} \int d\mathbf{x}' [H_{nsc}(\mathbf{x}'), H_F(\mathbf{x}) + H_{sc}(\mathbf{x})]. \end{aligned} \tag{58}$$

In a model with $H_{nsc} = 0$ the latter reduces to

$$\int e^{-i\mathbf{P}\mathbf{X}} \mathbf{I} e^{i\mathbf{P}\mathbf{X}} d\mathbf{X} = 0, \tag{59}$$

where

$$\mathbf{I} = \frac{1}{2} \int \mathbf{r} d\mathbf{r} [H_{sc}(\frac{1}{2}\mathbf{r}), H_{sc}(-\frac{1}{2}\mathbf{r})]. \tag{60}$$

By running again the way from Eq. (49) to Eqs. (59)-(60) we see that the nonlinear commutation (11)

$$[H, \mathbf{N}] = i\mathbf{P}$$

will take place once along with the Belinfante-type relation (56) the interaction density meets the condition

$$\int \mathbf{r} d\mathbf{r} [H_{sc}(\frac{1}{2}\mathbf{r}), H_{sc}(-\frac{1}{2}\mathbf{r})] = 0. \tag{61}$$

One should note that we have arrived to Eq. (56) being inside the Poincarè algebra itself without addressing the Nöther integrals, these stepping stones of the Lagrangian formalism. In the context, we would like to stress that the condition (61) is weaker compared to the constraint

$$[H_{sc}(\frac{1}{2}\mathbf{r}), H_{sc}(-\frac{1}{2}\mathbf{r})] = 0 \tag{62}$$

imposed for all $\mathbf{r}$ excepting, may be, the point $\mathbf{r} = 0$. But we recall it as a special case of the microcausality requirement that is realized in local field models. Beyond such models, as shown in Appendix B of (Shebeko & Frolov , 2011), Eqs. (56) and (49) may be incompatible. It makes us seek an alternative to assumption (56) in our attempts to meet Eq. (55).

At this point, we put $\mathbf{N}_I = \mathbf{N}_B + \mathbf{D}$ to get the relationship

$$[H_F, \mathbf{D}] = [\mathbf{N}_B + \mathbf{D}, H_{sc}] + [\mathbf{N}_F + \mathbf{N}_B + \mathbf{D}, H_{nsc}], \tag{63}$$

that replaces the commutator $[H, \mathbf{N}] = i\mathbf{P}$ and determines the displacement $\mathbf{D}$.

Further, assuming that the scalar density $H_{sc}(\mathbf{x})$ is of the first order in coupling constants involved and putting

$$H_{nsc}(\mathbf{x}) = \sum_{p=2}^{\infty} H_{nsc}^{(p)}(\mathbf{x}), \tag{64}$$

we will search the operator $\mathbf{D}$ in the form

$$\mathbf{D} = \sum_{p=2}^{\infty} \mathbf{D}^{(p)}, \tag{65}$$

i.e., as a perturbation expansion in powers of the interaction H_{sc}. Here the label (p) denotes the pth order in these constants. By the way, one should keep in mind that the terms in the r.h.s. of Eq. (64) are usually associated with perturbation series for mass and vertex counterterms. Evidently, their incorporation may affect the corresponding higher-order contributions with $p \geq 2$ to the boost. Therefore, to comprise different situations of practical interest let us consider field models in which $H_{nsc}(\mathbf{x}) = V_{nsc}(\mathbf{x}) + V_{ren}(\mathbf{x})$ with a nonscalar interaction $V_{nsc} = \int V_{nsc}(\mathbf{x}) d\mathbf{x}$ and some "renormalization" contribution $V_{ren} = \int V_{ren}(\mathbf{x}) d\mathbf{x}$. The latter may be scalar or not. Of course, such a division of $H_{nsc}(\mathbf{x})$ can be done at the beginning in Eq. (43). But the scheme presented here seems to us more flexible.

By substituting the expansions (64) and (65) into Eq. (63) we get the chain of relations

$$[H_F, \mathbf{D}^{(2)}] = [\mathbf{N}_F, H_{nsc}^{(2)}] + [\mathbf{N}_B, H_{sc}], \quad (66)$$

$$[H_F, \mathbf{D}^{(3)}] = [\mathbf{N}_F, H_{nsc}^{(3)}] + [\mathbf{D}^{(2)}, H_{sc}] + [\mathbf{N}_B, H_{nsc}^{(2)}], \quad (67)$$

$$[H_F, \mathbf{D}^{(p)}] = [\mathbf{N}_F, H_{nsc}^{(p)}] + [\mathbf{N}_B, H_{nsc}^{(p-1)}] + [\mathbf{D}^{(p-1)}, H_{sc}] + [\mathbf{D}, H_{nsc}]^{(p)}, \quad (68)$$
$$(p = 4, 5, \ldots)$$

for a recursive finding of the operators $\mathbf{D}^{(p)}$ $(p = 2, 3, ...)$.

Further, after such substitutions into the commutators

$$[P_k, N_j] = i\delta_{kj}H, \quad [J_k, N_j] = i\varepsilon_{kjl}N_l, \quad [N_k, N_j] = -i\varepsilon_{kjl}J_l$$

we deduce, respectively, the following relations:

$$[P_k, D_j^{(p)}] = i\delta_{kj}H_{nsc}^{(p)} \quad (p = 2, 3, ...) \quad (69)$$

from

$$[P_k, D_j] = i\delta_{kj}H_{nsc}, \quad (70)$$

$$[J_k, D_j^{(p)}] = i\varepsilon_{kjl}D_l^{(p)} \quad (71)$$

from

$$[J_k, D_j] = i\varepsilon_{kjl}D_l \quad (72)$$

and

$$[N_{Fk}, N_{Bj}] + [N_{Bk}, N_{Fj}] = 0, \quad (73)$$

The remaining Poincaré commutations are fulfilled once one deals with any rotationally and translationally invariant theory.

Now, keeping in mind an elegant method by Chandler (Chandler, 2003), we invoke on the property (see (Friedrichs, 1953)) of a formal solution Y of the equation

$$[H_F, Y] = X \quad (74)$$

to be any linear functional $F(X)$ of a given operator $X \neq 0$. In other words, it means that

$$[H_F, F(X)] = X \quad (75)$$

with $F(\lambda_1 X_1 + \lambda_2 X_2) = \lambda_1 F(X_1) + \lambda_2 F(X_2)$, where λ_1 and λ_2 are arbitrary c-numbers. In addition, one can see that

$$[H_F, F(X)] = F([H_F, X]). \quad (76)$$

Moreover, it turns out that

$$[\mathbf{P}, F(X)] = F([\mathbf{P}, X]), \quad (77)$$

$$[\mathbf{J}, F(X)] = F([\mathbf{J}, X]), \quad (78)$$

$$[\mathbf{N}_F, F(X)] = F([\mathbf{N}_F, X]) + iF(F([\mathbf{P}, X])). \quad (79)$$

In order to prove the relations let us employ the Jacobi identity

$$[A,[B,C]]+[C,[A,B]]+[B,[C,A]]=0 \tag{80}$$

and write

$$[\mathcal{O},[H_F,F(X)]]=-[F(X),[\mathcal{O},H_F]]+[H_F,[\mathcal{O},F(X)]]$$

with some operator $\mathcal{O}$. Then

$$[\mathcal{O},F(X)]=F([\mathcal{O},X])+F([F(X),[\mathcal{O},H_F]]). \tag{81}$$

Of course, to be more constructive one needs to have a definite realization of the functional $F(X)$. In this connection, we will use the representation

$$Y=-i\lim_{\eta\to 0+}\int_0^\infty X(t)e^{-\eta t}dt \tag{82}$$

of the operator Y that enters the equation (74). The existence proof for such a solution is sufficiently delicate (see discussion in Appendix A of Ref. (Shebeko & Shirokov, 2001)).

3.3 Application to a nonlocal field model

We will show how the method proposed works in combination with introducing certain cutoff (vertex) functions that makes an initial local model be nonlocal. In spite of our consideration may be extended to more realistic models its main idea becomes transparent for a simple system of "scalar nucleons" (more precisely, charged spinless bosons) and neutral scalar bosons with the interaction density $H_I(\mathbf{x}) = V_{loc}(\mathbf{x}) + V_{ren}(\mathbf{x})$ (cf. (Glöckle & Müller, 1981; Shirokov, 2002)):

$$V_{loc}(\mathbf{x}) = g\varphi_s(\mathbf{x}) : \psi_b^\dagger(\mathbf{x})\psi_b(\mathbf{x}) : \tag{83}$$

and

$$V_{ren}(\mathbf{x}) = \delta\mu_s : \varphi_s^2(\mathbf{x}) : +\delta\mu_b : \psi_b^\dagger(\mathbf{x})\psi_b(\mathbf{x}) : \tag{84}$$

with the mass shifts $\delta\mu_s = \frac{1}{2}(\mu_{0s}^2-\mu_s^2)$, $\delta\mu_b = \frac{1}{2}(\mu_{0b}^2-\mu_b^2)$. In order to regard a nonlocal extension of this local model let us substitute the expansions

$$\varphi_s(\mathbf{x}) = [2(2\pi)^3]^{-1/2}\int\frac{d\mathbf{k}}{\omega_\mathbf{k}}[a(k)+a^\dagger(k_-)]e^{i\mathbf{kx}},\quad \psi_b(\mathbf{x}) = [2(2\pi)^3]^{-1/2}\int\frac{d\mathbf{p}}{E_\mathbf{p}}[b(p)+d^\dagger(p_-)]e^{i\mathbf{px}}$$

into Eqs. (83) and (84) to get

$$V_{loc}(\mathbf{x}) = \frac{g}{2[2(2\pi)^3]^{1/2}}\int\frac{d\mathbf{p}'}{E_{\mathbf{p}'}}\int\frac{d\mathbf{p}}{E_\mathbf{p}}\int\frac{d\mathbf{k}}{\omega_\mathbf{k}}e^{-i\mathbf{p}'\mathbf{x}+i\mathbf{px}+i\mathbf{kx}}$$

$$\times\, a(k) : [b^\dagger(p')b(p)+b^\dagger(p')d^\dagger(p_-)+d(p'_-)b(p)+d(p'_-)d^\dagger(p_-)] : +H.c. \tag{85}$$

and $V_{ren}(\mathbf{x}) = \delta\mu_s(\mathbf{x}) + \delta\mu_b(\mathbf{x})$ with

$$\delta\mu_s(\mathbf{x}) = \frac{\delta\mu_s}{2(2\pi)^3}\int\frac{d\mathbf{k}'}{\omega_{\mathbf{k}'}}\int\frac{d\mathbf{k}}{\omega_\mathbf{k}} : [a(k')+a^\dagger(k'_-)]e^{i\mathbf{k}'\mathbf{x}+i\mathbf{kx}}[a(k)+a^\dagger(k_-)] :, \tag{86}$$

$$\delta\mu_b(\mathbf{x}) = \frac{\delta\mu_b}{2(2\pi)^3}\int\frac{d\mathbf{p}'}{E_{\mathbf{p}'}}\int\frac{d\mathbf{p}}{E_{\mathbf{p}}} : [b^\dagger(p') + d(p'_-)]e^{-i\mathbf{p}'\mathbf{x}+i\mathbf{p}\mathbf{x}}[b(p) + d^\dagger(p_-)] : . \tag{87}$$

It is implied that the operators $a(a^\dagger)$, $b(b^\dagger)$ and $d(d^\dagger)$ meet the commutation relations

$$[a(k), a^\dagger(k')] = k_0\delta(\mathbf{k} - \mathbf{k}'), \tag{88}$$

$$[b(p), b^\dagger(p')] = [d(p), d^\dagger(p')] = p_0\delta(\mathbf{p} - \mathbf{p}') \tag{89}$$

with all the remaining ones being zero.

The interaction operator itself $H_I = \int H_I(\mathbf{x})d\mathbf{x} = V_{loc} + V_{ren}$ with

$$V_{loc} = \int V_{nloc}(\mathbf{x})d\mathbf{x} = \frac{g}{2[2(2\pi)^3]^{1/2}}\int\frac{d\mathbf{p}'}{E_{\mathbf{p}'}}\int\frac{d\mathbf{p}}{E_{\mathbf{p}}}\int\frac{d\mathbf{k}}{\omega_{\mathbf{k}}}\delta(\mathbf{p}' - \mathbf{p} - \mathbf{k})$$

$$\times\, a(k) : [b^\dagger(p')b(p) + b^\dagger(p')d^\dagger(p_-) + d(p'_-)b(p) + d(p'_-)d^\dagger(p_-)] : +H.c., \tag{90}$$

$$V_{ren} = \int[\delta\mu_s(\mathbf{x}) + \delta\mu_b(\mathbf{x})]d\mathbf{x}. \tag{91}$$

Let us consider its nonlocal extension

$$H_I = V_{nloc} + M_s + M_b, \tag{92}$$

where in accordance with the representation (3) we introduce the following normally-ordered structures:

$$V_{nloc} = \int V_{nloc}(\mathbf{x})d\mathbf{x} = \int\frac{d\mathbf{p}'}{E_{\mathbf{p}'}}\int\frac{d\mathbf{p}}{E_{\mathbf{p}}}\int\frac{d\mathbf{k}}{\omega_{\mathbf{k}}}$$

$$\times\{\delta(\mathbf{p}' - \mathbf{p} - \mathbf{k})g_{11}(p', p, k)b^\dagger(p')b(p) + \delta(\mathbf{p}' + \mathbf{p} - \mathbf{k})g_{12}(p', p, k)b^\dagger(p')d^\dagger(p)$$

$$+\delta(\mathbf{p}' + \mathbf{p} + \mathbf{k})g_{21}(p', p, k)d(p')b(p)$$

$$+\,\delta(\mathbf{p}' - \mathbf{p} - \mathbf{k})g_{22}(p', p, k)d^\dagger(p')d(p)\}a(k) + H.c. \tag{93}$$

Furthermore, the creation/destruction operators have the transformation properties like (14). For example,

$$U_F(\Lambda)a(k)U_F^{-1}(\Lambda) = a(\Lambda k). \tag{94}$$

Therefore, in the Dirac picture

$$U_F(\Lambda)V_{loc}(x)U_F^{-1}(\Lambda) = V_{loc}(\Lambda x), \tag{95}$$

i.e., the interaction density $V_{loc}(x)$ is a Lorentz scalar.

For our nonlocal model we will retain the property assuming that

$$U_F(\Lambda)V_{nloc}(x)U_F^{-1}(\Lambda) = V_{nloc}(\Lambda x). \tag{96}$$

It is readily seen that this relation holds if the coefficients $g_{\varepsilon'\varepsilon}$ meet the condition

$$g_{\varepsilon'\varepsilon}(\Lambda p', \Lambda p, \Lambda k) = g_{\varepsilon'\varepsilon}(p', p, k). \tag{97}$$

On the mass shells with $p'^2 = p^2 = \mu_b^2$ and $k^2 = \mu_s^2$ the latter means that the functions $g_{\varepsilon'\varepsilon}(p',p,k)$ can depend only upon the invariants $p'p$, $p'k$ and pk.

The transition from V_{loc} to V_{nloc} can be interpreted as an endeavor to regularize the theory. In the context, the introduction of some cutoff functions $g_{\varepsilon'\varepsilon}$ in momentum space is aimed at removing ultraviolet divergences typical of local field models with interactions like expression (83).

An associated exploration carried out in (Shebeko & Frolov , 2011) with covariant cutoffs

$$g_{\varepsilon'\varepsilon}(p',p,k) = v_{\varepsilon'\varepsilon}([k + (-1)^{\varepsilon'} p' - (-1)^{\varepsilon} p][k - (-1)^{\varepsilon'} p' + (-1)^{\varepsilon} p]) \tag{98}$$

has allowed us to evaluate the lowest-order correction $\mathbf{D}^{(2)}$ to the Belinfante operator and get the leading-order analytic expressions for the coefficients in the "mass renormalization" terms :

$$M_s = \int \frac{d\mathbf{k}}{\omega_{\mathbf{k}}^2} \{m_1(k)a^\dagger(k)a(k) + m_2(k)[a^\dagger(k)a^\dagger(k_-) + a(k)a(k_-)]\}, \tag{99}$$

$$M_b = \int \frac{d\mathbf{p}}{E_{\mathbf{p}}^2} \{m_{11}(p)b^\dagger(p)b(p) + m_{12}(p)b^\dagger(p)d^\dagger(p_-) + m_{21}(p)b(p)d(p_-) + m_{22}(p)d^\dagger(p)d(p)\}. \tag{100}$$

4. The method of unitary clothing transformations in action

As shown in (Shebeko & Shirokov, 2001), the Belinfante ansatz turns out to be useful when constructing the Lorentz boosts in the CPR, viz., the generator $\mathbf{N} \equiv \mathbf{N}(\alpha)$, being a function of the primary operators $\{\alpha\}$ (such as $a^\dagger(a)$, $b^\dagger(b)$ and $d^\dagger(d)$ for the examples regarded above) in the BPR, is expressed through the corresponding operators $\{\alpha_c\}$ for particle creation and annihilation in the CPR. The transition $\{\alpha\} \Longrightarrow \{\alpha_c\}$ is implemented via the special unitary transformations $W(\alpha) = W(\alpha_c)$, viz.,

$$\alpha = W(\alpha_c)\alpha_c W^\dagger(\alpha_c). \tag{101}$$

These transformations satisfy certain physical requirements:

i) The physical vacuum (the H lowest eigenstate) must coincide with a new no–particle state Ω, i.e., the state that obeys the equations

$$a_c(\vec{k})\,|\Omega\rangle = b_c(\vec{p},\mu)\,|\Omega\rangle = d_c(\vec{p},\mu)\,|\Omega\rangle = 0, \;\; \forall\, \vec{k},\, \vec{p},\, \mu \tag{102}$$

$$\langle\Omega|\Omega\rangle = 1.$$

ii) New one-particle states $|\vec{k}\rangle_c \equiv a_c^\dagger(\vec{k})\Omega$ etc. are the H eigenvectors as well.

$$K(\alpha_c)|\vec{k}\rangle_c = K_F(\alpha_c)|\vec{k}\rangle_c = \omega_k|\vec{k}\rangle_c \tag{103}$$

$$K_I(\alpha_c)|\vec{k}\rangle_c = 0 \tag{104}$$

iii) The spectrum of indices that enumerate the new operators must be the same as that for the bare ones .

iv) The new operators α_c satisfy the same commutation rules as do their bare counterparts α that is provided via the link (101) with a unitary operator W to be obtained as in (Shebeko & Shirokov, 2001).

4.1 The Hamiltonian and other generators of the Poincaré group in the clothed-particle representation

A key point of the clothing procedure exposed in (Shebeko & Shirokov, 2001) is to remove the so-called bad terms from the Hamiltonian

$$H \equiv H(\alpha) = H_F(\alpha) + H_I(\alpha) = W(\alpha_c)H(\alpha_c)W^{\dagger}(\alpha_c) \equiv K(\alpha_c), \tag{105}$$

more exactly, from a primary interaction $V(\alpha)$ in $H_I(\alpha) = V(\alpha) + V_{ren}(\alpha)$. For example, these terms $b_c^{\dagger}b_c a_c^{\dagger}$, $b_c^{\dagger}d_c^{\dagger}a_c$, $b_c^{\dagger}d_c^{\dagger}a_c^{\dagger}$, $d_c d_c^{\dagger}a_c^{\dagger}$ enter $V(\alpha_c)$ determined by Eq. (90) after the replacement of the bare operators in it by the clothed ones. These terms should be removed together with their Hermitian conjugate counterterms! to retain the hermiticity of the similarity transformation (105). In general, such terms prevent the physical vacuum $|\Omega\rangle$ (the H lowest eigenstate) and the one-clothed-particle states $|n\rangle_c = a_c^{\dagger}(n)|\Omega\rangle$ to be the H eigenvectors for all n included. Here creation operators $a_c^{\dagger}(n)$ are clothed counterparts of those operators $a^{\dagger}(n)$ that are contained in expansion (2). The bad terms occur every time when any normally ordered product

$$a^{\dagger}(1')a^{\dagger}(2')...a^{\dagger}(n'_C)a(n_A)...a(2)a(1)$$

of the class [C.A] embodies, at least, one substructure which belongs to one of the classes $[k.0]$ $(k = 1, 2, ...)$ and $[k.1]$ $(k = 0, 1, ...)$.

Strictly speaking such a departure point should be specified and sometimes modified. Indeed, by trying to meet the requirements i) and ii) we, at first sight, leave out of consideration such undesirable terms in $V_{ren}(\alpha)$. Nevertheless, it is not accidental since the renormalization contribution is canceled in the course of the procedure itself that is some attractive feature of the UCT method as a whole (see below). In addition, it has turned out (Dubovyk & Shebeko, 2010) that the nonscalar contribution (the second integral in the r.h.s. of Eq. (48)) to the operator $V_V(\alpha)$ is canceled too when eliminating bad terms only from its scalar part (in fact, the first integral in the r.h.s. of Eq. (48)). Keeping this in mind, when handling the division

$$H_I(\alpha) = \int H_I(\mathbf{x})d\mathbf{x} = H_{sc}(\alpha) + H_{nsc}(\alpha), \tag{106}$$

we assume $H_{sc}(\alpha) = V_{bad}(\alpha) + V_{good}(\alpha)$ to remove the bad part V_{bad} from the similarity transformation

$$K(\alpha_c) = W(\alpha_c)[H_F(\alpha_c) + H_I(\alpha_c)]W^{\dagger}(\alpha_c)$$
$$= W(\alpha_c)[H_F(\alpha_c) + V_{bad}(\alpha_c) + V_{good}(\alpha_c) + H_{nsc}(\alpha_c)]W^{\dagger}(\alpha_c). \tag{107}$$

Remind that term "good", as an antithesis of "bad", is applied here to those operators (e.g., of the class [k.2] with $k \geq 2$) which destroy both the no-clothed-particle state Ω and the one-clothed-particle states.

For unitary transformation (UCT) $W = \exp R$ with $R = -R^{\dagger}$ it is implied that we will eliminate the bad terms V_{bad} in the r.h.s. of

$$K(\alpha_c) = H_F(\alpha_c) + V_{bad}(\alpha_c) + [R, H_F] + [R, V_{bad}] + \frac{1}{2}[R,[R,H_F]]$$

$$+ \frac{1}{2}[R,[R,V_{bad}]] + ... + e^R V_{good} e^{-R} + e^R H_{nsc} e^{-R} \tag{108}$$

(cf. Eq. (2.19) in (Shebeko & Shirokov, 2001)) by requiring that

$$[H_F, R] = V_{bad} \tag{109}$$

for the operator R of interest.

One should note that unlike the original clothing procedure exposed in (Shebeko & Shirokov, 2001), (Korda et al., 2007) we eliminate here the bad terms only from H_{sc} interaction in spite of such terms can appear in the nonscalar interaction as well. This preference is relied upon the previous experience (Dubovyk & Shebeko, 2010) when applying the method of UCTs in the theory of nucleon-nucleon scattering. Now we get the division

$$H = K(\alpha_c) = K_F + K_I \tag{110}$$

with a new free part $K_F = H_F(\alpha_c) \sim a_c^{\dagger} a_c$ and interaction

$$K_I = V_{good}(\alpha_c) + H_{nsc}(\alpha_c) + [R, V_{good}]$$

$$+ \frac{1}{2}[R, V_{bad}] + [R, H_{nsc}] + \frac{1}{3}[R,[R,V_{bad}]] + ..., \tag{111}$$

where the r.h.s. involves along with good terms other bad terms to be removed via subsequent UCTs described in Subsec. 2.4 of (Shebeko & Shirokov, 2001) and Sec. 3 of (Korda et al., 2007).

In parallel, we have

$$\mathbf{N} \equiv \mathbf{N}(\alpha) = \mathbf{N}_F(\alpha) + \mathbf{N}_I(\alpha) = W(\alpha_c)\mathbf{N}(\alpha_c)W^{\dagger}(\alpha_c) \equiv \mathbf{B}(\alpha_c) \tag{112}$$

or

$$\mathbf{B}(\alpha_c) = \mathbf{N}_F(\alpha_c) + \mathbf{N}_I(\alpha_c) + [R, \mathbf{N}_F] + [R, \mathbf{N}_I] + ..., \tag{113}$$

where accordingly the division

$$\mathbf{N}_I = \mathbf{N}_B + \mathbf{D}, \tag{114}$$

$$\mathbf{N}_B = -\int \mathbf{x} H_{sc}(\mathbf{x}) d\mathbf{x} = \mathbf{N}_{bad} + \mathbf{N}_{good},$$

Eq. (113) can be rewritten as

$$\mathbf{B}(\alpha_c) = \mathbf{N}_F(\alpha_c) + \mathbf{N}_{bad}(\alpha_c) + [R, \mathbf{N}_F] + [R, \mathbf{N}_{bad}] + \frac{1}{2}[R,[R,\mathbf{N}_F]]$$

$$+ \frac{1}{2}[R,[R,\mathbf{N}_{bad}]] + ... + e^R \mathbf{N}_{good} e^{-R} + e^R \mathbf{D} e^{-R}. \tag{115}$$

But it turns out (see the proof of Eq. (3.26) in (Shebeko & Shirokov, 2001)) that if R meets the condition (109), then

$$[\mathbf{N}_F, R] = \mathbf{N}_{bad} = -\int \mathbf{x} V_{bad}(\mathbf{x}) d\mathbf{x} \tag{116}$$

so the boost generators in the CPR can be written likely Eq. (110),

$$\mathbf{N} = \mathbf{B}(\alpha_c) = \mathbf{B}_F + \mathbf{B}_I, \tag{117}$$

where $\mathbf{B}_F = \mathbf{N}_F(\alpha_c)$ is the boost operator for noninteracting clothed particles while $\mathbf{B}_I$ includes the contributions induced by interactions between them

$$\mathbf{B}_I = \mathbf{N}_{good}(\alpha_c) + \mathbf{D}(\alpha_c) + [R, \mathbf{N}_{good}]$$

$$+ \frac{1}{2}[R, \mathbf{N}_{bad}] + [R, \mathbf{D}] + \frac{1}{3}[R, [R, \mathbf{N}_{bad}]] + ... \tag{118}$$

One should note that in formulae (111) and (118) we are focused upon the R-commutations with the first-eliminated interaction V_{bad}. As shown in (Shebeko & Shirokov, 2001), the brackets, on the one hand, yield new interactions responsible for different physical processes and, on the other hand, cancel (as a recipe!) the mass and other counterterms that stem from $H_{nsc}(\alpha_c)$ and $\mathbf{D}(\alpha_c)$.

But at this place we will come back to our model with $V_{bad} = V_{nloc}$, $V_{good} = 0$ and $R = R_{nloc}$ to calculate the simplest commutator $[R_{nloc}, V_{nloc}]$ in which accordingly condition (109) the clothing operator R_{nloc} is determined by

$$[H_F, R_{nloc}] = V_{nloc}. \tag{119}$$

From the equation it follows (cf. Appendix A in (Shebeko & Shirokov, 2001)) that its solution can be given by

$$R_{nloc} = \int \frac{d\mathbf{k}}{\omega_{\mathbf{k}}} : F_b^{\dagger} R(k) F_b : a(k) - H.c. = \mathcal{R}_{nloc} - \mathcal{R}_{nloc}^{\dagger}. \tag{120}$$

with the row $F_b^{\dagger} = \left[b(p), d^{\dagger}(p)\right]$ and the column F_b (cf. Eq.(A.8) in (Shebeko & Shirokov, 2001)). The matrix $R(k)$ is composed of the elements

$$R_{\varepsilon'\varepsilon}(p', p, k) = -\frac{\tilde{g}_{\varepsilon'\varepsilon}(p', p, k)}{\omega_{\mathbf{k}} + (-1)^{\varepsilon'} E_{\mathbf{p}'} - (-1)^{\varepsilon} E_{\mathbf{p}}} \delta(\mathbf{k} + (-1)^{\varepsilon'} \mathbf{p}' - (-1)^{\varepsilon} \mathbf{p}). \tag{121}$$

$$(\varepsilon', \varepsilon = 1, 2)$$

Such a solution is valid if $\mu_s < 2\mu_b$. In other words, under such an inequality the operator R_{nloc} has the same structure as V_{nloc} itself. Then, all we need is to evaluate the commutator $[R_{nloc}, V_{nloc}]$.

For example, our calculations result in the boson-boson interaction operator

$$\frac{1}{2}[R_{nloc}, V_{nloc}](bb \to bb) = -\frac{1}{4} \int \frac{d\mathbf{p}_2'}{E_{\mathbf{p}_2'}} \int \frac{d\mathbf{p}_2}{E_{\mathbf{p}_2}} \int \frac{d\mathbf{p}_1'}{E_{\mathbf{p}_1'}} \int \frac{d\mathbf{p}_1}{E_{\mathbf{p}_1}} \delta(\mathbf{p}_1' + \mathbf{p}_2' - \mathbf{p}_1 - \mathbf{p}_2)$$

$$\times g_{11}(p_1', p_1, k) g_{11}(p_2', p_2, k)$$

$$\times \left\{ \frac{1}{(p_1 - p_1')^2 - \mu_s^2} + \frac{1}{(p_2 - p_2')^2 - \mu_s^2} \right\} b_c^\dagger(p_2') b_c^\dagger(p_1') b_c(p_2) b_c(p_1) \qquad (122)$$

with $\mathbf{k} = \mathbf{p}_1' - \mathbf{p}_1$ and the respective contribution to $\mathbf{B}_I$,

$$\frac{1}{2}[R_{nloc}, \mathbf{N}_B](bb \to bb)$$

$$= \frac{i}{4} \int \frac{d\mathbf{p}_2'}{E_{\mathbf{p}_2'}} \int \frac{d\mathbf{p}_2}{E_{\mathbf{p}_2}} \int \frac{d\mathbf{p}_1'}{E_{\mathbf{p}_1'}} \int \frac{d\mathbf{p}_1}{E_{\mathbf{p}_1}} \frac{\partial}{\partial \mathbf{p}_1'} \delta(\mathbf{p}_1' + \mathbf{p}_2' - \mathbf{p}_1 - \mathbf{p}_2)$$

$$\times g_{11}(p_1', p_1, k) g_{11}(p_2', p_2, k)$$

$$\times \left\{ \frac{1}{(p_1 - p_1')^2 - \mu_s^2} + \frac{1}{(p_2 - p_2')^2 - \mu_s^2} \right\} b_c^\dagger(p_2') b_c^\dagger(p_1') b_c(p_2) b_c(p_1) \qquad (123)$$

In Eqs. (122) and (123) we encounter a covariant (Feynman-like) "propagator"

$$\frac{1}{2} \left\{ \frac{1}{(p_1 - p_1')^2 - \mu_s^2} + \frac{1}{(p_2 - p_2')^2 - \mu_s^2} \right\}, \qquad (124)$$

which on the energy shell

$$E_{\mathbf{p}_1} + E_{\mathbf{p}_1} = E_{\mathbf{p}_1'} + E_{\mathbf{p}_2'} \qquad (125)$$

is converted into the genuine Feynman propagator for the corresponding S matrix (cf. the first results in (Shebeko & Shirokov, 2001)).

4.2 Relativistic interactions between clothed particles in meson-nucleon systems

Following the same scenario one can derive analytical expressions for separate contributions to the operator

$$K_I \sim a_c^\dagger b_c^\dagger a_c b_c(\pi N \to \pi N) + b_c^\dagger b_c^\dagger b_c b_c(NN \to NN) + d_c^\dagger d_c^\dagger d_c d_c(\bar{N}\bar{N} \to \bar{N}\bar{N})$$
$$+ b_c^\dagger b_c^\dagger b_c^\dagger b_c b_c b_c(NNN \to NNN) + \ldots + [a_c^\dagger a_c^\dagger b_c d_c + H.c.](N\bar{N} \leftrightarrow 2\pi) + \ldots$$
$$+ [a_c^\dagger b_c^\dagger b_c^\dagger b_c b_c + H.c.](NN \leftrightarrow \pi NN) + \ldots \qquad (126)$$

and, in particular, the operator

$$K_I^{(2)} = K(NN \to NN) + K(\bar{N}\bar{N} \to \bar{N}\bar{N}) + K(N\bar{N} \to N\bar{N}) + K(bN \to bN) + K(b\bar{N} \to b\bar{N})$$
$$+ K(bb' \to N\bar{N}) + K(N\bar{N} \to bb') \qquad (127)$$

if one starts with the interactions by Eqs. (50)-(52). It has been done in (Shebeko & Shirokov, 2001) and (Korda et al., 2007) so many technical details of those derivations can be found

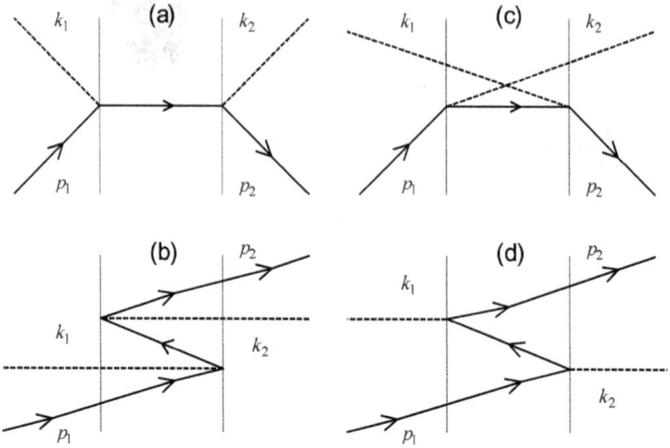

Fig. 1. Different contributions to the πN quasipotential.

therein. Note only that the normal ordering is an essential element of them. We will confine ourselves to the two examples that originate from the corresponding commutator $\frac{1}{2}[R, V(bad)]$.

4.2.1 Pion-nucleon interaction operator

As a result, we show the contribution

$$K(\pi N \to \pi N) = \int d\vec{p}_1 d\vec{p}_2 d\vec{k}_1 d\vec{k}_2 \, V_{\pi N}(\vec{k}_2, \vec{p}_2; \vec{k}_1, \vec{p}_1) a_c^\dagger(\vec{k}_2) b_c^\dagger(\vec{p}_2) a_c(\vec{k}_1) b_c(\vec{p}_1), \tag{128}$$

with the following covariant (Feynman-like) form:

$$V_{\pi N}(\vec{k}_2, \vec{p}_2; \vec{k}_1, \vec{p}_1) = \frac{g^2}{2(2\pi)^3} \frac{m}{\sqrt{\omega_{\vec{k}_1} \omega_{\vec{k}_2} E_{\vec{p}_1} E_{\vec{p}_2}}} \delta(\vec{p}_1 + \vec{k}_1 - \vec{p}_2 - \vec{k}_2)$$

$$\bar{u}(\vec{p}_2) \left\{ \frac{1}{2} \left[\frac{1}{\hat{p}_1 + \hat{k}_1 + m} + \frac{1}{\hat{p}_2 + \hat{k}_2 + m} \right] \right.$$

$$\left. + \frac{1}{2} \left[\frac{1}{\hat{p}_1 - \hat{k}_2 + m} + \frac{1}{\hat{p}_2 - \hat{k}_1 + m} \right] \right\} u(\vec{p}_1)$$

For brevity, the spin and isospin indices have been omitted.

The corresponding πN quasipotential in momentum space is determined by

$$\tilde{V}_{\pi N}(\vec{k}_2, \vec{p}_2; \vec{k}_1, \vec{p}_1) = \left\langle a_c^\dagger(\vec{k}_2) b_c^\dagger(\vec{p}_2) \Omega | K(\pi N \to \pi N) | a_c^\dagger(\vec{k}_1) b_c^\dagger(\vec{p}_1) \Omega \right\rangle \tag{129}$$

Graphs in Fig. 1 are topologically equivalent to the well-known time-ordered Feynman diagrams. However, in Schrödinger picture used here, where all events are related to one and the same instant $t = 0$, such an analogy could be misleading: line directions in Fig. 1 are given with the sole scope to discriminate between nucleon and antinucleon states. Moreover, the energy conservation is not assumed in constructing this and other quasipotentials. Indeed, the coefficients in front of $a_c^\dagger b_c^\dagger a_c b_c$ generally do not fulfill the on-energy-shell condition

$$E_{\vec{p}_1} + \omega_{\vec{k}_1} = E_{\vec{p}_2} + \omega_{\vec{k}_2},$$

In this connection, the "left" four-vector s_1 is not necessarily equal to the "right" Mandelstam vector $s_2 = p_2 + k_2$.

4.2.2 Nucleon-nucleon interaction operator

Accordingly (Dubovyk & Shebeko, 2010) the operator of interest has the following structure:

$$K(NN \to NN) = \sum_b K_b(NN \to NN),$$

$$K_b(NN \quad \to \quad NN) \quad = \quad \sum\!\!\!\!\!\!\int d\vec{p}_1'\, d\vec{p}_2'\, d\vec{p}_1\, d\vec{p}_2 V_b(1',2';1,2) b_c^\dagger(1') b_c^\dagger(2') b_c(1) b_c(2), \quad (130)$$

where, for example, the c–number matrix V_b in the second order in the PS coupling is given by

$$V_{ps}(1',2';1,2) = \frac{1}{(2\pi)^3} \frac{m^2}{\sqrt{E_{\vec{p}_1'} E_{\vec{p}_2'} E_{\vec{p}_1} E_{\vec{p}_2}}} \delta\left(\vec{p}_1' + \vec{p}_2' - \vec{p}_1 - \vec{p}_2\right) v_{ps}(1',2';1,2), \quad (131)$$

$$v_{ps}(1',2';1,2) = \frac{g_{ps}^2}{2} \bar{u}(\vec{p}_1')\gamma_5 u(\vec{p}_1) \frac{1}{(p_1 - p_1')^2 - m_{ps}^2} \bar{u}(\vec{p}_2')\gamma_5 u(\vec{p}_2), \quad (132)$$

omitting again the discrete quantum numbers. Here m_{ps} the mass of the clothed pion (its physical value).

Corresponding relativistic and properly symmetrized NN quasipotential

$$\tilde{V}_{NN}(\vec{p}_1', \vec{p}_2'; \vec{p}_1, \vec{p}_2) = \left\langle b_c^\dagger(\vec{p}_1') b_c^\dagger(\vec{p}_2')\Omega \mid K_{NN} \mid b_c^\dagger(\vec{p}_1) b_c^\dagger(\vec{p}_2)\Omega \right\rangle$$

can be written as

$$\tilde{V}_{NN}(\vec{p}_1', \vec{p}_2'; \vec{p}_1, \vec{p}_2) = \frac{1}{2} \frac{g_{ps}^2}{(2\pi)^3} \frac{m^2}{2\sqrt{E_{\vec{p}_1} E_{\vec{p}_2} E_{\vec{p}_1'} E_{\vec{p}_2'}}} \delta(\vec{p}_1' + \vec{p}_2' - \vec{p}_1 - \vec{p}_2)$$

$$\times\, \bar{u}(\vec{p}_1')\gamma_5 u(\vec{p}_1) \frac{1}{2} \left\{ \frac{1}{(p_1 - p_1')^2 - m_{ps}^2} \right.$$

$$\left. + \frac{1}{(p_2 - p_2')^2 - m_{ps}^2} \right\} \bar{u}(\vec{p}_2')\gamma_5 u(\vec{p}_2) - (1 \leftrightarrow 2). \quad (*)$$

Meson		Bonn B	UCT
π	$g_\pi^2/4\pi$	14.4	13.395
	Λ_π	1700	2500
	m_π	138.03	138.03
η	$g_\eta^2/4\pi$	3	5.0
	Λ_η	1500	1219
	m_η	548.8	548.8
ρ	$g_\rho^2/4\pi$	0.9	1.2
	Λ_ρ	1850	1593.0
	f_ρ/g_ρ	6.1	6.1
	m_ρ	769	769
ω	$g_\omega^2/4\pi$	24.5	17.349
	Λ_ω	1850	2494
	m_ω	782.6	782.6
δ	$g_\delta^2/4\pi$	2.488	5.0
	Λ_δ	2000	2169
	m_δ	983	983
$\sigma, T=0, T=1$	$g_\sigma^2/4\pi$	18.3773, 8.9437	22.015, 5.514
	Λ_σ	2000, 1900	1200, 2500
	m_σ	720, 550	691.78, 510.62

Table 1. The best-fit parameters for the two models. The column *Potential B* (*UCT*) taken from Table A.1 in (Machleidt, 1989) (obtained by least squares fitting the OBEP values in Table 1 of that survey). All masses are in MeV.

Distinctive feature of potential (*) is the presence of covariant (Feynman-like) "propagator",

$$\frac{1}{2}\left\{\frac{1}{(p_1-p_1')^2-\mu^2}+\frac{1}{(p_2-p_2')^2-\mu^2}\right\}.$$

On the energy shell for NN scattering, that is

$$E_i \equiv E_{\vec{p}_1}+E_{\vec{p}_2}=E_{\vec{p}_1'}+E_{\vec{p}_2'}\equiv E_f,$$

this expression is converted into the genuine Feynman propagator.

A little part of our numerical results with the best-fit values of the coupling constants g_b and cutoff parameters Λ_b in the meson-nucleon vertices are compared with those by the Bonn group (Machleidt, 1989) in Table 1 and Fig. 2. They labeled by abbreviation UCT have been obtained by solving the partial Lippmann-Schwinger equations (coupled and uncoupled) for the R-matrix of the nucleon-nucleon scattering. Details are in (Dubovyk & Shebeko, 2010).

4.3 Deuteron properties

Besides, we would like to outline the basic elements of another our exploration that is in progress. It is the case, where relying upon the available experience of relativistic calculations of the deuteron static moments and the deuteron form factors (see reviews (Bondarenko et al., 2002) and (Garcon & Van Orden, 2002) and refs. therein) one has to deal with the matrix elements $\langle \mathbf{P}', M'|J^\mu(0)|\mathbf{P}=0, M\rangle$ (to be definite in the laboratory frame). Here the operator

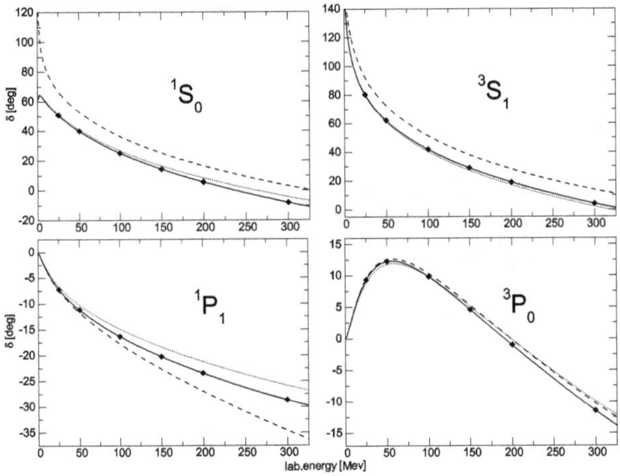

Fig. 2. Neutron-proton phase parameters plotted versus nucleon kinetic energy in lab. system. Solid curves calculated for Potential B. Dashed (dotted) - for UCT potential with *Potential B* (*UCT*) parameters from Table 1. The rhombs show original OBEP results.

$J^\mu(0)$ is the Nöther current density $J^\mu(x)$ at $x = 0$, sandwiched between the eigenstates of a "strong" field Hamiltonian H (cf., discussion in Sec. 5 of lecture (Shebeko & Shirokov, 2000)). In the CPR with $H = K(\alpha_c)$ (Eq. (110)) and $\mathbf{N} = \mathbf{B}(\alpha_c)$ (Eq. (112)) the deuteron state $|\mathbf{P} = 0, M\rangle$ ($|\mathbf{P}' = \mathbf{q}, M'\rangle$) in the rest (the frame moving with the velocity $\mathbf{v} = \mathbf{q}/m_d$) meets the eigenvalue equation

$$P^\mu|\mathbf{P}, M\rangle = P_d^\mu|\mathbf{P}, M\rangle \tag{133}$$

with the three-momentum transfer $\mathbf{q}$, four-momentum $P_d^\mu = (E_d, \mathbf{P})$, $E_d = \sqrt{\mathbf{P}^2 + m_d^2}$, $m_d = m_p + m_n - \varepsilon_d$ and the deuteron binding energy $\varepsilon_d > 0$.

We know that such observables as the charge, magnetic and quadrupole moments of the deuteron can be expressed through the matrix elements in question (e.g., within the Bethe-Salpeter (BS) formalism (Bondarenko et al., 2002)) by introducing the corresponding covariant form factors. With the aid of cumbersome numerical methods the latter have been evaluated in terms of the Mandelstam current sandwiched between the deuteron BS amplitudes.

Unlike this, following Shebeko & Shirokov (2000) , we consider the expansion in the R-commutators

$$J^\mu(0) = WJ_c^\mu(0)W^\dagger = J_c^\mu(0) + [R, J_c^\mu(0)] + \frac{1}{2}[R, [R, J_c^\mu(0)]] + ..., \tag{134}$$

where $J_c^\mu(0)$ is the initial current in which the bare operators $\{\alpha\}$ are replaced by the clothed ones $\{\alpha_c\}$. Decomposition (134) involves one-body, two-body and more complicated interaction currents, if one uses the terminology customary in the theory of meson exchange currents. Further, to the approximation

$$K_I = K(NN \to NN) \sim b_c^\dagger b_c^\dagger b_c b_c \tag{135}$$

and

$$\mathbf{B}_I = \mathbf{B}(NN \to NN) \sim b_c^\dagger b_c^\dagger b_c b_c \tag{136}$$

(see, respectively, (122) and (123)) the eigenvalue problem (133) becomes simpler so its solution acquires the form

$$|\mathbf{P}, M\rangle = \int d\mathbf{p}_1 \int d\mathbf{p}_2 D_M([\mathbf{P}]; \mathbf{p}_1\mu_1; \mathbf{p}_2\mu_2) b_c^\dagger(\mathbf{p}_1\mu_1) b_c^\dagger(\mathbf{p}_2\mu_2)|\Omega\rangle. \tag{137}$$

In this connection, let us recall the relation

$$|\mathbf{q}, M\rangle = exp[i\vec{\beta}\mathbf{B}(\alpha_c)]|\mathbf{0}, M\rangle \tag{138}$$

with $\vec{\beta} = \beta\mathbf{n}$, $\mathbf{n} = \mathbf{n}/n$ and $\tanh\beta = v$, that takes place owing to the property

$$e^{i\vec{\beta}\mathbf{B}} p^\mu e^{-i\vec{\beta}\mathbf{B}} = p^\nu L_\nu^\mu(\vec{\beta}), \tag{139}$$

where $L(\vec{\beta})$ is the matrix of the corresponding Lorentz transformation. Note also that the label $M = (\pm 1, 0)$ denotes the eigenvalue of the third component of the total (field) angular-momentum operator in the deuteron center-of-mass (details can be found in (Dubovyk & Shebeko, 2010)). The c-coefficients D_M in Eq. (137) are calculated by solving the homogeneous Lippmann-Schwinger equation with the quasipotentials taken from (Dubovyk & Shebeko, 2010) (see formulae (67)-(69) therein). Numerical results can be obtained either using the angular-momentum decomposition (as in (Dubovyk & Shebeko, 2010)) or without it.

Several our results are shown in Table 2 and Fig. 3.

Parameter	Bonn B	UCT	Experiment
a_s (fm)	-23.71	-23.57	-23.748±0.010
r_s (fm)	2.71	2.65	2.75±0.05
a_t (fm)	5.426	5.44	5.419±0.007
r_t (fm)	1.761	1.79	1.754±0.008
ε_d (MeV)	2.223	2.224	2.224575
P_D (%)	4.99	4.89	

Table 2. Deuteron and low-energy parameters. The experimental values are from Table 4.2 of Ref. (Machleidt, 1989).

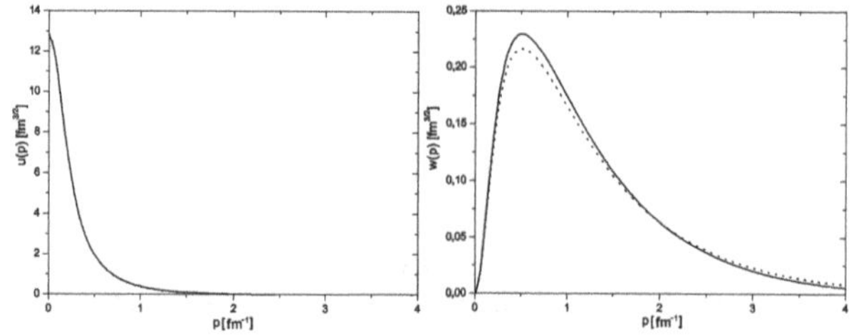

Fig. 3. Deuteron wave functions $\psi_0^d(p) = u(p)$ and $\psi_2^d(p) = w(p)$. Solid(dotted) curves for Bonn Potential B (UCT) potential.

In its turn, the operator (134) being between the two-clothed-nucleon states contributes as

$$\eta_c J^\mu(0)\eta_c = J^\mu_{one-body} + J^\mu_{two-body}, \tag{140}$$

where the operator

$$J^\mu_{one-body} = \int d\mathbf{p}' d\mathbf{p} F^\mu_{p,n}(\mathbf{p}',\mathbf{p}) b_c^\dagger(\mathbf{p}) b_c(\mathbf{p}) \tag{141}$$

with

$$F^\mu_{p,n}(\mathbf{p}',\mathbf{p}) = e\bar{u}(\mathbf{p}')F_1^{p,n}[(p'-p)^2]\gamma^\mu + i\sigma^{\mu\nu}(p'-p)_\nu F_2^{p,n}[(p'-p)^2]u(\mathbf{p}) \tag{142}$$

that describes the virtual photon interaction with the clothed proton (neutron)[5].

Its appearance follows from the observation, in which the primary Nöther current operator, being between the physical (clothed) states $|\Psi_N\rangle = b_c^\dagger|\Omega\rangle$, yields the usual on-mass-shell expression

$$\langle \Psi_{p,n}(\mathbf{p}')|J^\mu(0)|\Psi_{p,n}(\mathbf{p})\rangle = F^\mu_{p,n}(\mathbf{p}',\mathbf{p})$$

in terms of the Dirac and Pauli nucleon form factors.[6]

By keeping in the r.h.s. of Eq. (140) only the one-body contribution we arrive to certain off-energy-shell extrapolation of the so-called relativistic impulse approximation (RIA) in the theory of e.m. interactions with nuclei (bound systems). In a recent work by Dubovyk and Shebeko the deuteron magnetic and quadrupole moments have been calculated to be submitted to Few Body Systems.

Of course, the RIA results should be corrected including more complex mechanisms of e-d scattering, that are contained in

$$J^\mu_{two-body} = \int d\mathbf{p}_1' d\mathbf{p}_2' d\mathbf{p}_1 d\mathbf{p}_2 F^\mu_{MEC}(\mathbf{p}_1',\mathbf{p}_2';\mathbf{p}_1,\mathbf{p}_2) b_c^\dagger(\mathbf{p}_1') b_c^\dagger(\mathbf{p}_1') b_c(\mathbf{p}_1) b_c(\mathbf{p}_2). \tag{143}$$

Analytic (approximate) expressions for the coefficients F^μ_{MEC} stem from the R-commutators (beginning with the third one) in the expansion (134), which, first, belong to the class [2.2], as in Eq. (141), and, second, depend on even numbers of mesons involved. It requires a separate consideration aimed at finding a new family of meson exchange currents, as we hope not only for the e-d scattering.

At last, one should note that, as before, we prefer to handle the explicitly gauge-independent (GI) representation of photonuclear reaction amplitudes with one-photon absorption or emission (Levchuk & Shebeko, 1993). That representation is an extension of the Siegert theorem, in which, the amplitude of interest is expressed through the Fourier transforms of electric (magnetic) field strengths and the generalized electric (magnetic) dipole moments of hadronic system. It allows us to retain the GI in the course of inevitably approximate calculations.

[5] In Eqs. (140) η_c is the projection operator on the subspace $\mathcal{H}_{2N} \in \mathcal{H}$ spanned on the two-clothed-nucleon states $|2N\rangle = b_c^\dagger b_c^\dagger|\Omega\rangle$

[6] Of course, all nucleon polarization labels are implied here together with necessary summations over them in Eq. (141) and so on

5. Summary

We propose a constructive way of ensuring the RI in QFT with cutoffs in momentum space. In contrast to the traditional approach, where the generators of Π are determined as the Nöther integrals of the energy-momentum tensor density , we do not utilize the Lagrangian formalism so fruitful in case of local field models. Our purpose is to find these generators as elements of the Lie algebra of Π starting from the total Hamiltonian whose interaction density in the Dirac picture includes a Lorentz-scalar part $H_{sc}(x)$. Respectively, the algebraic aspect of the RI as a whole for the present exploration with the so-called instant form of relativistic dynamics is of paramount importance.

In the context, using purely algebraic means the boost generators can be decomposed into the Belinfante operator built of H_{sc} and the operator which accumulates the chain of recursive relations in the second and higher orders in H_{nsc}. Thereby, it becomes clear that the Poincaré commutations are not fulfilled if the Hamiltonian does not contain some additional ingredients, which we call the mass renormalization terms, though beyond local field models such a terminology looks rather conventional. The UCT method enables us to determine the corresponding operators including their nonlocal extensions that satisfy the requirements of special relativity and preserve certain continuity with local QFTs.

We see that our approach is sufficiently flexible being applied not merely to local field models including ones with derivative couplings and spins $j \geq 1$. Within the approach all interactions constructed are responsible for physical (not virtual) processes in a given system of interacting fields. Such interactions are Hermitian and energy independent including the off-energy-shell and recoil effects (the latter in all orders of the $1/c^2$ - expansion). In particular, we have managed to build up a new family such interactions in the system of $\pi-$, $\eta-$, $\rho-$, $\omega-$, $\delta-$ and σ - mesons and nucleons. Besides, the interaction operators for processes of the type $NN \rightarrow NN$, $\pi N \rightarrow \pi N$, and $NN \leftrightarrow \pi NN$ are derived on one and the same physical footing.

After constructing the interaction operators in the CPR we express the conventional S matrix through the clothed-particle interactions and states that simplifies the initial field-theoretical task. It becomes possible owing to the isomorphism between the α-algebra with the bare vacuum and the α_c-algebra with the physical vacuum when, first, the requirement iii) is fulfilled and, second, the R - generators of unitary clothing transformations in the Dirac picture come to zero in the distant past and future (see our talk in Durham (Shebeko, 2004)).

In the course of our current work we are trying to understand to what extent the deuteron quenching in flight affects the deuteron electromagnetic form factors. In our opinion, the exposed approach has promising prospects, e.g., in the theory of decaying states (after evident refinements), certainly in quantum electrodynamics and, we believe, in quantum chromodynamics. Such endeavors are under way.

At last, we offer not only a fresh look at constructing the interactions in question but also a nonstandard renormalization procedure in relativistic quantum field theory. In this context, let us remind the prophetic words by Dirac (Dirac, 1963): "I am inclined to suspect that renormalization theory is something that will not survive in the future, and the remarkable agreement between theory and experiment should be looked on as a fluke".

6. Acknowledgements

This exposition is dedicated to the memory of Mikhail Shirokov, the excellent scientist whose contribution to the subfield is difficult to be overestimated. I am very grateful to Luciano Canton, Evgenyi Dubovik, Pavel Frolov and Vladimir Korda for our fruitful collaboration.

7. References

Albeverio, S. (1973). Scattering theory in a model of quantum fields, *I. J. Math. Phys.,* 18, 1800-1816.

Arnous, E.; Heitler W. ; Takahashi Y. (1960). On a convergent non-local field theory, *Nuovo. Cim.,* 16, 671-682.

Bakker, B.L.G. (2001). Forms of relativistic dynamics, *In:Lecture Notes in Physics*, Latal, H. and Schweiger, W. (ed.), Springer-Verlag, pp. 1-54.

Belinfante, F.J. (1940). On the covariant derivative of tensor-undors, *Physica*, 7, 305-324.

Bondarenko, S. G.; Burov, V. V.; Molochkov, A. V.; Smirnov, G. I.; Toki, H. (2002). The Bethe-Salpeter approach with the separable interaction for the deuteron, *Prog. Part. Nucl. Phys.*, 48, 449-535.

Chandler, C. (2003). Relativistic scattering theory with particle creation. Contribution to the 17th International IUPAP Conference on Few Body Physics (Durham, USA) and private communication.

Dirac, P.A.M. (1949). Forms of relativistic dynamics, *Rev. Mod. Phys.,* 21, 392-399.

Dirac, P.A.M. (1963). The evolution of the physicist's picture of nature, *Sci. Amer.*, 208, 45-53.

Dubovyk, E.A.; Shebeko, A.V. (2010). The method of unitary clothing transformations in the theory of nucleon-nucleon scattering, *Few-Body Syst.,* 48, 109-142.

Eckmann, J.-P. (1970). A model with persistent vacuum, *Commun. Math. Phys.,* 18, 247-264.

Efimov, G.V. (1985). Problems of the quantum theory of nonlocal interactions, *Nauka, Moscow, (in Russian).*

Foldy, L.L. (1961). Relativistic particle systems with interactions, *Phys. Rev.,* 122, 275-288.

Friar, J.L. (1975). Relativistic effects on the wave function of a moving system, *Phys. Rev.* C 12, 695-698.

Friedrichs, K. (1953). Mathematical Aspects of the Quantum Theory of Field, *Interscience, New York.*

Glöckle, W.; Müller, L. (1981). Relativistic theory of interacting particles, *Phys. Rev.* C 23, 1183-1195.

Garçon, M.; Van Orden, J.W. (2002). The deuteron: structure and form factors. *Adv. Nucl. Phys.*, 26, 293-378.

Greenberg, O.; Schweber, S. (1958). Clothed particle operators in simple models of quantum field theory, *Nuovo Cim.,* 8, 378-406.

Guillot, J.C.; Jaus, W.; O'Raifeartaigh, L. (1965). A survey of the Heitler-Arnous non-local field theory, *Proc. R.I.A.,* 64, 93-105.

Heinzl, T. (2001). Light-cone quantization foundations and applications, *In:Lecture Notes in Physics*, Latal, H. and Schweiger, W. (ed.), Springer-Verlag, pp. 55-78.

Keister, B.D.; Polyzou, W.N. (1991). Relativistic Hamiltonian dynamics in nuclear and particle physics, *Adv. Nucl. Phys.,* 20, 225-479.

Kita, H. (1966). A non-trivial example of a relativistic quantum theory of particles without divergence difficulties, *Prog. Theor. Phys.,* 35, 934-959.

Kita, H. (1968). Another convergent, relativistic model theory of interacting particles. A relativistic version of a modified Lee model, *Prog. Theor. Phys.*, 39, 1333-1360.

Kobayashi, M.; Sato, T.; Ohtsubo, H. (1997). Effective interactions for mesons and baryons in nuclei, *Progr. Theor. Phys.*, 98, 927-941.

Korda, V.Yu.; Shebeko, A.V. (2004). Clothed particle representation in quantum field theory: Mass renormalization, *Phys. Rev. D* 70, 085011.

Korda, V.Yu.; Canton, L.; Shebeko, A.V. (2007). Relativistic interactions for the meson-two-nucleon system in the clothed-particle unitary representation, *Ann. Phys.*, 322, 736-768.

Krüger, A.; Glöckle, W. (1999). Approach towards N-nucleon effective generators of the Poincaré group derived from a field theory, *Phys. Rev.* C 59, 31919-1929.

Levchuk, L.G.; Shebeko, A.V. (1993). On a generalization of Siegert's theorem. A corrected result, *Phys. At. Nuclei*, 56, 227-229.

Machleidt, R. (1989). The meson theory of nuclear forces and nuclear structure, *Adv. Nucl. Phys.*, 19, 189-376.

Melde, T.; Canton, L.; Plessas, W. (2009). Structure of meson-baryon interaction vertices, *Phys. Rev. Lett.*, 102, 132002.

Nelson, E. (1964). Interaction of nonrelativistic particles with a quantized scalar field, *J. Math. Phys.*, 5, 1190-1197.

Okubo, S. (1954). Diagonalization of Hamiltonian and Tamm-Dancoff equation, *Prog. Theor. Phys.*, 12, 603-608.

Schweber, S.S. (1961). An Introduction to Relativistic Quantum Field Theory, *Row, Peterson, New York.*

Schwinger, J. (1962). Non-abelian gauge fields. Relativistic invariance, *Phys. Rev.*, 127, 324-330.

Shebeko, A.V. (1990). Local analog of the Siegert theorem for covariant description of electromagnetic interactions with nuclei. *Sov. J. Nucl. Phys.*, 52, 970-975.

Shebeko, A.V. (2004). The S matrix in the method of unitary clothing transformations, *Nucl. Phys. A* 737, 252-254.

Shebeko, A.V. and Shirokov, M.I. (2000).Clothing procedure in relativistic quantum field theory and its applications to description of electromagnetic interactions with nuclei (bound systems), *Progr. Part. Nucl. Phys.*, 44, 75-86.

Shebeko, A.V. and Shirokov, M.I. (2001). Unitary transformations in quantum field theory and bound states, *Phys. Part. Nuclei* , 32, 31-95.

Shebeko, A.; Frolov, P. (2011). A possible way for constructing the generators of the Poincaré group in quantum field theory, *to appear in : Few Body Systems*

Shirokov, M.I. (2002). Relativistic nonlocal quantum field theory, *Int. J. Theor. Phys.*, 41, 1027-1041.

Stefanovich, E. (2001). Quantum field theories without infinities, *Ann. Phys.*, 292, 139.

Weinberg, S. (1995). The Quantum Theory of Fields, *University Press, Cambridge*, Vol.1.

Wentzel, G. (1949). Quantum Theory of Fields, *Interscience, New York.*

3

Ab Initio Hamiltonian Approach to Light-Front Quantum Field Theory

James P. Vary
Department of Physics and Astronomy, Iowa State University, Ames, Iowa USA

1. Introduction

Non-perturbative solutions of quantum field theory represent opportunities and challenges that span particle physics and nuclear physics. Increasingly, it is also gaining attention in condensed matter physics. Fundamental understanding of, among others, the phase structure of strongly interacting systems, the spin structure of the proton, the neutron electromagnetic form factor, and the generalized parton distributions of the baryons should emerge from results derived from a non-perturbative light-front Hamiltonian approach. The light-front Hamiltonian quantized within a basis function approach as described here offers a promising avenue that capitalizes on theoretical and computational achievements in quantum many-body theory over the past decade.

By way of background, one notes that Hamiltonian light-front field theory in a discretized momentum basis (1) and in transverse lattice approaches (2; 3) have shown significant promise. I outline here a Hamiltonian basis function approach following Refs. (4–10) that exploits recent advances in solving the non-relativistic strongly interacting nuclear many-body problem (11; 12). There are many issues faced in common - i.e. how to (1) define the Hamiltonian; (2) renormalize for the available finite spaces while preserving all symmetries; (3) solve for eigenvalues and eigenvectors; (4) evaluate experimental observables; and, (5) take the continuum limit.

I begin with a brief overview of recent advances in solving light nuclei with realistic nucleon-nucleon (NN) and three-nucleon (NNN) interactions using *ab initio* no-core methods. After reviewing some advances with two-dimensional theories, I outline a basis function approach suitable for light front gauge theories including the issues of renormalization/regularization. I present an introduction to the approach for cavity-mode QED, to systems in the absence of an external cavity and I discuss its extension to QCD.

2. No Core Shell Model (NCSM) and No Core Full Configuration (NCFC) methods

To solve for the properties of self-bound strongly interacting systems, such as nuclei, with realistic Hamiltonians, one faces immense theoretical and computational challenges. Recently, *ab initio* approaches have been developed that treat all the nucleons on an equal footing, preserve all the underlying symmetries and converge to the exact result given sufficient computational effort. The basis function approach (11; 12) is one of several methods shown to be successful. The primary advantages are its flexibility for choosing the Hamiltonian, the method of renormalization/regularization and the basis space. These advantages support the adoption of the basis function approach in light-front quantum field theory.

Refs. (11; 13–18) and (12; 19; 20) provide examples of the recent advances in the *ab initio* NCSM and NCFC, respectively. The NCSM adopts a renormalization method that provides an effective interaction dependent on the chosen many-body basis space (e.g. on the harmonic oscillator length scale) and on its cutoff (N_{max} below). The NCFC either retains the un-renormalized interaction or adopts a basis-space independent renormalization so that the exact results are obtained either by using a sufficiently large basis space or by extrapolation to the infinite matrix limit. Recent results for the NCSM employ realistic nucleon-nucleon (NN) and three-nucleon (NNN) interactions derived from chiral effective field theory to solve nuclei with Atomic Numbers 10-13 (15) and Atomic Number 14 (17). For an overview of the NCSM including applications to reactions and to effective interactions with a core, see Ref. (18). Recent results for the NCFC feature a realistic NN interaction that is sufficiently soft that binding energies and spectra from a sequence of finite matrix solutions may be extrapolated to the infinite matrix limit (20). Experimental binding energies, spectra, magnetic moments and Gamow-Teller transition rates are well-reproduced in both the NCSM and NCFC approaches. Convergence of long range observables such as the RMS radius and the electric quadrupole are more challenging since they are sensitive to the exponential tails of the nuclear wavefunctions.

It is important to note two recent analytical and technical advances. First, non-perturbative renormalization has been developed to accompany these basis-space methods and their success is impressive. Several schemes have emerged and current research focuses on understanding of the scheme-dependence of convergence rates. Among the many issues to consider, I note that different observables converge at different rates (19) even within a fixed scheme. Second, large scale calculations are performed on leadership-class parallel computers to solve for the low-lying eigenstates and eigenvectors and to evaluate a suite of experimental observables. Low-lying solutions for matrices of basis-space dimension 10-billion on 215,000 cores with a 5-hour run is the current record. However, one expects these limits to continue growing as the techniques are evolving rapidly (16) and the computers are also growing dramatically. Matrices with dimensions in the several tens of billions will soon be solvable with strong interaction Hamiltonians. Note, however, that it is not simply the matrix dimension that controls the level of the computational challenge but a set of issues that includes the sparsity of the Hamiltonian matrix (which depends dramatically on whether NNN interactions are employed), the density of the eigenvalue spectrum, the range of excitation energies desired, etc.

In a NCSM or NCFC application, one adopts a 3D harmonic oscillator (HO) with HO energy ω (using $\hbar = 1$ units) for all the particles in the nucleus, treats the neutrons and protons independently, and generates a many-fermion basis space that includes the lowest oscillator configurations as well as all those generated by allowing up to N_{max} oscillator quanta of excitations. The single-particle states specify the orbital angular momentum projection and the basis is referred to as the m-scheme basis. For the NCSM one also selects a renormalization scheme linked to the basis truncation while in the NCFC the renormalization is either absent or of a type that retains the infinite matrix problem. In the NCFC case (12), one either proceeds to a sufficiently large basis that converged results are obtained (if that is computationally feasible) or extrapolates to the continuum limit as I now illustrate.

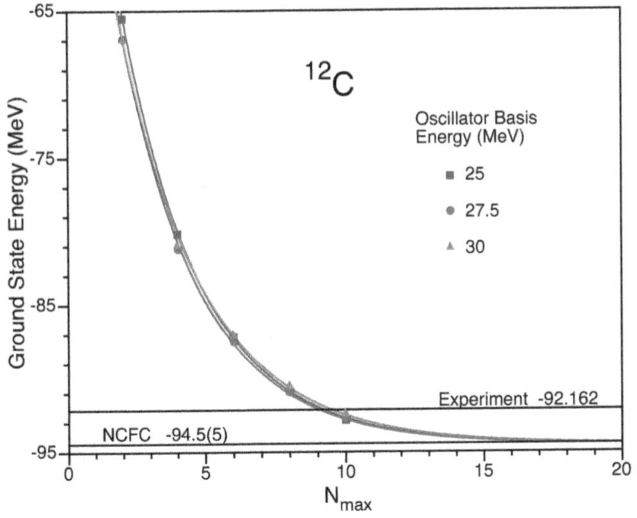

Fig. 1. (Color online) Calculated ground state (gs) energy of ^{12}C for $N_{max} = 2-10$ (symbols) at selected values of ω indicated in the legend. For each ω, the results are fit to an exponential plus a constant, the asymptote, constrained to be the same for all ω(12). Horizontal lines indicate the experimental gs and the NCFC result (uncertainty = 0.5 MeV).

I show in Fig. 1 results for the ground state (gs) of ^{12}C as a function of N_{max} obtained with a realistic NN interaction, JISP16 (14). The smooth curves portray fits that achieve asymptotic independence of N_{max} and ω. The NCFC gs energy (the common asymptote) of -94.5 MeV indicates $\sim$ 3% overbinding. The assessed uncertainty in the NCFC result is 0.5 MeV indicated in parenthesis in the figure. The largest ^{12}C calculations correspond to $N_{max} = 10$, with a matrix dimension near 8 billion. $N_{max} = 12$ produces a matrix dimension near 81 billion which we hope to solve in the future.

In order to further illustrate the successes of the *ab initio* NCSM, I display in Fig. 2 the natural-parity excitation spectra of four nuclei in the middle of the $0p-$shell with both the NN and the NN+NNN effective interactions from χEFT (15). Overall, the NNN interaction contributes significantly to improve theory in comparison with experiment. This is especially well-demonstrated in the odd mass nuclei for the lowest, few excited states. The case of the g.s. spin of ^{10}B and its sensitivity to the presence of the NNN interaction is clearly evident. The results of numerous *ab initio* NCSM applications not only show good convergence with regard to increasing size of the basis space but also have reproduced known properties of 0p-shell nuclei (nuclei up to ^{16}O) as well as explained existing puzzles and made predictions of, as yet, unexplained nuclear phenomena. I cite another prominent example to illustrate this point.

We recently evaluated the Gamow-Teller (GT) matrix element for the beta decay of ^{14}C, including the effect of chiral NNN forces (17). These investigations showed that the very long lifetime for ^{14}C arises from a cancellation between 0p-shell NN-and NNN-interaction contributions to the GT matrix element, as shown in Figure 3. The net result is a GT matrix

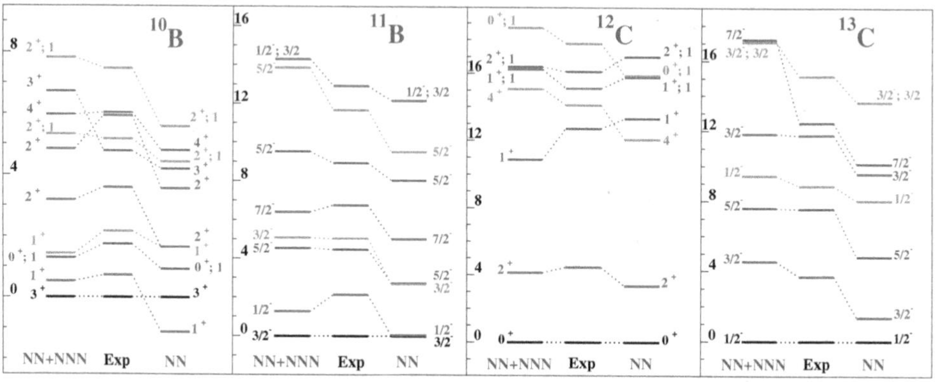

Fig. 2. States dominated by $0p$-shell configurations for ^{10}B, ^{11}B, ^{12}C, and ^{13}C calculated at $N_{\mathrm{max}} = 6$ using $\hbar\Omega = 15$ MeV (14 MeV for ^{10}B). Most of the eigenstates are isospin T=0 or 1/2, the isospin label is explicitly shown only for states with T=1 or 3/2. The excitation energy scales are in MeV (adopted from Ref (15)).

element close to zero (final point of the green curve in the lower half of Fig. 3) which is far more consistent with the 5730 year halflife of ^{14}C. The same calculations show that including the NNN-interactions also bring the binding energies of ^{14}C and ^{14}N into closer agreement with experiment. These A=14 beta decay results were obtained in the largest basis space achieved to date with NNN interactions, $N_{\mathrm{max}} = 8$, or approximately one billion m-scheme configurations.

Other noteworthy results include calculations for ^{12}C explaining the measured ^{12}C B(M1) transition from the g.s. to the $(1^{+},1)$ state at 15.11 MeV and showing more than a factor of 2 enhancement arising from the NNN interaction (13). Neutrino elastic and inelastic cross sections on ^{12}C were shown to be similarly sensitive to the NNN interaction and their contributions significantly improve agreement with experiment (13). Working in collaboration with experimentalists, we uncovered a puzzle in the GT-excited state strengths in A=14 nuclei (21). Its resolution may lie in the role of intruder-state admixtures, but this will require further work.

In addition to numerous successful applications to spectra and electroweak transitions in light nuclei, major efforts are underway to develop extensions to *ab initio* nuclear reactions(18). Key motivations include the goal to further refine our understanding of the fundamental strong interactions among the constituent nucleons and to provide, at the same time, accurate predictions of crucial reaction rates for nuclear astrophysics.

An *ab initio* approach to nuclear reactions based on the NCSM requires a precise treatment of the wave-function asymptotics and the coupling to the continuum. These requirements have led to a new approach, the *ab initio* NCSM/RGM (22; 23), capable of simultaneously describing both bound and scattering states in light nuclei, by combining the resonating-group method (RGM) (24) with the *ab initio* NCSM. The RGM is a microscopic cluster technique based on the use of A-nucleon Hamiltonians, with fully anti-symmetric many-body wave functions built assuming that the nucleons are grouped into clusters. By combining the NCSM with the RGM, one complements the ability of the RGM to deal with scattering and reactions with the utilization of realistic interactions and a consistent *ab initio* microscopic description of

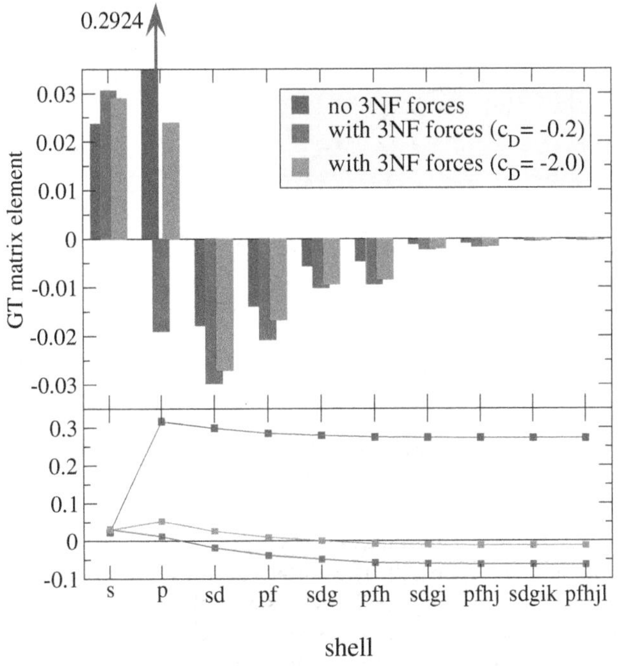

Fig. 3. (Color online) Contributions to the ^{14}C beta decay matrix element as a function of the 3D harmonic oscillator shell in the basis space when the nuclear structure is described by the χEFT interaction (adopted from Ref. (17)). Top panel displays the contributions with (two right bars, the red and green, of each triplet) and without (leftmost bar, the blue bar, of each triplet) the NNN force at $N_{\max} = 8$. Contributions are summed within each shell to yield a total for that shell. The bottom panel displays the running sum of the GT contributions over the shells with the same color coding scheme. Two reasonable choices for coupling constants (red and green components of the histogram and lines) in the NNN-interaction lead to similar strong suppression of the GT matrix element. Note, in particular, the order-of-magnitude suppression of the $0p$-shell contributions arising from the NNN force.

the nucleonic clusters, while preserving important symmetries, including the Pauli exclusion principle and translational invariance.

3. Light-front Hamiltonian field theory

It has long been known that light-front Hamiltonian quantum field theory has similarities with non-relativistic quantum many-body theory and this has prompted applications with established non-relativistic many-body methods (see Ref. (1) for a review). These applications include theories in 1+1, 2+1 and 3+1 dimensions. Several of my efforts in 1+1 dimensions, in collaboration with others, have focused on developing an understanding of how one detects and characterizes phase transition phenomena in the Hamiltonian approach. To this end, I list the following developments:

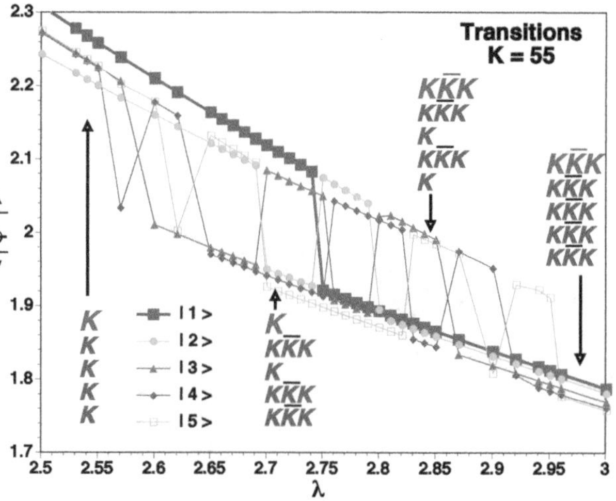

Fig. 4. Expectation value of the square of the scalar field as a function of the coupling constant λ at light-front harmonic resolution K=55 for the lowest five excitations of two dimensional ϕ^4 in the broken phase (27). The pattern of transitions correspond to 5 states falling with increasing λ and crossing the 5 lowest states, thus replacing them and becoming the new 5 lowest states. At selected values of λ, the character of the lowest states is indicated on the figure with the top level of each column signifying the nature of the lowest state. Successive excited states are signified by the labels proceeding down the column. The letter "K" represents "kink" while "$K\bar{K}K$" represents "kink-antikink-kink".

1. identification and characterization of the quantum kink solutions in the broken symmetry phase of two dimensional ϕ^4 including the extraction of the vacuum energy and kink mass that compare well with classical and semi - classical results (25);
2. detailed investigation of the strong coupling region of the topological sector of the two-dimensional ϕ^4 theory demonstrating that low-lying states with periodic boundary conditions above the transition coupling are dominantly kink-antikink coherent states (26);
3. switching to anti-periodic boundary conditions in the strong coupling region of the topological sector of the two-dimensional ϕ^4 theory and demonstrating that low-lying states above the critical coupling are dominantly kink-antikink-kink states as well as presenting evidence for the onset of kink condensation(27). Fig. 4 presents the detailed transition of the lowest 5 mass eigenstates in the broken phase from kink to kink-antikink-kink structure over a narrow range in the coupling. Increasing the resolution K shrinks the range in coupling over which the transitions occur.

More recently, full-fledged applications to gauge theories in 3+1 dimensions have appeared along with roadmaps for addressing QCD. A brief summary of some of the major developments in 3+1 dimensional Hamiltonian light front field theory includes the solutions of:

1. light-front QED wave equations for the electron plus electron-photon system (28–30)
2. simplified gauge theories with a transverse lattice (2; 3; 31)

3. Hamiltonian QED for the electron plus electron-photon system in a trap with a basis function approach (4; 7; 8) that I discuss in the next section.
4. Hamiltonian QED for the electron plus electron-photon system without an external trap that I also discuss in the next section(9; 10)

These successes open pathways for ambitious research programs to evaluate non-perturbative amplitudes and to address the multitude of experimental phenomena that are conveniently evaluated in a light-front quantized approach. As one important example, consider the deeply virtual Compton scattering (DVCS) process which provides the opportunity to study the 3-dimensional coordinate space structure of the hadrons. Recent efforts with model 3+1 dimensional light-front amplitudes (32) have shown that the Fourier spectra of DVCS should reveal telltale diffractive patterns indicating detailed properties of the coordinate space structure.

Additional applications include the non-perturbative regime of QED that future experiments with ultra-strong pulsed lasers will explore, for example, looking for non-perturbative lepton pair production (33–35). Yet another application resides with the strong time-dependent QED fields generated in relativistic heavy-ion collisions where puzzling excesses of electron-positron pairs have been observed (36; 37).

4. Basis light-front quantization applied to QED

We define our light-front coordinates as $x^{\pm} = x^0 \pm x^3$, $x^{\perp} = (x^1, x^2)$, where the variable x^+ is light-front time and x^- is the longitudinal coordinate. We adopt $x^+ = 0$, the "null plane", for our quantization surface. Here we adopt basis states for each constituent that consist of transverse 2D harmonic oscillator (HO) states combined with discretized longitudinal modes, plane waves, satisfying selected boundary conditions. This basis function approach follows Refs. (4–6). Note that the choice of basis functions is arbitrary except for the standard conditions of orthonormality and completeness. Adoption of this particular basis is consistent with recent developments in AdS/QCD correspondence with QCD (38; 39).

The HO states are characterized by a principal quantum number n, orbital quantum number m, and HO energy. Here we adopt the convention that Ω represents both the energy of the transverse HO trap and the basis representation when the trap is present (i.e we match the basis to the trap potential). To signal that the trap is absent we use ω to represent the frequency choice for the basis.

Working in momentum space, it is convenient to write the 2D oscillator as a function of the dimensionless variable $\rho = |p^{\perp}|/\sqrt{M_0\Omega}$, and M_0 has units of mass. The orthonormalized HO wave functions in polar coordinates (ρ, φ) are then given in terms of the generalized Laguerre polynomials, $L_n^{|m|}(\rho^2)$, by

$$\begin{aligned} \Phi_{nm}(\rho, \varphi) &= \langle \rho\varphi | nm \rangle \\ &= \sqrt{\frac{2\pi}{M_0\Omega}} \sqrt{\frac{2n!}{(|m|+n)!}} e^{im\varphi} \rho^{|m|} e^{-\rho^2/2} L_n^{|m|}(\rho^2), \end{aligned} \tag{1}$$

with HO eigenvalues $E_{n,m} = (2n + |m| + 1)\Omega$. The HO wavefunctions have the same analytic structure in both coordinate and momentum space, a feature reminiscent of a plane-wave basis.

The longitudinal modes, ψ_k, in our basis are defined for $-L \leq x^- \leq L$ with periodic boundary conditions for the photon and antiperiodic boundary conditions for the electron:

$$\psi_k(x^-) = \frac{1}{\sqrt{2L}} e^{i \frac{\pi}{L} k x^-}, \tag{2}$$

where $k = 1, 2, 3, ...$ for periodic boundary conditions (we neglect the zero mode) and $k = \frac{1}{2}, \frac{3}{2}, \frac{5}{2}, ...$ for antiperiodic boundary conditions. The full 3D single-particle basis state is defined by the product form

$$\Psi_{k,n,m}(x^-, \rho, \varphi) = \psi_k(x^-)\Phi_{n,m}(\rho, \varphi). \tag{3}$$

For illustrative purposes, we select a transverse mode with $n = 1, m = 0$ joined together with the $k = \frac{1}{2}$ longitudinal antiperiodic boundary condition mode of Eq. 2 and display slices of the real part of this 3-D basis function at selected longitudinal coordinates, x^- in Fig. 5. For comparison, we present a second example with box boundary conditions for the longitudinal mode in Fig. 6. Our purpose in presenting both Figs. 5 and 6 is to suggest the richness, flexibility and economy of texture available for solutions in a basis function approach.

Next, we introduce the total invariant mass-squared M^2 for the low-lying physical states in terms of a Hamiltonian H times a dimensionless integer for the total light-front momentum K

$$M^2 + P_\perp P_\perp \to M^2 + const = P^+ P^- = KH \tag{4}$$

where we absorb the constant into M^2. For simplicity, the transverse functions for both the electron and the photon were taken as eigenmodes of the external trap in our initial application (7) which we discuss here (below, we present results with the external trap removed). The noninteracting Hamiltonian $H_0 = 2M_0 P_c^-$ for this system with a trap is then defined by the sum of the occupied modes i in each many-parton state:

$$H_0 = \frac{2M_0\Omega}{K} \sum_i \frac{2n_i + |m_i| + 1 + \bar{m}_i^2/(2M_0\Omega)}{x_i}, \tag{5}$$

where $\bar{m}_i$ is the mass of the parton i. The photon mass is set to zero throughout this work and the electron mass $\bar{m}_e$ is set at the physical mass 0.511 MeV in our nonrenormalized calculations. We also set $M_0 = \bar{m}_e$.

The light-front QED Hamiltonian interaction terms we need are the electron to electron-photon vertex, given as

$$V_{e \to e\gamma} = g \int dx_+ d^2x_\perp \overline{\Psi}(x)\gamma^\mu \Psi(x) A_\mu(x) \Big|_{x^+=0}, \tag{6}$$

and the instantaneous electron-photon interaction,

$$V_{e\gamma \to e\gamma} = \frac{g^2}{2} \int dx_+ d^2x_\perp \, \overline{\Psi}\gamma^\mu A_\mu \, \frac{\gamma^+}{i\partial^+} \, (\gamma^\nu A_\nu \Psi) \Big|_{x^+=0}, \tag{7}$$

where the coupling constant $g^2 = 4\pi\alpha$, and α is the fine structure constant. The nonspinflip vertex terms of Eq.(6) are $\propto M_0\Omega$, whereas spinflip terms are $\propto \sqrt{M_0\Omega} m_e$. Selecting the initial state electron helicity in the single electron sector always as "up" the process $e \to e\gamma$ is nonzero

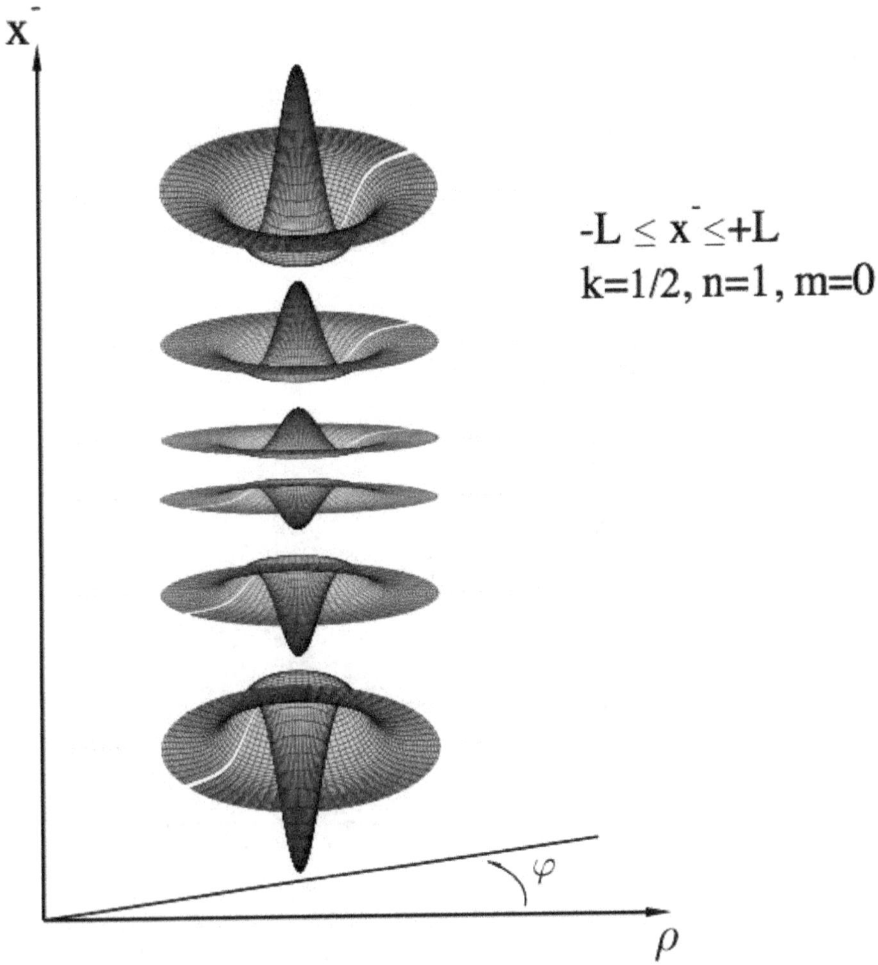

Fig. 5. (color online) Transverse sections of the real part of a 3-D basis function involving a 2-D harmonic oscillator and a longitudinal mode of Eq. 2 with antiperiodic boundary conditions. The quantum numbers for this basis function are given in the legend. The basis function is shown for the full range $-L \leq x^- \leq L$ (adapted from Ref. (4)).

for three out of eight helicity combinations, and the process $e\gamma \to e\gamma$ is nonzero only with all four spin projections aligned (two out of 16 combinations), resulting in a sparse matrix.

We implement a symmetry constraint for the basis by fixing the total angular momentum projection $J_z = M + S = \frac{1}{2}$, where $M = \sum_i m_i$ is the total azimuthal quantum number, and $S = \sum_i s_i$ the total spin projection along the x^- direction. For cutoffs, we select the total light-front momentum, K, and the maximum total quanta allowed in the transverse mode of

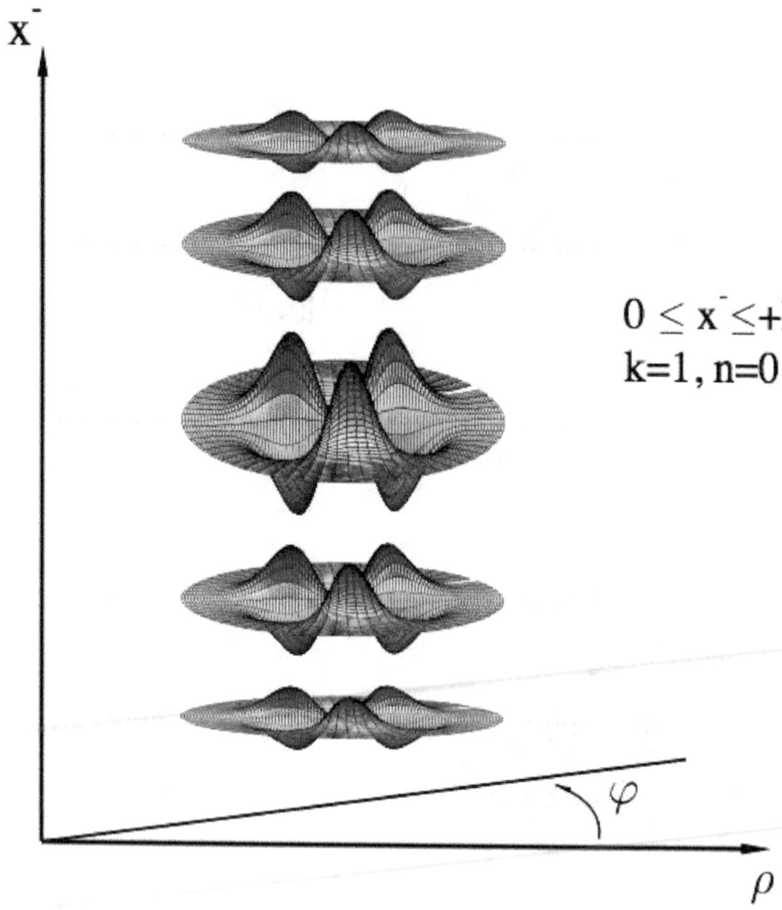

Fig. 6. (color online) Transverse sections of a 3-D basis function involving a 2-D harmonic oscillator and a longitudinal mode with box boundary conditions (wavefunction vanishes at ±L). The quantum numbers for this basis function are given in the legend. The basis function is shown for positive values of x^- and is antisymmetric with respect to $x^- = 0$ (adapted from Ref. (4)).

each one or two-parton state, N_{max}, such that

$$\sum_i x_i = 1 = \frac{1}{K}\sum_i k_i, \tag{8}$$

$$\sum_i 2n_i + |m_i| + 1 \leq N_{max}, \tag{9}$$

where, for example, k_i defines the longitudinal modes of Eq.(2) for the ith parton. Equation (8) signifies total light-front momentum conservation written in terms of boost-invariant momentum fractions, x_i. Since we employ a mix of boundary conditions and all states have

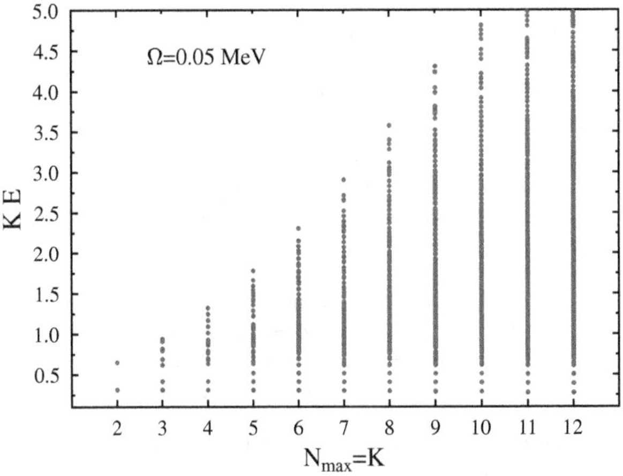

Fig. 7. (color online). Eigenvalues (multiplied by K) for a nonrenormalized light-front QED Hamiltonian for an electron in an external trap with $\Omega = 0.05$ MeV which includes the electron-photon vertex and the instantaneous electron-photon interaction. The cutoffs for the basis space dimensions are selected such that K increases simultaneously with the N_{max}.

half-integer total K, we will quote K values rounded downwards for convenience, except when the precise value is required.

In our approach, the HO parameters Ω, M_0, the electron mass m_e, and the total longitudinal momentum K appear as prefactors for the matrix elements in the Hamiltonian. Therefore, we can rather straightforwardly vary the size of the Hamiltonian matrix by keeping N_{max} fixed, and changing K alone. This facilitates examination of the convergence rates at each value of N_{max}.

In our initial applications, we focus on QED and consider a system including only $|e\rangle$ and $|e\gamma\rangle$ sectors in a transverse scalar harmonic trap (7) and, more recently, in the absence of the external trap (10; 40). Both of these setups, once the Fock space is extended, will be useful for addressing a range of strong field QED problems such as electron-positron pair production in relativistic heavy-ion collisions and with ultra-strong pulsed lasers planned for the future. We adopt the sector dependent non-perturbative renormalization scheme (41).

In Fig. 7 we show the eigenvalues (multiplied by K) for a nonrenormalized light-front QED Hamiltonian given in Eqs.(5,6,7), with fixed $\Omega = 0.05$ MeV and simultaneously increasing K and N_{max}. The resulting dimension of the Hamiltonian matrix increases rapidly. For $N_{max} = K = 2, 10$, and 20, the dimensions of the corresponding symmetric $d \times d$ matrices are $d =$ 2, 1670, and 26 990, respectively.

The number of the single electron basis states, considering all the symmetries, increases slowly with increasing $N_{max} = K$ cutoff. For $N_{max} = K = 2, 10$, and 20 the number of single electron basis states is $1, 5$, and 10, respectively. Our lowest-lying eigenvalue corresponds to a solution dominated by the electron with $n = m = 0$. The ordering of excited states, due to significant interaction mixing, does not always follow the highly degenerate unperturbed spectrum of Eq.(5). States dominated by spin-flipped electron-photon components are evident in the solutions. Nevertheless, the lowest-lying eigenvalues appear with nearly harmonic separations in Fig. 7 as would be expected at the coupling of QED. The multiplicity of the higher eigenstates increases rapidly with increasing $N_{max} = K$ and the states exhibit stronger mixing with other states than the lowest-lying states. In principle, the electron-photon basis states interact directly with each other in leading order through the instantaneous electron-photon interaction, but numerically the effect of this interaction is very weak, and thus does not contribute significantly to the mixing. Even though we work within a Fock-space approach, our numerical results should approximate the lowest order perturbative QED results for sufficiently weak external field.

In the most recent application to QED, we still retain the truncated basis including only $|e\rangle$ and $|e\gamma\rangle$ sectors as in Ref. (7). However, we introduce major extensions and improvements. For a more complete description, I refer to the paper by Zhao, et al. (10) and to a separate paper (40). Here, I simply list a few of the key extensions and improvements.

1. In order to expand the range of applications, we extend the application of BLFQ to a free space system by omitting the external transverse trap.

2. In order to improve computational efficiency and numerical precision, we replace numerical integrations previously used in Ref. (7) to evaluate matrix elements of QED interaction vertices with newly-developed analytic methods.

3. To achieve improved convergence, we allow the HO basis length scale to be fixed separately in each Fock sector which allows a more efficient treatment of the transverse center-of-momentum degree of freedom.

4. We correct the evaluation of the anomalous magnetic moment a_e and a factor appearing in the vertex matrix elements. These corrections go in opposing directions for the previously evaluated a_e in an external trap (7) and updated results will be provided in a separate paper (40).

5. Results for electron anomalous magnetic moment a_e

With the methods and improvements summarized in Secs. 3 and 4, we evaluate and diagonalize the light-front QED Hamiltonian in $|e\rangle$ and $|e\gamma\rangle$ sectors without the external transverse trap and evaluate a_e from the resulting light-front amplitude for the lowest mass eigenstate.

In Ref. (7) the electron anomalous magnetic moment was approximated (based on non-relativistic quantum mechanics) by the squared modulus of the helicity-flip (for the constituent electron) components of the eigenstates. The precise definition of the electron anomalous magnetic moment in relativistic QED is a_e, the electron Pauli form factor F_2 evaluated at momentum transfer $q^2 \to 0$ (42),

$$a_e \equiv \frac{g-2}{2} = \lim_{q^2 \to 0} F_2(q^2). \tag{10}$$

In BLFQ the a_e can be calculated by sandwiching the operator corresponding to $F_2(0)$ with the solution for the ground state for the electron with opposing helicities,

$$
\begin{aligned}
a_e \equiv \frac{g-2}{2} &= \langle \Psi_e^{\downarrow} | F_2(0) | \Psi_e^{\uparrow} \rangle \\
&= \sum_{i',i} \langle \Psi_e^{\downarrow} | e\gamma, i' \rangle \langle e\gamma, i' | F_2(0) | e\gamma, i \rangle \langle e\gamma, i | \Psi_e^{\uparrow} \rangle.
\end{aligned} \tag{11}
$$

Here $\langle e\gamma, i' | F_2(0) | e\gamma, i \rangle$ is the matrix element of the Pauli form factor in the BLFQ basis. The $\langle e\gamma, i | \Psi_e^{\uparrow(\downarrow)} \rangle$ is the wavefunction of a physical electron with helicity up (down) in the $|e\gamma\rangle$ sector (the only sector contributing to a_e in our truncated basis). The i denotes a complete set of quantum numbers. Although Eq.(11) involves two electron eigenstates with opposite helicities, in practice one needs only to solve for one of them and infer the other by exploiting the parity symmetry in light-front QED (43). The explicit expression for $\langle e\gamma, i' | F_2(0) | e\gamma, i \rangle$ and the exact relation between $\langle e\gamma, i | \Psi_e^{\uparrow} \rangle$ and $\langle e\gamma, i | \Psi_e^{\downarrow} \rangle$ will be reported in a later work (40).

In this work without the external trap we reduce the QED coupling constant α by a factor of 10^4 in order to reduce higher order effects and facilitate comparison with a_e from perturbation theory (44). In addition, we omit the instantaneous electron exchange vertex for the same reason.

We define our basis space with total longitudinal momentum K=80 which we found adequate for the present application but will be extended in the future. In fact, the results presented in Ref. (10) already extend the basis to K=160. Furthermore, we use 2D HO single-particle states with frequencies ω ranging from 0.01MeV to 1.4MeV. These ω's bracket the electron mass m_e=0.511 MeV, the only scale-setting parameter in the QED Hamiltonian. At each ω we calculate a_e with N_{max} in the range of 10 to 118 to map out its convergence behavior with increasing N_{max}. Larger N_{max} translates to a larger basis with higher effective ultraviolet cutoff and lower effective infrared cutoff in the transverse plane. We expect that, with increasing N_{max}, the results more closely approximate the Schwinger result. The rate of convergence may be different for different ω's, depending on a_e's sensitivity to the effective cutoffs of the basis space. Our results agree with this expectation and approach the Schwinger result uniformly as N_{max} increases with increments of 4.

In Fig. 8 I present the results evaluated with ω=0.1MeV. For comparison, see the results in Ref. (10) at ω=0.02MeV and ω=0.5MeV. For each ω the results exhibit a simple pattern with increasing N_{max}: the results with even $N_{\text{max}}/2$ are systematically larger than those with odd $N_{\text{max}}/2$ so that the former and the latter separate into two individual groups. Within each group the results define a trend which is understandable by analysis of the perturbative calculation in light-front QED (10; 39; 40). Other features of these results are similarly understandable (10; 40).

The data points in Fig. 8 appear to define straight lines as a function of $1/\sqrt{N_{\text{max}}}$ as can be seen by the linear fits to all the points shown (solid lines). We can therefore easily extrapolate to the limit of no basis truncation ($N_{\text{max}} \to \infty$) where we expect to recover the Schwinger result. Indeed as seen in Fig. 8 the lines converge close to the Schwinger result in this limit. Their intercepts at $1/N_{\text{max}}$=0 are: 0.1131(1.0%) and 0.1133(1.4%) for even $N_{\text{max}}/2$ and odd $N_{\text{max}}/2$, respectively. The percentages in the parenthesis are their corresponding relative deviation from the Schwinger result, $\frac{a_e}{e^2} = \frac{\alpha}{2\pi e^2} = \frac{1}{8\pi^2} \approx 0.012665$.

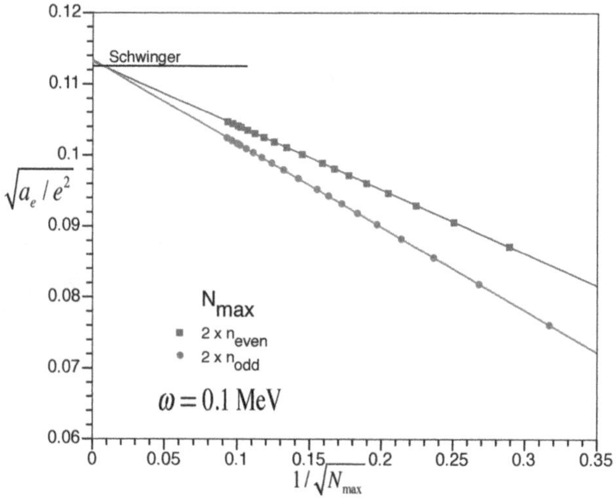

Fig. 8. (Color online) Anomalous magnetic moment of the electron calculated in BLFQ compared to the Schwinger result (44). The vertical axis is the square root of anomalous magnetic moment normalized to electron charge, e, so the Schwinger value is $\sqrt{\frac{1}{8\pi^2}} = 0.11254$. The horizontal axis is the square root of the reciprocal of N_{max}. Symbols are for the BLFQ results. Squares: even $N_{max}/2$; circles: odd $N_{max}/2$. The HO frequency for the basis is 0.1 MeV as indicated in the legend. The lines are linear extrapolations of BLFQ results based on all the points shown which span $N_{max} = 10 - 118$

What is not so apparent from a visual inspection of Fig. 8 is the fact that the extrapolated values come closer to the Schwinger result if one limits the linear fit to results for only the larger values of N_{max}. For example, if the linear fit is performed for $N_{max} \geq 64$ the extrapolated values improve to 0.1129(0.7%) and 0.1130(0.9%) for even $N_{max}/2$ and odd $N_{max}/2$, respectively. Continuing this avenue of investigation, if the linear fit is performed only for results with $N_{max} \geq 100$ the extrapolated values improve to 0.1128(0.4%) and 0.1129(0.6%) for even $N_{max}/2$ and odd $N_{max}/2$, respectively. This is an encouraging sign of expected systematic improvement with increasing N_{max}.

What is also important to note is that these results are systematically improvable. We will extend the calculations to larger K and N_{max} values to further improve accuracy and reduce extrapolation uncertainties. That is, we will evaluate additional results in regions where they are expected to scale more accurately as a function of $\sqrt{\frac{1}{N_{max}}}$. In order to compare with the perturbative result for a_e with the rescaling as shown in Fig. 8 (i.e. to achieve results for $\frac{a_e}{e^2}$) it is also advantageous to further decrease the fine structure constant below $10^{-4}\alpha$, the value for the results presented here.

6. Conclusion

The recent history of light-front Hamiltonian field theory features many advances that pave the way for non-perturbative solutions of gauge theories. The goal is to evaluate the light-front amplitudes for strongly interacting composite systems and predict experimental observables. High precision tests of the Standard Model may be envisioned as well as applications to theories beyond the Standard Model.

We can extend the BLFQ approach to QCD by implementing the SU(3) color degree of freedom for each parton - 3 colors for each fermion and 8 for each boson. We have investigated two methods for implementing the global color singlet constraint and we illustrate the resulting multiplicity of color configurations for each space-spin configuration in Fig. 9. In the first case, we follow Ref. (45) by constraining all color components to have zero color projection and adding a Lagrange multiplier term to the Hamiltonian to select global color singlet eigenstates. This produces the upper curves in each panel of Fig. 9. In the second case, we restrict the basis space to global color singlets (4–6; 46). The second method produces the lower curves in each panel of Fig. 9 and shows a factor of 30-40 lower many-parton basis space dimension at the cost of increased computation time for matrix elements. Either implementation provides an exact treatment of the global color symmetry constraint but the use of the second method provides overall more efficient use of computational resources. Nevertheless, the computational requirements of this approach are substantial, and we foresee extensive use of leadership-class computers to obtain practical results.

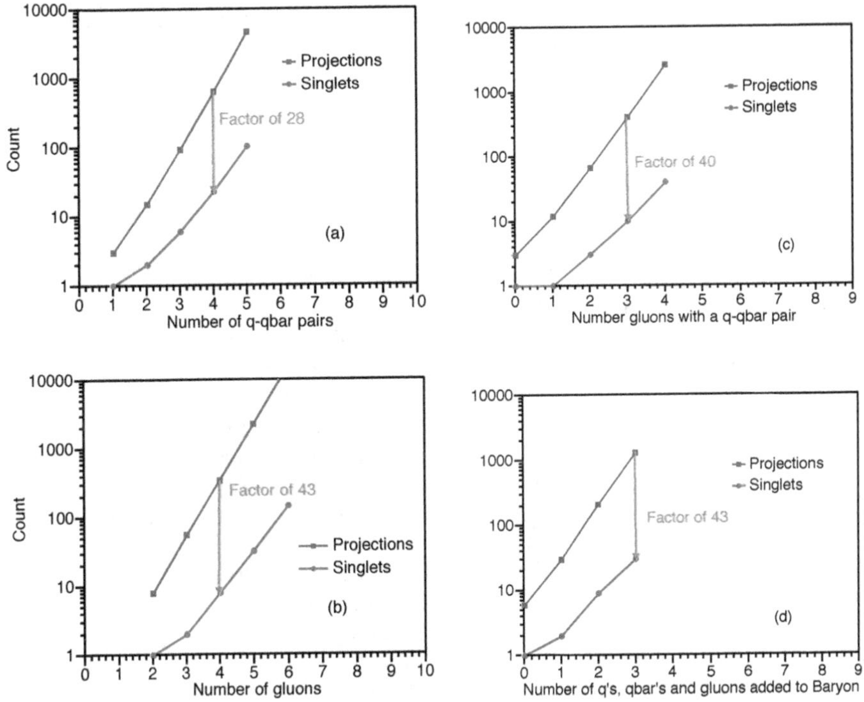

Fig. 9. (color online) Number of color space states that apply to each space-spin configuration of selected multi-parton states for two methods of enumerating the color basis states. The upper curves are counts of all color configurations with zero color projection. The lower curves are counts of global color singlets (adapted from Ref. (4)).

I would like to close by mentioning that we are extending the QED application in several directions. One specific goal is to include the capability for treating strong time-dependent laser pulses to address non-perturbative QED processes (35). In addition, we are launching an initial effort to evaluate the properties of charmonium in a BLFQ treatment of QCD with a first application to the heavy-quarkonia sector leading to predictions for the hybrid states (states dominated by q-qbar-glue configurations).

7. Acknowledgments

The author thanks all his collaborators on the cited publications as well as K. Tuchin, J. Hiller, S. Chabysheva, V. Karmanov, A. Ilderton, Y. Li and P. Wiecki for fruitful discussions. Computational resources were provided by the National Energy Research Scientific Computing Center (NERSC), which is supported by the Office of Science of the U.S. Department of Energy under Contract No. DE-AC02-05CH11231. This work was supported in part by US DOE Grants DE-FG02-87ER40371 and DE-FC02-09ER41582 (UNEDF SciDAC Collaboration). This work was also supported in part by US NSF grant 0904782.

8. References

[1] H. C. Pauli and S. J. Brodsky, *Solving Field Theory In One Space One Time Dimension, Phys. Rev. D* 32, (1985)1993; S. J. Brodsky, H. C. Pauli and S. S. Pinsky, *Quantum Chromodynamics and Other Field Theories on the Light Cone, Phys. Reports* 301 (1998) 299 [*hep-ph/9705477*].

[2] D. Chakrabarti, A. Harindranath and J. P. Vary, *A Study of q-qbar States in Transverse Lattice QCD Using Alternative Fermion Formulations, Phys. Rev. D* 69 , (2004) 034502 [*hep-ph/0309317*].

[3] D. Grunewald, E. M. Ilgenfritz, E. V. Prokhvatilov and H. J. Pirner, *Formulating Light Cone QCD on the Lattice, Phys. Rev. D* 77 (2008) 014512.

[4] J. P. Vary, H. Honkanen, J. Li, P. Maris, S. J. Brodsky, A. Harindranath, G. F. de Teramond, P. Sternberg, E. G. Ng and C. Yang, *Hamiltonian light-front field theory in a basis function approach,* Phys. Rev. C 81,(2010) 035205.

[5] J. P. Vary, H. Honkanen, Jun Li, P. Maris, S. J. Brodsky, A. Harindranath, G. F. de Teramond, P. Sternberg, E. G. Ng, C. Yang, *Hamiltonian light-front field theory within an AdS/QCD basis, Nucl. Phys. B* 199, 64 (2010).

[6] J. P. Vary, H. Honkanen, Jun Li, P. Maris, A. M. Shirokov, S. J. Brodsky, A. Harindranath, G. F. de Teramond, P. Sternberg, E. G. Ng, C. Yang and M. Sosonkina, *Ab initio Hamiltonian approach to light nuclei and to quantum field theory, Pramana Jnl of Phys.* 75, 39 (2010).

[7] H. Honkanen, P. Maris, J. P. Vary and S. J. Brodsky, *Electron in a transverse harmonic cavity,'* Phys. Rev. Lett. 106, 061603 (2011).

[8] J. P. Vary, *Non-Perturbative Hamiltonian Light-Front Field Theory: Progress and Prospects, Proceedings of Science,* PoS(LC2010)001 (available online).

[9] J. P. Vary, *Hamiltonian Light-Front Field Theory: Recent Progress and Tantalizing Prospects,* Few-Body Systems (to appear), 2012; arXiv:1110.1071.

[10] X. Zhao, H. Honkanen, P. Maris, J. P. Vary and S. J. Brodsky *Electron Anomalous Magnetic Moment in Basis Light-Front Quantization Approach,* Few-Body Systems (to appear), 2012; arXiv:1110.0553.

[11] P. Navrátil, J. P. Vary and B. R. Barrett, *Properties of 12-C in the ab-initio Nuclear Shell Model, Phys. Rev. Lett.* 84 (2000) 5728; *Large-basis ab-initio No-core Shell Model and its application to 12-C, Phys. Rev. C* 62 (2000) 054311.

[12] P. Maris, J. P. Vary and A. M. Shirokov, *Ab Initio no-core full configuration calculations of light nuclei*, *Phys. Rev. C.* 79 (2009) 014308 [*nucl-th/0808.3420*].
[13] A. C. Hayes, P. Navrátil and J. P. Vary, *Neutrino-12C Scattering in the ab initio Shell Model with a Realistic Three-Body Interaction Phys. Rev. Lett.* 91, 012502 (2003); nucl-th/0305072.
[14] A. M. Shirokov, J. P. Vary, A. I. Mazur and T. A. Weber, *Realistic Nuclear Hamiltonian: Ab exitu approach*, *Phys. Letts. B* 644 (2007) 33 [*nucl-th/0512105*].
[15] P. Navratil, V. G. Gueorguiev, J. P. Vary, W. E. Ormand and A. Nogga, *Structure of A = 10-13 nuclei with two- plus three-nucleon interactions from chiral effective field theory*, *Phys. Rev. Lett.* 99 (2007)042501 [*nucl-th/0701038*].
[16] P. Maris, M. Sosonkina, J. P. Vary, E. G. Ng and C. Yang, *Scaling of ab-initio nuclear physics calculations on multicore computer architectures*, International Conference on Computer Science, ICCS 2010, *Procedia Computer Science* 1 (2010) 97.
[17] P. Maris, J. P. Vary, P. Navratil, W. E. Ormand, H. Nam, D. J. Dean, *Origin of the anomalous long lifetime of 14C*, Phys. Rev. Lett. 106, 202502(2011).
[18] B. R. Barrett, P. Navratil and J. P. Vary, *Ab initio No Core Shell Model*, Nuclear Physics News International, 21, 5 (2011).
[19] S. K. Bogner, R. J. Furnstahl, P. Maris, R. J. Perry, A. Schwenk and J. P. Vary, *Convergence in the no-core shell model with low-momentum two-nucleon interactions*, *Nucl. Phys. A* 801, (2008) 21[*nucl-th/0708.3754*].
[20] P. Maris, A. M. Shirokov and J. P. Vary, *Ab initio nuclear structure simulations: the speculative* ^{14}F *nucleus*, *Phys. Rev. C* 81 (2010) 021301(R).
[21] A. Negret, et al., *Gamow-Teller Strengths in the A = 14 Multiplet: A Challenge to the Shell Model*, *Phys. Rev. Lett.* 97, 062502 (2006).
[22] S. Quaglioni and P. Navrátil, *Phys. Rev. Lett.* 101, 092501 (2008).
[23] S. Quaglioni and P. Navrátil, *Phys. Rev. C* 79, 044606 (2009).
[24] K. Wildermuth and Y. C. Tang, *A unified theory of the nucleus* (Vieweg, 1977, Braunschweig).
[25] D. Chakrabarti, A. Harindranath, L. Martinovic and J. P. Vary, *Kinks in discrete light cone quantization*, *Phys. Lett. B* 582, (2004) 196 [arXiv:hep-th/0309263].
[26] D. Chakrabarti, A. Harindranath, L. Martinovic, G. B. Pivovarov and J. P. Vary, *Ab initio results for the broken phase of scalar light front field theory*, *Phys. Lett. B* 617, (2005) 92 [arXiv:hep-th/0310290].
[27] D. Chakrabarti, A. Harindranath and J. P. Vary, *A transition in the spectrum of the topological sector of phi**4(2) theory at strong coupling*, *Phys. Rev.* D 71, (2005) 125012 [arXiv:hep-th/0504094].
[28] S. J. Brodsky, V. A. Franke, J. R. Hiller, G. McCartor, S. A. Paston and E. V. Prokhvatilov, *A nonperturbative calculation of the electron's magnetic moment*, *Nucl. Phys. B* 703 (2004) 333;
[29] S. S. Chabysheva and J. R. Hiller, *A nonperturbative calculation of the electron's magnetic moment with truncation extended to two photons*, *Phys. Rev. D* 81 (2010) 074030;
[30] S. S. Chabysheva and J. R. Hiller, *On the nonperturbative solution of Pauli–Villars-regulated light-front QED: A comparison of the sector-dependent and standard parameterizations*, Annals Phys. 325 (2010) 2435.
[31] S. Dalley and B. van de Sande, *Transverse lattice calculation of the pion light-cone wavefunctions*, *Phys. Rev. D* 67 (2003) 114507, [arXiv:hep-ph/0212086].
[32] S. J. Brodsky, D. Chakrabarti, A. Harindranath, A. Mukherjee and J. P. Vary, *Hadron optics: Diffraction patterns in deeply virtual Compton scattering*, *Phys. Lett. B* 641, (2006) 440 [arXiv:hep-ph/0604262]; S. J. Brodsky, D. Chakrabarti, A. Harindranath, A. Mukherjee

and J. P. Vary, *Hadron optics in three-dimensional invariant coordinate space from deeply virtual Compton scattering, Phys. Rev. D* 75, 014003 (2007) [arXiv:hep-ph/0611159].

[33] M. Ruf, G. R. Mocken, C. Muller, K. Z. Hatsagortsyan and C. H. Keitel, *Pair production in laser fields oscillating in space and time,* Phys. Rev. Lett. 102 (2009) 080402.

[34] C. K. Dumlu and G. V. Dunne, *The Stokes Phenomenon and Schwinger Vacuum Pair Production in Time-Dependent Laser Pulses,* Phys. Rev. Lett. 104 (2010) 250402,

[35] A. Ilderton, *Trident pair production in strong laser pulses,* Phys. Rev. Lett. 106, 020404 (2011).

[36] A. Adare *et al.* [PHENIX Collaboration], *Detailed Measurement Of The E^+E^- Pair Continuum In $P+P$ And Au+Au Collisions At* $\sqrt{s_{nn}} = 200$ *Gev And Implications For Direct Photon Production,* Phys. Rev. C 81 (2010) 034911.

[37] K. Tuchin, *Photon decay in strong magnetic field in heavy-ion collisions,* Phys. Rev. C83, 017901 (2011).

[38] G. F. de Teramond and S. J. Brodsky, *Light-Front Holography: A First Approximation to QCD,* Phys. Rev. Lett. 102, 081601 (2009).

[39] S. J. Brodsky and G. F. de Teramond, *Light-Front hadron dynamics and AdS/CFT correspondence, Phys. Lett. B* 582 (2004) 211 [*hep-th/0310227*]; A. Karch, E. Katz, D. T. Son and M. A. Stephanov, *Linear confinement and AdS/QCD, Phys. Rev. D* 74, (2006)015005.

[40] X. Zhao, H. Honkanen, P. Maris, J. P. Vary and S. J. Brodsky, *Electron in a transverse harmonic cavity,* in preparation.

[41] V. A. Karmanov, J.-F. Mathiot and A. V. Smirnov, *Systematic renormalization scheme in light-front dynamics with Fock space truncation, Phys. Rev. D* 77 (2008) 085028 [*hep-th/0801.4507*]

[42] S. J. Brodsky and S. D. Drell, *The Anomalous Magnetic Moment and Limits on Fermion Substructure,* Phys. Rev. D 22, 2236 (1980).

[43] S. J. Brodsky, S. Gardner and D. S. Hwang, *Discrete symmetries on the light front and a general relation connecting nucleon electric dipole and anomalous magnetic moments,* Phys. Rev. D 73, 036007 (2006).

[44] J. S. Schwinger, *On Quantum Electrodynamics And The Magnetic Moment Of The Electron,* Phys. Rev. 73 (1948) 416.

[45] R. J. Lloyd and J. P. Vary, *All-charm Tetraquarks, Phys. Rev. D* 70 (2004) 014009 [*hep-ph/0311179*].

[46] Jun Li, *Light front Hamiltonian and its application in QCD* Ph. D. thesis, Iowa State University (2009), unpublished.

Part 2

Applications to Low-Energy Physics

4

Quantum Field Theory of Exciton Correlations and Entanglement in Semiconductor Structures

M. Erementchouk, V. Turkowski and Michael N. Leuenberger
NanoScience Technology Center and Department of Physics, University of Central Florida USA

1. Introduction

The usual ideology when dealing with many-body systems using second quantization is to treat elementary excitations, say electrons, as the main entities determining the physical properties of the system: the states are enumerated by population numbers, the response is described using Green's functions defined in terms of the elementary excitation, and so on. The situation changes drastically when interaction allows for the formation of bound states. The whole view of the system has to be revisited. Perhaps the most famous example is given by superconductors. In the normal state the material fits the canonical universal description when, roughly speaking, most of the properties are explained (at least qualitatively) by the position of the Fermi level. In the superconducting state, however, one has complete reconstruction of the ground state. Now properties of the material are defined by Cooper pairs with behavior qualitatively different from that of individual electrons. For one the exclusion principle has significantly diminished effect, so that pairs can even condense. Such change in the character of elementary excitations leads to significant consequences: resistivity drops practically to zero.

In the present chapter we review the basic approaches to treating such new states in semiconductors, where elementary excitations, electrons and holes, have opposite charge and the Coulomb attraction leads to formation of excitons, bound states of electron-hole pairs. Excitons are the major factor determining the semiconductor optical response below the fundamental absorption edge: they are responsible for resonant absorption at these frequencies, where in the absence of excitons the material would be transparent. Therefore the problem of main interest addressed in the present chapter is the interaction with an external electromagnetic field tuned in resonance with interband transitions.

In Section 2 we present the general description of semiconductor optical response based on the perturbative treatment of light-matter interaction. We develop a diagrammatical representation of the perturbation series which we use to discuss the optical response of initially unperturbed semiconductor and semiconductor where dark excitons form Bose-Einstein condensate. The essential component of such a description is the solid knowledge of the dynamics of semiconductor many-body excitations. In Section 3 the application of time-dependent density functional theory (TDDFT) to the problem of exciton dynamics is reviewed. TDDFT is a rapidly growing area of research, which shows a great potential to be used in studies of the dynamics of real systems, including ultrafast processes. We review our recent results on studies of the bound states, excitons and biexcitons, using this approach. Also, we discuss some nonlinear effects in the excitonic systems obtained by means of the TDDFT approach.

2. Semiconductor optical response

2.1 Dynamics of semiconductor excitations

In the most general setup the problem of combined time evolution of semiconductor excitations and electromagnetic field requires considering very complex Hamiltonians of electrons in a periodic lattice and coupled to quantized field, whose dynamics, in turn, may be affected by the spatial variation of the dielectric function. However, the physics of phenomena of our main interest, as will be seen, is quite rich on its own. Therefore, in order to demonstrate the major effects we consider the simplest case of a semiconductor with well separated spherical bands excited by a classical electromagnetic field. An example of such a material is GaAs, where the approximation of spherical bands proved to be good. Keeping in mind the applications of techniques discussed below for GaAs we consider the case when the states in the valence band are characterized by the projections $\sigma = \pm 3/2$ of angular momentum (heavy holes).

We will be interested in interband optical transitions, when the external electromagnetic field has its frequency tuned close to the value of the gap separating the conduction and valence bands. Due to the high main frequency the vector potential can be presented in a form convenient for adopting the rotating wave approximation $\mathbf{A}(\mathbf{x},t) = \mathbf{A}_\Omega(\mathbf{x},t)e^{-i\Omega t} + \mathbf{A}^*_\Omega(\mathbf{x},t)e^{i\Omega t}$ with relatively slowly changing amplitude $\mathbf{A}_\Omega(\mathbf{x},t)$.

The dynamics of excitations is described by the semiconductor-light Hamiltonian $\mathcal{H} = \mathcal{H}_{SC} + \mathcal{H}_F + \mathcal{H}_{exc}$, which is composed of the Hamiltonians of the nonperturbed semiconductor $\mathcal{H}_{SC}$, the free electromagnetic field $\mathcal{H}_E$ and the light-matter interaction $\mathcal{H}_{exc}$. We will concentrate on interband optical transitions so that the frequency of the relevant photon modes is close to the value of the gap (throughout the chapter we use units with $\hbar = 1$). If the external field is not too strong such an interaction is well described by the rotating wave approximation that takes into account only resonant transitions. Adopting this approximation and using the coordinate representation we have in the rotating frame

$$\begin{aligned}
\mathcal{H}_{SC} = &\int d\mathbf{x} \left[\sum_s c_s^\dagger(\mathbf{x}) \left(\epsilon_c - \frac{\nabla^2}{2m_e} - \frac{1}{2}\Omega \right) c_s(\mathbf{x}) - \sum_\sigma v_\sigma^\dagger(\mathbf{x}) \left(\epsilon_v + \frac{\nabla^2}{2m_h} - \frac{1}{2}\Omega \right) v_\sigma(\mathbf{x}) \right] \\
&+ \frac{1}{2} \int d\mathbf{x}_1 d\mathbf{x}_2 \sum_{s_1,s_2} c_{s_1}^\dagger(\mathbf{x}_1) c_{s_2}^\dagger(\mathbf{x}_2) V(\mathbf{x}_1 - \mathbf{x}_2) c_{s_2}(\mathbf{x}_2) c_{s_1}(\mathbf{x}_1) \\
&+ \frac{1}{2} \int d\mathbf{x}_1 d\mathbf{x}_2 \sum_{\sigma_1,\sigma_2} v_{\sigma_1}^\dagger(\mathbf{x}_1) v_{\sigma_2}^\dagger(\mathbf{x}_2) V(\mathbf{x}_1 - \mathbf{x}_2) v_{\sigma_2}(\mathbf{x}_2) v_{\sigma_1}(\mathbf{x}_1) \\
&- \int d\mathbf{x}_1 d\mathbf{x}_2 \sum_{s,\sigma} c_s^\dagger(\mathbf{x}_1) v_\sigma^\dagger(\mathbf{x}_2) V(\mathbf{x}_1 - \mathbf{x}_2) v_\sigma(\mathbf{x}_2) c_s(\mathbf{x}_1).
\end{aligned} \tag{1}$$

Here we have introduced $c_s^\dagger(\mathbf{x})$ and $v_\sigma(\mathbf{x})$, operators creating electron with spin s and hole in spin state σ, respectively, at point $\mathbf{x}$. These are fermion operators satisfying the canonical anticommutation relations $\{c_{s_1}(\mathbf{x}_1), c_{s_2}(\mathbf{x}_2)\} = \{c_{s_1}^\dagger(\mathbf{x}_1), c_{s_2}^\dagger(\mathbf{x}_2)\} = \{v_{\sigma_1}(\mathbf{x}_1), v_{\sigma_2}(\mathbf{x}_2)\} =$

$\{v^\dagger_{\sigma_1}(\mathbf{x}_1), v^\dagger_{\sigma_2}(\mathbf{x}_2)\} = 0$ and

$$\left\{c_{s_1}(\mathbf{x}_1), c^\dagger_{s_2}(\mathbf{x}_2)\right\} = \delta_{s_1,s_2}\delta(\mathbf{x}_1 - \mathbf{x}_2), \qquad \left\{v_{\sigma_1}(\mathbf{x}_1), v^\dagger_{\sigma_2}(\mathbf{x}_2)\right\} = \delta_{\sigma_1,\sigma_2}\delta(\mathbf{x}_1 - \mathbf{x}_2). \tag{2}$$

The first line in Eq. (1) represents the Hamiltonian of non-interacting electrons and holes with $m_{e,h}$ being the respective masses and ϵ_c and ϵ_v denoting the positions of the bottom of the conduction band and of the top of the valence band, respectively, so that $\epsilon_c - \epsilon_v$ is the gap. The last three lines describe the electrostatic interaction with the Coulomb potential $V(\mathbf{x})$.

The Hamiltonian of interaction with external electromagnetic field is

$$\mathcal{H}_{exc} = \int d\mathbf{x} \sum_{s,\sigma} \left[\mathbf{A}_\Omega(\mathbf{x},t)\cdot \mathbf{d}_{s,\sigma} v^\dagger_\sigma(\mathbf{x}) c^\dagger_s(\mathbf{x}) + \mathbf{A}^*_\Omega(\mathbf{x},t)\cdot \mathbf{d}_{\sigma,s} c_s(\mathbf{x}) v_\sigma(\mathbf{x})\right], \tag{3}$$

where $\mathbf{d}_{s,\sigma} = \mathbf{d}^*_{\sigma,s} = -i\langle s|\nabla|\sigma\rangle e/m_0$, with m_0 the electron mass in vacuum, quantifies coupling between the respective states in conduction and valence bands. Interaction of light with the semiconductor occurs through absorption and emission of electron-hole pairs thus specifying the quantity of main interest.

Because the main technical tool used in this part is perturbation theory, it is convenient to incorporate the time dependence into the Heisenberg representation of quantum operators $\widetilde{\mathcal{O}} = e^{i\mathcal{H}_{SC}t}\mathcal{O}e^{-i\mathcal{H}_{SC}t}$, so that $i\partial\mathcal{O}/\partial t = [\mathcal{H}_{SC}, \mathcal{O}]$. For electron-hole pair operator the equation of motion has the form

$$i\frac{\partial}{\partial t} v^\dagger_\sigma(\mathbf{x}_2)c^\dagger_s(\mathbf{x}_1) = -\widehat{L}_{s,\sigma}(\mathbf{x}_1,\mathbf{x}_2) v^\dagger_\sigma(\mathbf{x}_2)c^\dagger_s(\mathbf{x}_1) - v^\dagger_\sigma(\mathbf{x}_2)c^\dagger_s(\mathbf{x}_1)\left[\mathcal{U}(\mathbf{x}_1) - \mathcal{U}(\mathbf{x}_2)\right], \tag{4}$$

where operator $\widehat{L}_{s,\sigma}(\mathbf{x}_1,\mathbf{x}_2)$ describes the one-pair dynamics

$$\widehat{L}_{s,\sigma}(\mathbf{x}_1,\mathbf{x}_2) = \epsilon_c - \epsilon_v - \Omega - \frac{1}{2m_e}\nabla_1^2 - \frac{1}{2m_h}\nabla_2^2 - V(\mathbf{x}_1 - \mathbf{x}_2) \tag{5}$$

and $\mathcal{U}(\mathbf{x})$ accounts for interaction of the pair with surrounding charges

$$\mathcal{U}(\mathbf{x}) = \int d\mathbf{x}'\, V(\mathbf{x}-\mathbf{x}')\left[\sum_\sigma v^\dagger_\sigma(\mathbf{x}')v_\sigma(\mathbf{x}') - \sum_s c^\dagger_s(\mathbf{x}')c_s(\mathbf{x}')\right]. \tag{6}$$

In the case when its contribution vanishes (for instance, when operators in Eq. (4) act on vacuum state, i.e. state with empty conduction band and filled valence band) the semiconductor dynamics becomes very simple and the structure of excitations is determined by the spectral decomposition

$$\widehat{L}_{s,\sigma}(\mathbf{x}_1,\mathbf{x}_2)f(\mathbf{x}_1,\mathbf{x}_2) = \sum_\mu E_\mu \phi^*_\mu(\mathbf{x}_1,\mathbf{x}_2)\int d\mathbf{x}'_1 d\mathbf{x}'_2\, \phi_\mu(\mathbf{x}'_1,\mathbf{x}'_2) f(\mathbf{x}'_1,\mathbf{x}'_2), \tag{7}$$

where the formal summation over μ implies summing over the discrete quantum numbers and integrating over continuous ones. In Eq. (7) E_μ and ϕ_μ are, respectively, the eigenvalues and the eigenfunctions, $\widehat{L}\phi_\mu = E_\mu\phi_\mu$. The operator $\widehat{L}$ is self-adjoint and, hence, its eigenvalues are real and the eigenfunctions form a complete orthonormal set. Convoluting both sides of Eq. (4) (taken with $\mathcal{U} = 0$) with ϕ_μ we find that operators

$$\mathcal{B}^\dagger_\mu = \int d\mathbf{x}_1 d\mathbf{x}_2\, \phi_\mu(\mathbf{x}_1,\mathbf{x}_2) v^\dagger_\sigma(\mathbf{x}_2) c^\dagger_s(\mathbf{x}_1) \tag{8}$$

acting on vacuum create eigenstates of the semiconductor Hamiltonian $\mathcal{H}_{SC}\mathcal{B}_\mu^\dagger|0\rangle = E_\mu\mathcal{B}_\mu^\dagger|0\rangle$. Such electron-hole states are called excitons and, respectively, $\mathcal{B}_\mu^\dagger$ are called exciton operators. As follows from Eq. (7) excitons are characterized by a set of quantum numbers describing the solutions of the respective Schrödinger equation. For convenience we also include the spin variables into this set, so that $\mu = \{s_\mu, \sigma_\mu, n_\mu, \ell_\mu, \mathbf{K}_\mu\}$, where $\mathbf{K}_\mu$ is the momentum of the exciton center of mass, n_μ is the principal quantum number, and ℓ_μ is orbital momentum.

Among the full variety of exciton states not all of them play an equally important role in the dynamics of a semiconductor interacting with external electromagnetic field. In order to see this we employ the fact that wave-functions ϕ_μ form a complete set and obtain the relation between electron-hole and exciton operators

$$v_{\sigma_\mu}^\dagger(\mathbf{x}_2)c_{s_\mu}^\dagger(\mathbf{x}_1) = \sum_{\mu|s_\mu,\sigma_\mu} B_\mu^\dagger\phi_\mu^*(\mathbf{x}_1,\mathbf{x}_2), \tag{9}$$

where the sum is taken for fixed values of electron and hole spins. This relation allows us to express $\mathcal{H}_{exc}$ in terms of the exciton operators

$$\mathcal{H}_{exc} = \sum_\mu \left(A_\mu^*(t)\mathcal{B}_\mu + A_\mu(t)\mathcal{B}_\mu^\dagger\right), \tag{10}$$

where $A_\mu(t)$ are the projections of the external field onto the respective exciton mode

$$A_\mu(t) = \int d\mathbf{x}\,\mathbf{A}_\Omega(\mathbf{x},t)\cdot\mathbf{d}_{s_\mu,\sigma_\mu}\phi_\mu^*(\mathbf{x},\mathbf{x}). \tag{11}$$

As a result only states with $\ell = 0$ are directly coupled with the electromagnetic field. Among four possible heavy-hole exciton states two of them are dark because the respective transitions between valence and conduction bands are dipole-forbidden (Ivchenko, 2005). This results in a life-time of dark excitons that is significantly greater than the life time of bright states. Additionally, the interaction of bright excitons with light raises their energy compared to dark excitons (Combescot & Leuenberger, 2009). These circumstances are very important from the perspective of exciton Bose-Einstein condensation as will be discussed below.

2.2 Dark excitons

A simple way to see why the dark exciton states have lower energies than the bright exciton states (Combescot & Leuenberger, 2009) is to rearrange the electron-hole exchange scattering diagram as shown in Fig. 1. Then it becomes obvious that the electron-hole exchange corresponds to the exchange of a virtual photon between electron-hole pairs. Since only bright excitons interact with photons, it is only possible for bright excitons to exchange virtual photons among each other. This interaction pushes the energies of the bright excitons above the energies of the dark excitons. The reason for adding energy can be intuitively understood by comparing direct and exchange Coulomb interactions for electrons; i.e. the direct interaction among electrons leads to repulsion, corresponding to adding energy, while the exchange interaction leads to attraction, corresponding to reducing energy. In the case of the Coulomb interaction between electrons and holes, this situation is completely reversed; i.e. the direct interaction between electrons and holes leads to attraction, corresponding to reducing energy, while the exchange interaction leads to repulsion, corresponding to adding energy. It is this exchange-based repulsion which lets the bright exciton energies be higher than the dark exciton energies.

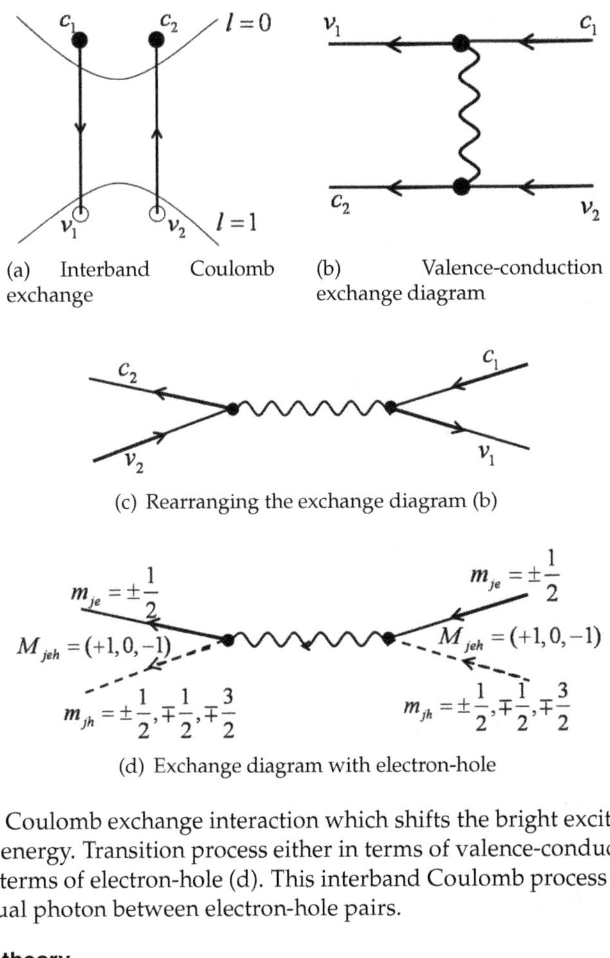

(a) Interband Coulomb exchange

(b) Valence-conduction exchange diagram

(c) Rearranging the exchange diagram (b)

(d) Exchange diagram with electron-hole

Fig. 1. Interband Coulomb exchange interaction which shifts the bright exciton energy above the dark exciton energy. Transition process either in terms of valence-conduction electrons (b) and (c), or in terms of electron-hole (d). This interband Coulomb process is nothing but an exchange of virtual photon between electron-hole pairs.

2.3 Perturbation theory

Coupled operator equations of motion of semiconductor excitations and electromagnetic field, turn out to be too complex for a detailed analysis and, therefore, a reliable approximation scheme must be applied. While such scheme can be worked out at the level of the equations of motion it is convenient to start from the Hamiltonian formulation of the problem and to use the standard approach developed in the quantum field theory on the ground of interaction representation.

The equation of motion of the external electromagnetic field driven by semiconductor excitations has the form of classical wave equation

$$\frac{1}{c^2}\ddot{\mathbf{A}}(\mathbf{x},t) - \nabla^2\mathbf{A}(\mathbf{x},t) = -\mu_0\mathbf{J}(\mathbf{x},t) \tag{12}$$

with a source

$$\mathbf{J}(\mathbf{x},t) = \sum_{\mu} \mathbf{d}_{\sigma_\mu, s_\mu} \phi_\mu(\mathbf{x},\mathbf{x}) \langle \mathcal{B}_\mu \rangle . \tag{13}$$

This shows that the effect of material excitations is described by the exciton polarization

$$P_\mu(t) = \langle \mathcal{B}_\mu \rangle = \mathrm{Tr}\left[\rho(t)\mathcal{B}_\mu\right], \tag{14}$$

where $\rho(t)$ is the density matrix at instant t. Specifying the semiconductor state by the density matrix allows one to address the most general situation. We, however, start from the special case when the system at all times is in some pure state $|\Psi(t)\rangle$, so that $P_\mu(t) = \langle\Psi(t)|\, \mathcal{B}_\mu\, |\Psi(t)\rangle$. Next, we treat the interaction with the electromagnetic field, $\mathcal{H}_{exc}$ as a perturbation and follow the standard prescription. First, we account the nonperturbed dynamics by introducing $\left|\widetilde{\Psi}(t)\right\rangle = \exp\left[i\mathcal{H}_{SC}t\right]|\Psi(t)\rangle$, which satisfies $i\partial\left|\widetilde{\Psi}(t)\right\rangle/\partial t = \widetilde{\mathcal{H}}_{exc}(t)\left|\widetilde{\Psi}(t)\right\rangle$, where $\widetilde{\mathcal{H}}_{exc}(t)$ is the Hamiltonian of light-matter interaction in the Heisenberg picture

$$\widetilde{\mathcal{H}}_{exc}(t) = e^{i\mathcal{H}_{SC}t}\mathcal{H}_{exc}e^{-i\mathcal{H}_{SC}t}. \tag{15}$$

Iterating the equation of motion we find

$$\left|\widetilde{\Psi}(t)\right\rangle = \mathcal{S}(t)\left|\widetilde{\Psi}(0)\right\rangle = \mathcal{S}(t)\,|\Psi(0)\rangle, \tag{16}$$

where

$$S(t) = 1 + \sum_{n=1}^{\infty}(-i)^n \int\cdots\int_{t_1\le\ldots\le t_n\le t} dt_1\ldots dt_n\, \widetilde{\mathcal{H}}_{exc}(t_1)\ldots\widetilde{\mathcal{H}}_{exc}(t_n). \tag{17}$$

In what follows we will need the explicit form of such expansion. For a compact notation, however, it is convenient to introduce the time ordering operator $\mathcal{T}_+$, so that one has $\mathcal{S}(t) = \mathcal{T}_+\exp\left\{-i\int_0^t dt'\widetilde{\mathcal{H}}_{exc}(t')\right\}$. Following the same line of arguments we can also derive $\left\langle\widetilde{\Psi}(t)\right| = \left\langle\widetilde{\Psi}(0)\right|\mathcal{S}^\dagger(t)$, where $\mathcal{S}^\dagger(t) = \mathcal{T}_-\exp\left\{i\int_0^t dt'\widetilde{\mathcal{H}}_{exc}(t')\right\}$. Thus we obtain for the exciton polarization (and, actually, for any observable)

$$P_\mu(t) = \left\langle\widetilde{\Psi}(t)\right| e^{i\mathcal{H}_{SC}t}\mathcal{B}e^{-i\mathcal{H}_{SC}t}\left|\widetilde{\Psi}(t)\right\rangle = \langle\Psi(0)|\,\mathcal{S}^\dagger(t)\widetilde{\mathcal{B}}_\mu(t)\mathcal{S}(t)\,|\Psi(0)\rangle . \tag{18}$$

Finally, the perturbational series for $P_\mu(t)$ is obtained substituting instead of $\mathcal{S}(t)$ and $\mathcal{S}^\dagger(t)$ their expansions following Eq. (17).

This was the general consideration, which is applicable for arbitrary system and perturbation. For the problem of our main interest, however, the fact of great importance is that exciton operators entering $\mathcal{H}_{exc}$ change the total number of electrons and holes. In particular, if the initial state $|\Psi(0)\rangle$ is characterized by a definite number of particles this form of perturbation implies that only terms with matching numbers of exciton creation and annihilation operators would make nonzero contribution. At the same time a lot of terms enter even low orders of the perturbation series [as illustrated by Eq. (21) below]. Therefore, for analysis of the series it is convenient to represent terms graphically using as building blocks

$$= \pm iA_\mu^*(t)\widetilde{\mathcal{B}}_\mu(t),$$

$$= \pm iA_\mu(t)\widetilde{\mathcal{B}}_\mu^\dagger(t). \tag{19}$$

The raising (lowering) line corresponds to the exciton annihilation (creation) operator in the interaction representation at particular instant and a hollow vertex attached to the line corresponds to the external field taken at the same instant. Thus the order of the diagram is determined by the number of vertices. Integration over time is shown by filled vertex and taking the diagonal matrix element is indicated by a horizontal line, for example

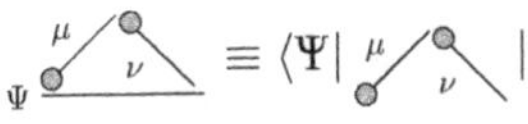

$$\equiv \langle \Psi | \quad |\Psi\rangle = -\int_0^t dt_2 \int_0^{t_2} dt_1 A^*_\mu(t_1) A_\nu(t_2) \left\langle \widetilde{\mathcal{B}}_\mu(t_1) \widetilde{\mathcal{B}}^\dagger_\nu(t_2) \right\rangle_\Psi . \tag{20}$$

While writing down an expression corresponding to particular term in the perturbational series for $P_\mu(t)$ one should take into account that elements to the left and to the right from $\mathcal{B}_\mu$ originate from expansions for $\mathcal{S}^\dagger$ and $\mathcal{S}$, respectively, which have factors i and $-i$ in correspondence rule (19). The diagrammatical representation of the first few terms of the perturbational series for $P_\mu(t)$ has the form

$$P_\mu(t) =$$

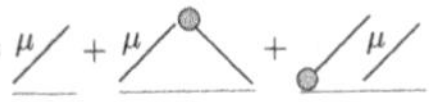

$$+ \ldots = \left\langle \widetilde{\mathcal{B}}_\mu(t) \right\rangle_\Psi - \\ - i \int_0^t dt' \left\langle \widetilde{\mathcal{B}}_\mu(t) \widetilde{\mathcal{B}}^\dagger_\nu(t') \right\rangle_\Psi A_\nu(t') + i \int_0^t dt' \left\langle \widetilde{\mathcal{B}}_\nu(t') \widetilde{\mathcal{B}}_\mu(t) \right\rangle_\Psi A^*_\nu(t') + \ldots . \tag{21}$$

2.4 Linear and nonlinear responses of initially unperturbed semiconductor

We illustrate the application of this analysis by considering the optical response of a non-perturbed semiconductor (Erementchouk & Leuenberger, 2010b; Ostreich et al., 1998), that is when the initial state is vacuum $|0\rangle$ (empty conduction band and filled valence band). We will emphasize this fact by using the dashed horizontal (vacuum) line in diagrams. It is seen that only diagrams starting and ending at the vacuum line and above it provide non-zero contributions into the series. The total number of lines in this case is even and, hence, the total number of vertices is odd implying that we have only odd orders of the perturbation theory in this case. In particular among the diagrams shown in Eq. (21) only the second diagram survives yielding the polarization of linear response

$$P^{(1)}_\mu(t) =$$

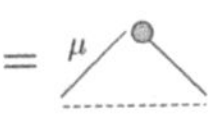

$$= -i \sum_\nu \int_0^t dt' \left\langle \mathcal{B}_\mu e^{-i\mathcal{H}_{SC}(t-t')} \mathcal{B}^\dagger_\nu \right\rangle_0 A_\mu(\tau). \tag{22}$$

For the following consideration it is constructive to analyze this expression in somewhat excessive details. We introduce the vacuum exciton propagator $\Phi^{(0)}_{\mu,\nu}(\tau) = \left\langle \mathcal{B}_\mu e^{-i\mathcal{H}_{SC}\tau} \mathcal{B}^\dagger_\nu \right\rangle_0$ and differentiating with respect to τ we find that it satisfies the dynamical equation

$$\dot{\Phi}^{(0)}_{\mu,\nu}(\tau) = -i \left\langle \mathcal{B}_\mu e^{-i\mathcal{H}_{SC}\tau} \left[\mathcal{H}_{SC}, \mathcal{B}^\dagger_\nu \right] \right\rangle_0 = -iE_\nu \Phi^{(0)}_{\mu,\nu}(\tau) \tag{23}$$

with the initial value $\Phi^{(0)}_{\mu,\nu}(0) = \delta_{\mu,\nu}$, where we have taken into account that $\mathcal{B}_\mu \mathcal{B}^\dagger_\nu |0\rangle = \delta_{\mu,\nu} |0\rangle$ following from the commutation relation for the exciton operators

$$\left[\mathcal{B}_\mu, \mathcal{B}^\dagger_\nu \right] = \delta_{\mu,\nu} - \mathcal{C}_{\mu,\nu}. \tag{24}$$

Operators $\mathcal{C}_{\mu,\nu}$ describe the deviation of excitons from bosons

$$\mathcal{C}_{\mu,\nu} = \int d\mathbf{x}_1 d\mathbf{x}_2 d\mathbf{x}'_1 d\mathbf{x}'_2 \phi^*_\mu(\mathbf{x}_1, \mathbf{x}_2) \phi_\nu(\mathbf{x}'_1, \mathbf{x}'_2) \\ \times \left[v^\dagger_{\sigma_\nu}(\mathbf{x}'_2) v_{\sigma_\mu}(\mathbf{x}_2) \delta_{s_\mu, s_\nu} \delta(\mathbf{x}_1 - \mathbf{x}'_1) + c^\dagger_{s_\nu}(\mathbf{x}'_1) c_{s_\mu}(\mathbf{x}_1) \delta_{\sigma_\mu, \sigma_\nu} \delta(\mathbf{x}_2 - \mathbf{x}'_2) \right]. \tag{25}$$

Thus we find

$$P^{(1)}(t) = -i \int_0^t dt' e^{-iE_\mu (t-t')} A_\mu(t'). \tag{26}$$

Simple but important feature of the linear response is that wave vectors containing in the external excitation directly transferred to the linear exciton polarization. For example, if $\mathbf{A}(\mathbf{x}) \propto e^{i\mathbf{K}\cdot\mathbf{x}}$ then $P_\mu^{(1)} \propto \delta(\mathbf{K}_\mu - \mathbf{K})$.

In the next (third) order the exciton polarization is given by the following diagrams

$$P_\mu^{(3)}(t) =$$

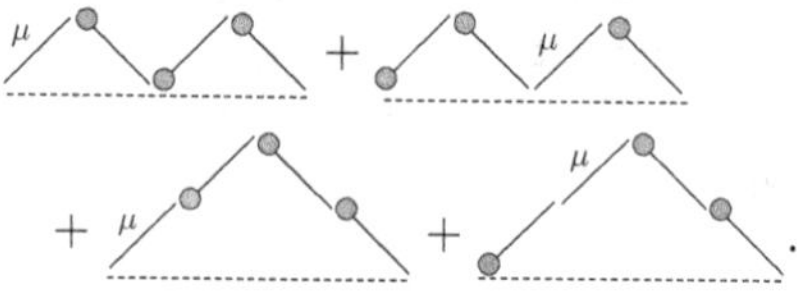

(27)

In order to present this expression in more familiar form we differentiate Eq. (27) with respect to t remembering that we also need to differentiate the upper limits of the respective integrals. This yields

$$\dot{P}_\mu^{(3)}(t) = -i\omega_\mu P_\mu^{(3)}(t) +$$

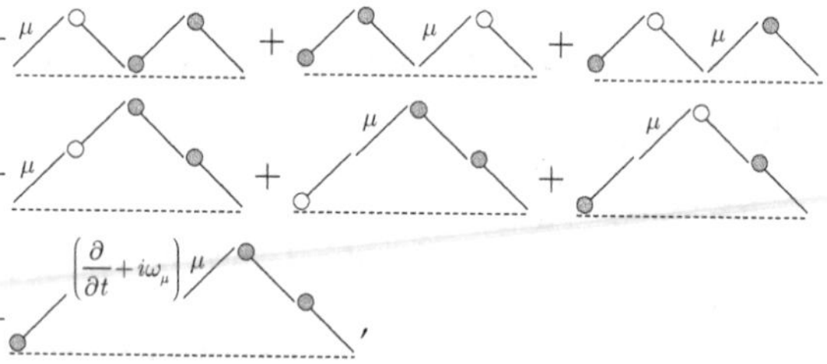

(28)

where the elements with the hollow vertices are taken at the instant t and the respective diagrams describe the modification of the instantaneous effect of the electromagnetic field and thus account for the phase-space filling effect. It can be seen that fifth and fourth diagram cancel each other by virtue of $[\mathcal{B}_\mu, \mathcal{B}_\nu] = 0$. The first and the second diagrams combine together yielding $-iA_\mu(t)\sum_\nu |P_\nu^{(1)}(t)|^2$. This term is canceled by the commutator appearing after combining the third and the sixth diagrams. Thus the phase-space filling effect is described by

$$K_\mu(t) = i \sum_{\kappa,\lambda,\nu} \left\langle \mathcal{B}_\lambda \mathcal{C}_{\mu,\nu} \mathcal{B}_\kappa^\dagger \right\rangle_0 A_\nu(t) P_\lambda^{(1)^*}(t) P_\kappa^{(1)}(t). \tag{29}$$

The effect of the Coulomb interaction is described by the last term in Eq. (28), which has the form

$$M_\mu(t) = \sum_{\kappa,\lambda,\nu} P_\lambda^{(1)^*}(t) \int_0^t dt_3 \int_0^{t_3} dt_1\, A_\nu(t_3) A_\kappa(t_1) \left\langle \mathcal{D}_{\lambda,\mu} e^{-i\mathcal{H}_{SC}t} \widetilde{\mathcal{B}}_\nu^\dagger(t_3) \widetilde{\mathcal{B}}_\kappa^\dagger(t_1) \right\rangle_0, \tag{30}$$

where we have introduced $\mathcal{D}_{\lambda,\mu} = \left[\mathcal{B}_\lambda, \left[\mathcal{B}_\mu, \mathcal{H}_{SC}\right]\right]$. This operator can be presented as

$$\begin{aligned} \mathcal{D}_{\lambda,\mu} = \int d\mathbf{x}_1 d\mathbf{x}_2 d\mathbf{x}_1' d\mathbf{x}_2' \mathcal{B}_\lambda(\mathbf{x}_1, \mathbf{x}_2) \mathcal{B}_\mu(\mathbf{x}_1', \mathbf{x}_2') \times \\ \left[V(\mathbf{x}_1 - \mathbf{x}_1') + V(\mathbf{x}_2 - \mathbf{x}_2') - V(\mathbf{x}_1 - \mathbf{x}_2') - V(\mathbf{x}_1' - \mathbf{x}_2)\right], \end{aligned} \tag{31}$$

where we have denoted $\mathcal{B}_\mu(\mathbf{x}_1,\mathbf{x}_2) = \phi_\mu(\mathbf{x}_1,\mathbf{x}_2)c_{s_\mu}(\mathbf{x}_1)v_{\sigma_\mu}(\mathbf{x}_2)$. Thus operator $\mathcal{D}_{\lambda,\mu}$ describes the Coulomb interaction between excitons λ and μ. Taking into account the symmetry with respect to $\nu \leftrightarrow \kappa$ we can rewrite $M_\mu(t)$ in terms of polarizations of linear response resulting in

$$M_\mu(t) = \frac{1}{2}\sum_{\kappa,\lambda,\nu} P_\lambda^{(1)^*}(t)\left[i\beta_{\lambda,\mu}^{\nu,\kappa}P_\kappa^{(1)}(t)P_\nu^{(1)}(t) + \int_0^t dt'\, F_{\lambda,\mu}^{\nu,\kappa}(t-t')P_\kappa^{(1)}(t')P_\nu^{(1)}(t')\right], \quad (32)$$

where $\beta_{\lambda,\mu}^{\nu,\kappa} = \left\langle \mathcal{D}_{\lambda,\mu}\mathcal{B}_\nu^\dagger\mathcal{B}_\kappa^\dagger\right\rangle_0$ and $F_{\lambda,\mu}^{\nu,\kappa} = \left\langle \mathcal{D}_{\lambda,\mu}e^{-i\mathcal{H}_{SC}t}\mathcal{D}_{\nu,\kappa}^\dagger\right\rangle_0$. These coefficients contain typical average of the form $\langle \mathcal{B}_\lambda\mathcal{B}_\mu\mathcal{B}_\kappa^\dagger\mathcal{B}_\nu^\dagger\rangle$, which determines the spin selection rules governing different optical processes. In order to derive them we present this average in terms of the electron and hole operators and rearrange operators to have $\langle vvv^\dagger v^\dagger\rangle_0\langle ccc^\dagger c^\dagger\rangle_0$ and then expand each term using the Wick theorem for fermions. It produces four terms, which are represented by the diagrams in Fig. 2. The points on the upper line of a diagram represent the spin states of electrons in the conduction band and the points on the lower line stand for the spin states of holes in the valence band. For example, the anticommutator $\{v_{\sigma_\kappa}^\dagger, v_{\sigma_\mu}\} \propto \delta_{\sigma_\kappa,\sigma_\mu}$ requires the equality of the respective hole spins in the valence band. We denote this equality by connecting the vertices κ and μ on the lower line by the arc.

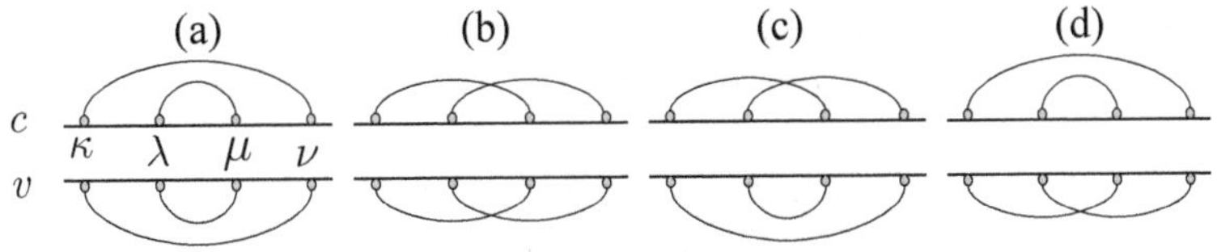

Fig. 2. The spin diagrams corresponding to non-zero terms in $\langle \mathcal{B}_\lambda\mathcal{B}_\mu\mathcal{B}_\kappa^\dagger\mathcal{B}_\nu^\dagger\rangle$. Upper and lower lines correspond to electron and hole spins, respectively. Arcs connecting two vertices denote equal spins. The diagrams (a) and (b) lead to helicity selection rules $\propto \delta_{\kappa,\sigma_\nu}\delta_{\sigma_\lambda,\sigma_\mu}$ and $\propto \delta_{\sigma_\kappa,\sigma_\mu}\delta_{\sigma_\lambda,\sigma_\nu}$, respectively. Diagrams (c) and (d) enter the average with minus sign and for bright excitons require the spin states in the conduction band to be the same.

Also, diagrams in Fig. 2 show which coordinates are identified by delta-functions appearing after anti-commutation and thus demonstrate how exciton wave functions are convoluted in such averages. Thus diagrams in Fig. 2a and 2b describe direct exciton scattering, while those shown in Fig. 2c and 2d take into account scattering with exchange by electron or hole.

The memory term in Eq. (32) accounts for the effect of exciton-exciton interaction. Unfortunately, an exact evolution of this integral is impossible (it is related to four-particle propagator) and one has to rely on approximation schemes. It should be noted, however, that the most significant effect of this interaction is when excitons form a bound state (possibly metastable). When the contribution of such a state is small or non-existent (e.g. there are no bound states in the spectrum of two copolarized excitons) one can employ a short memory approximation, which accounts for the effect of the nonlocal term as a modification of $\beta_{\lambda,\mu}^{\nu,\kappa}$.

The essential difference between linear and nonlinear responses is that the latter is a combination of several linear polarizations. As a result if the external excitation has components with different wave vectors the nonlinear polarization contains not only all of them but also their combinations. More precisely one can easily show that $P_\mu^{(3)} \propto \delta(\mathbf{K}_\mu + \mathbf{K}_\lambda - \mathbf{K}_\nu - \mathbf{K}_\kappa)$. The four momenta have to add up to zero and therefore this is called four-wave mixing response. Because the direction of the response is different from the direction of excitation this allows observing the effect of interaction that is not blurred by linear response or non-absorbed light. This makes four-wave mixing spectroscopy possible.

We would like to emphasize the generality of the derivation of the exciton optical response. For example, the specific form of the exciton states has not been used for the derivation of main formulas and therefore the same arguments can be repeated for excitons in arbitrary confinement potential. This allows using this approach for description of nonlinear optical response of disordered quantum wells (Erementchouk et al., 2011).

Another important feature of the diagrammatic representation is that it establishes the connection between nonlinear optical response and other phenomena which involve exciton dynamics. As an example we would like to discuss the problem of entanglement of photons interacting with a semiconductor quantum well. For simplicity we neglect the possible effect of variation of the dielectric function and assume that the states of the quantized electromagnetic field are plane waves so that the vector potential is presented as

$$\mathcal{A} = \frac{1}{(2\pi)^{3/2}} \sum_{\widehat{k}} (\epsilon)_{\widehat{k}} \frac{1}{\sqrt{2\omega_{\widehat{k}}}} e^{i\mathbf{k}\cdot\mathbf{x}} a^{\dagger}_{\widehat{k}} + \text{h.c.}, \tag{33}$$

where $\widehat{k} = \{\sigma, \mathbf{k}\}$ combines all photon quantum numbers, polarization and wave vector. Then the two-photon states are described by the density matrix, which in the interaction picture has the form

$$\rho^{\widehat{k}_1,\widehat{k}_2}_{\widehat{q}_1,\widehat{q}_2}(t) = \langle \Psi(t) | \, a^{\dagger}_{\widehat{q}_1} a^{\dagger}_{\widehat{q}_2} a_{\widehat{k}_1} a_{\widehat{k}_2} \, | \Psi(t) \rangle , \tag{34}$$

where $|\Psi(t)\rangle$ is the state of the semiconductor-photon system. As the first approximation it suffices to consider vacuum as the initial state of the semiconductor $|\Psi(0)\rangle = |0\rangle$ and to neglect the processes of photon re-absorption, which is justified if the photon lifetime within the quantum well is short (it should be noted that the situation may change in a cavity). In the lowest approximation the photon annihilation operators in Eq. (34) act on a two-photon state yielding the factorization $\rho^{\widehat{k}_1,\widehat{k}_2}_{\widehat{q}_1,\widehat{q}_2}(t) = \Psi^{*}_{\widehat{q}_1,\widehat{q}_2}(t)\Psi_{\widehat{k}_1,\widehat{k}_2}(t)$, where

$$\Psi_{\widehat{k}_1,\widehat{k}_2}(t) = \left\langle a_{\widehat{k}_1}(t) a_{\widehat{k}_2}(t) \mathcal{S}(t) \right\rangle_0 , \tag{35}$$

where $a_{\widehat{k}}(t)$ are the photon operators in the Heisenberg representation. The interaction Hamitonian in this case has the form $\mathcal{H}_{exc} = \sum_\mu \left(\mathcal{A}^{\dagger}_{\mu} \mathcal{B}_\mu + A_\mu(t) \mathcal{B}^{\dagger}_{\mu} \right)$, where

$$\mathcal{A}^{\dagger}_{\mu} = \frac{1}{(2\pi)^{3/2}} \sum_{\widehat{k}} \frac{1}{\sqrt{2\omega_{\widehat{k}}}} \mathbf{d}_{\sigma_\mu, s_\mu} \cdot \epsilon_{\widehat{k}} a^{\dagger}_{\widehat{k}} \int d\mathbf{x} \, \phi_\mu(\mathbf{x}, \mathbf{x}) e^{i\mathbf{k}\cdot\mathbf{x}}. \tag{36}$$

Thus the same diagrams as before can be drawn with the only difference that $\mathcal{B}$ line is accompanied with $\mathcal{A}^{\dagger}$. Quick analysis shows that only two diagrams contribute into Ψ

$$\Psi_{\widehat{k}_1,\widehat{k}_2}(t) =$$ 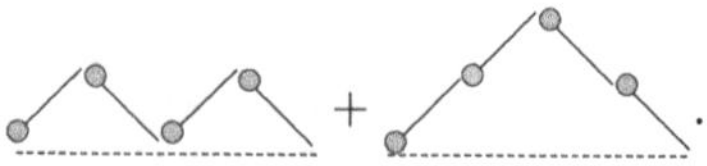 (37)

It is immediately seen that the first diagrams describes the emission of photons along the direction of external excitation and the emitted photons are disentangled: the polarization two-photon state is the direct product of the external excitation polarization. The two-exciton process represented by the second diagram, however, admits oblique emission with non-trivial dependence of entanglement of emitted photons on the direction of observation and the polarization state of the excitation field (Erementchouk & Leuenberger, 2010a) as is summarized in Fig. 3. The polarization of external excitation is described using Poincare sphere, that is the excitation is presented as a combination of left- and right-polarized components with amplitudes $A_- = e^{-i\chi/2}\sin(\beta/2)$ and $A_+ = e^{i\chi/2}\cos(\beta/2)$, respectively, where β is the polar angle on the Poincare sphere and χ is azimuthal angle ($\chi/2$ is the angle between the axis of the ellipse of polarization and the plane spanned by the wave vectors of emitted photons). Near the frequency of the heavy-hole exciton resonance entanglement may reach maximum $E_N = 1$ only in the case of linear polarization of the pump field and entanglement demonstrates interesting dependence on the orientation of the plane of polarization. Near the light-hole exciton resonance the most advantageous orientation of the ellipse of polarization is $\chi = 0$, however, the direction along which the most emitted photons are emitted strongly depends on the ellipticity of the external excitation.

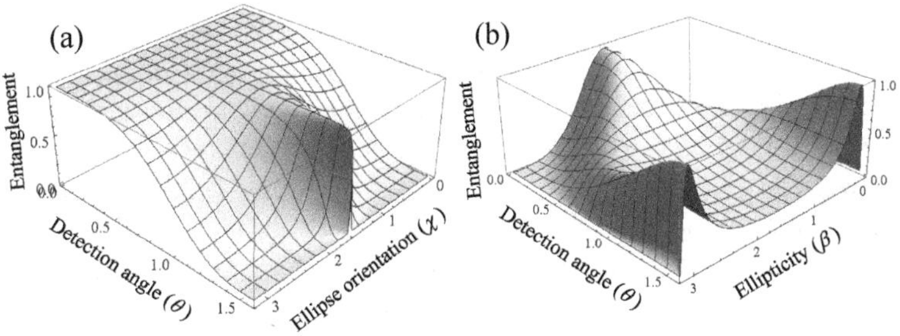

Fig. 3. Dependence of two-photon entanglement on detection angle and on polarization of external excitation in the vicinities of (a) heavy-hole and (b) light-hole exciton resonances.

2.5 Optical response of Bose-Einstein condensate

Bose-Einstein condensation (BEC) is a phenomenon when at non-zero temperature the majority of particles occupy only a few states. This is in striking contrast to a distribution prescribed by the classical theory, where it is governed by the Boltzmann exponent $\exp(-E/k_BT)$ and significant difference in occupations may be expected only when the energy levels are sufficiently far away from each other $\Delta E/k_BT \gg 1$. The Bose-Einstein condensation, as is well known from the standard textbook consideration of ideal Bose gases (see e.g. Chapter 12 in (Huang, 1987), where it is clearly shown how a condensate emerges during the transition to the thermodynamic limit), does not require such level separation and may as well appear in a system with continuous spectrum.

The effect of BEC is tightly connected to such highly unusual from the classical point of view phenomena as superconductivity and superfluidity, which enjoy detailed developed theories (Lifshitz & Pitaevskii, 2002) and still are inexhaustible sources of new questions. In contrast to superfluidity and superconductivity, for which experiments have taken the lead over theoretical considerations, experimental studies of BEC fall well behind the theory (Moskalenko & Snoke, 2000): being predicted in 1925 BEC was obtained in a laboratory only in 1995. The difficulty of observing BEC motivates the constant search for more optimal systems. With this regard the significant attention has been paid to excitons in semiconductors. The physics of transition of semiconductors into the condensate state is similar to the superconducting transition (in conventional superconductors). While the elementary excitations are electrons, at sufficiently low temperatures Cooper pairs of electrons play the important role, which are formed by phonon-mediated attraction. Pairs are no longer subject to the exclusion principle and, moreover, at low densities obey the boson commutation relation. Thus, they may undergo the transition into the BEC-state. With excitons in semiconductors a similar scenario may take place and may even be more favorable because the Coulomb interaction binds electron and hole instead of preventing them from forming a bound state as it is in superconductors.

The finite life time of excitons, however, makes it difficult to reach the condensate state — excitons decay before the equilibrium is established. Therefore, recently indirect excitons in coupled quantum wells became the object of special interest (Butov, Gossard & Chemla, 2002; Butov, Lai, Ivanov, Gossard & Chemla, 2002; Snoke et al., 2002). The electrons and holes are spatially separated in such a structure that leads to increased life-time. Recently another possibility, BE condensate formed by dark excitons, started to attract attention (Combescot & Leuenberger, 2009).

The obvious difficulty related to dark excitons is how to observe them. One of possible ways to test properties of dark excitons is to use indirect interband spectroscopy, which relies on dynamics of bright states modified by the presence of dark excitons.

The problem of the optical response can be approached along the same lines as in the previous sections. The Hamiltonian of light-matter interaction is treated as perturbation and using the interaction picture the exciton polarization of *bright* excitons $P = \langle \mathcal{B} \rangle$ is found in terms of $\mathcal{S}$ operator, which produces the perturbation theory. Before we apply this ideology we need to revisit the notion of averaging in formulas containing $\langle \ldots \rangle$. In Section 2.4 the initial state of semiconductor was taken to be vacuum. Here, however, we need to take into account that initially the system is in thermal equilibrium and therefore its state is given by density matrix ρ rather than by a vector of state. Thus for an operator $\mathcal{O}$ we need to consider

$$\langle \mathcal{O} \rangle = \mathrm{Tr}\left[\rho \mathcal{O}\right]. \tag{38}$$

In the case when the system has BE condensate this expression significantly simplifies because the main contribution results from the condensate states, which contain the macroscopic number of particles. This observation is formally expressed by the spectral decomposition for the density matrix $\rho = \sum_n w_n \left|\psi_n\right\rangle \left\langle \psi_n\right|$, where $\left|\psi_n\right\rangle$ are some orthogonal states and w_n are their weights. The ratio of the weights of non-condensate and condensate states is small, $w_{nc}/w_c \ll 1$, and thus the contribution of the respective terms in Eq. (38) can be neglected leaving us with

$$\langle \mathcal{O} \rangle = \sum_c w_c \left\langle \psi_c \right| \mathcal{Q} \left|\psi_c\right\rangle. \tag{39}$$

We model the condensate state by a coherent state, which is obtained from vacuum by Glauber's shift operator $\left|\psi_c\right\rangle = \mathcal{D}_c^\dagger(\lambda)\left|0\right\rangle$, where

$$\mathcal{D}_\sigma(\lambda) = \exp\left[\lambda\left(\mathcal{B}_c^\dagger - \mathcal{B}_c\right)\right] \tag{40}$$

with $\mathcal{B}_\sigma^\dagger = \int d\mathbf{x}_1 d\mathbf{x}_2\, \chi(\mathbf{x}_1, \mathbf{x}_2) v_\sigma^\dagger(\mathbf{x}_2) c_s^\dagger(\mathbf{x}_1)$. Here λ and function χ are parameters of the state and will be determined later. It should be noted, however, that translational symmetry of the system implies that $\chi(\mathbf{x}_1, \mathbf{x}_2)$ should posses the symmetry with respect to translations of both arguments $\chi(\mathbf{x}_1 + \mathbf{a}, \mathbf{x}_2 + \mathbf{a}) = e^{i\mathbf{P}\cdot\mathbf{a}}\chi(\mathbf{x}_1, \mathbf{x}_2)$ with some $\mathbf{P}$. Clearly $\mathbf{P}$ can be eliminated in a moving frame and, hence, among states created by $\mathcal{B}^\dagger_{\boldsymbol{\sigma}}$ the smallest energy would be of those with $\mathbf{P} = 0$. Thus we need to consider

$$\mathcal{B}^\dagger_{\boldsymbol{\sigma}} = \int d\mathbf{x}_1 d\mathbf{x}_2\, \chi(\mathbf{x}_1 - \mathbf{x}_2) v^\dagger_\sigma(\mathbf{x}_2) c^\dagger_s(\mathbf{x}_1). \tag{41}$$

The spin states of hole and electron entering the pair creation operator $\mathcal{B}^\dagger_{\boldsymbol{\sigma}}$ are denoted by $\boldsymbol{\sigma} = \{\sigma, s\}$ and are such that the pair does not interact with the electromagnetic field. As has been shown above, there are two such states with $\boldsymbol{\sigma}_- = \{3/2, -1/2\}$ and $\boldsymbol{\sigma}_+ = \{-3/2, 1/2\}$. In the absence of external static magnetic field the energy of these states is the same thus we naturally have fragmented condensate approximately described by the density matrix $\rho = (|\psi_+\rangle\langle\psi_+| + |\psi_-\rangle\langle\psi_-|)/2$, where $|\psi_\pm\rangle = \mathcal{D}^\dagger_{\boldsymbol{\sigma}_\pm}|0\rangle$. Using this approximation in Eq. (39) we obtain

$$\langle \mathcal{O}\rangle = \frac{1}{2}\left[\langle\psi_+|\,\mathcal{O}\,|\psi_+\rangle + \langle\psi_-|\,\mathcal{O}\,|\psi_-\rangle\right], \tag{42}$$

which reduces the problem of finding exciton polarization to the problem with pure initial state similar to analyzed in Section 2.4. The first and the second terms in Eq. (42) turn into each other under the inversion of spins in the condensate, which does not present any difficulty. Therefore, we consider in details only the first term. In order to simplify notations we denote the spin states of the condensate by simply σ and s and the complementary values by $\bar{\sigma}$ and $\bar{s}$. Thus for $|\psi_+\rangle$ we have $\sigma = -3/2$ and $s = 1/2$ while $\bar{\sigma} = 3/2$ and $\bar{s} = -1/2$. The bright excitons correspond to spins $\{\bar{\sigma}, s\}$ and $\{\sigma, \bar{s}\}$, while $\{\bar{\sigma}, \bar{s}\}$ are the spins in the another fraction (spanned by $|\psi_-\rangle$).

Applying the diagrammatic representation of the perturbation series (as illustrated in Eq. (21) one can immediately see that the series will contain only the same diagrams as in Section 2.4. In contrast, if the condensate was made of bright excitons then all diagrams shown in Eq. (21) would contribute. For example, the first diagram, without the external field, would describe the radiative decay of excitons in the condensate. For dark condensate, however, this diagram turns to zero because $\langle c_{\bar{s}}\rangle_{\psi_\pm} = \langle c_{\bar{\sigma}}\rangle_{\psi_\pm} = 0$. Thus one can see that only diagrams with matching number of creation and annihilation operators of electrons or holes not containing in the condensate are not vanishing and one need to keep only diagrams with the vacuum line, where vacuum is understood for electrons with spin $\bar{s}$ or for holes with spin $\bar{\sigma}$.

Unitarity of the shift operator $\mathcal{D}_+(\lambda)$ allows one to present the average as taken over vacuum $\langle\psi_+|\,\mathcal{O}\,|\psi_+\rangle = \langle\mathcal{O}(\lambda)\rangle_0$, where

$$\mathcal{O}(\lambda) = \mathcal{D}_+(\lambda)\mathcal{O}\mathcal{D}^\dagger_+(\lambda). \tag{43}$$

Only electron and hole operators with spins s and σ are affected by this transformation. In order to find how they transform it is convenient to use the momentum representation, e.g. $v_\sigma(\mathbf{x}) = (2\pi)^{-d/2}\int d\mathbf{k} v_\sigma(\mathbf{k})e^{i\mathbf{k}\cdot\mathbf{x}}$ with d being the dimensionality of the problem. In this representation we have

$$\mathcal{B}^\dagger_+ = \int d\mathbf{k}\, \chi(\mathbf{k}) v^\dagger_\sigma(\mathbf{k}) c^\dagger_s(-\mathbf{k}) \tag{44}$$

and thus

$$\frac{\partial}{\partial\lambda} v_\sigma(\mathbf{k};\lambda) = \mathcal{D}_+(\lambda)\left[\mathcal{B}^\dagger_+, v_\sigma(\mathbf{k})\right]\mathcal{D}^\dagger_+(\lambda) = -\chi(-\mathbf{k}) c^\dagger_s(-\mathbf{k};\lambda). \tag{45}$$

Solving the system of equations we find the transformation induced by the shift operator is equivalent to the Bogoliubov transformation

$$v_\sigma(\mathbf{k};\lambda) = v_\sigma(\mathbf{k})\alpha(\mathbf{k}) - c_s^\dagger(-\mathbf{k})\beta(-\mathbf{k}), \qquad c_s(\mathbf{k};\lambda) = c_s(\mathbf{k})\alpha(\mathbf{k}) + v_\sigma^\dagger(-\mathbf{k})\beta(-\mathbf{k}), \tag{46}$$

where

$$\alpha(\mathbf{k}) = \cos[\lambda\chi(\mathbf{k})], \qquad \beta(\mathbf{k}) = \sin[\lambda\chi(\mathbf{k})]. \tag{47}$$

In order to clarify the physical meaning of the transformation we present Eq. (46) in coordinate representation

$$v_\sigma(\mathbf{x};\lambda) = v_\sigma(\mathbf{x}) - \int d\mathbf{x}'\,\tilde{\alpha}(\mathbf{x}-\mathbf{x}')v_\sigma(\mathbf{x}') - \int d\mathbf{x}'\,\beta(\mathbf{x}-\mathbf{x}')c_s^\dagger(\mathbf{x}'), \tag{48}$$

where $\tilde{\alpha}(\mathbf{x})$ is the Fourier transform of $2\sin^2[\lambda\chi(\mathbf{k})/2]$.

Under transformation (43) the Heisenberg representation $e^{i\mathcal{H}_{SC}t}\mathcal{O}e^{-i\mathcal{H}_{SC}t}$ is mapped into the Heisenberg representation with transformed Hamiltonian $e^{i\mathcal{H}_{SC}(\lambda)t}\mathcal{O}(\lambda)e^{-i\mathcal{H}_{SC}(\lambda)t}$. Thus the transformation can be interpreted as a transition to new particles. The condition of the quadratic part of $\mathcal{H}_{SC}(\lambda)$ to be diagonal yields an equation with respect to function $\chi(\mathbf{k})$

$$\left[\widetilde{E}(\mathbf{k}) - 2\int d\mathbf{q}V(\mathbf{k}-\mathbf{q})\beta^2(\mathbf{q})\right]\alpha(\mathbf{k})\beta(\mathbf{k}) = \left[\alpha^2(\mathbf{k}) - \beta^2(\mathbf{k})\right]\int d\mathbf{q}V(\mathbf{k}-\mathbf{q})\alpha(\mathbf{q})\beta(\mathbf{q}), \tag{49}$$

where $\widetilde{E}(\mathbf{k}) = \epsilon_c - \epsilon_v - \Omega + \mathbf{k}^2/2m_e + \mathbf{k}^2/2m_h$. As the zeroth approximation we obtain $\alpha(\mathbf{k}) = 1$ and $\beta(\mathbf{k}) = c\phi(\mathbf{k})$, where $\phi(\mathbf{k}) = 2\sqrt{2r_B^3}/[\pi(r_B^2k^2+1)^2]$ is the Fourier transform of 1s-exciton states in 3d and r_B is the exciton Bohr radius. The constant c is found from the "normalization condition": the electron (or hole) density is equal to the condensate density

$$n_+ = \left\langle c_s^\dagger(\mathbf{x})c_s(\mathbf{x})\right\rangle_+ = \frac{1}{(2\pi)^3}\int d\mathbf{k}d\mathbf{q}e^{i\mathbf{x}\cdot(\mathbf{q}-\mathbf{k})}\left\langle c_s^\dagger(\mathbf{k};\lambda)c_s(\mathbf{q};\lambda)\right\rangle_0 = \frac{1}{(2\pi)^3}\int d\mathbf{k}\beta^2(\mathbf{k}), \tag{50}$$

that is $c^2 = n_+(2\pi)^3$. In the dilute regime, when the condition

$$\eta = 64\pi n_+ r_B^3 < 1 \tag{51}$$

holds we have $\beta(\mathbf{k}) \ll 1$ and hence $\lambda\chi(\mathbf{k}) \approx c\phi(\mathbf{k})$ thus completely defining the condensate state.

The major effect of the dark condensate on bright excitons is that the condensate changes the structure of excitons. In order to see this we consider the polarization of linear response given by Eq. (22) with transformed exciton operators. Let us consider for definiteness the case of right polarized external excitation, which is coupled to bright exciton with $\boldsymbol{\sigma} = \{\sigma, \bar{s}\} = \{-3/2, -1/2\}$. The time dependence of propagator $\Phi_{\mu,\nu}(\tau)$ is determined by

$$\begin{aligned}\mathcal{D}_+(\lambda)\left[\mathcal{H}_{SC}, v_\sigma^\dagger(\mathbf{k}_1)c_{\bar{s}}^\dagger(\mathbf{k}_2)\right]\mathcal{D}_+^\dagger(\lambda)\left|0\right\rangle &= -\int d\mathbf{q}L(\mathbf{k}_1,\mathbf{k}_2;\mathbf{q})v_\sigma^\dagger(\mathbf{k}_1-\mathbf{p})c_{\bar{s}}^\dagger(\mathbf{k}_2+\mathbf{p})\left|0\right\rangle \\ &+\int d\mathbf{p}d\mathbf{q}\,V(\mathbf{p})\left[v_\sigma^\dagger(\mathbf{k}_1-\mathbf{p})c_{\bar{s}}^\dagger(\mathbf{k}_2)\alpha(\mathbf{k}_1-\mathbf{p})\right. \\ &\left.-v_\sigma^\dagger(\mathbf{k}_1)c_{\bar{s}}^\dagger(\mathbf{k}_2-\mathbf{p})\alpha(\mathbf{k}_1)\right]v_\sigma^\dagger(\mathbf{q})c_{\bar{s}}^\dagger(\mathbf{p}-\mathbf{q})A(\mathbf{p},\mathbf{q})\left|0\right\rangle,\end{aligned} \tag{52}$$

where $A(\mathbf{p},\mathbf{q}) = \alpha(\mathbf{q})\beta(\mathbf{q}-\mathbf{p}) - \alpha(\mathbf{q}-\mathbf{p})\beta(\mathbf{q})$ and

$$L(\mathbf{k}_1,\mathbf{k}_2;\mathbf{q}) = \left[\epsilon_c - \epsilon_v - \Omega + \frac{\mathbf{k}_1^2}{2m_h} + \int d\mathbf{p} V(\mathbf{p})\beta(\mathbf{k}_1-\mathbf{p})A(\mathbf{p},\mathbf{k}_1) + \frac{\mathbf{k}_2^2}{2m_e}\right]\delta(\mathbf{q}) \\ - V(\mathbf{q})\left[1-\beta(\mathbf{k}_1)A(\mathbf{q},-\mathbf{k}_1)\right]. \quad (53)$$

The last term in Eq. (52) yields the modification of the energy through perturbation of the condensate. This term can be estimated to lead to relatively long beatings in the linear response and will be neglected in the following.

Similarly to the case of the dynamics of initially unperturbed semiconductor Eq. (52) provides the structure of elementary excitations given by the spectral decomposition of the kernel $\widetilde{L}(\mathbf{k}_1,\mathbf{k}_2;\mathbf{k}_1',\mathbf{k}_2') = \delta(\mathbf{k}_1+\mathbf{k}_2-\mathbf{k}_1'-\mathbf{k}_2')L(\mathbf{k}_1,\mathbf{k}_2;\mathbf{k}_1-\mathbf{k}_2)$: its eigenfunctions define the modified exciton operators, which then should be used in expansion (9). We present results for the lowest energy state in Fig. 4.

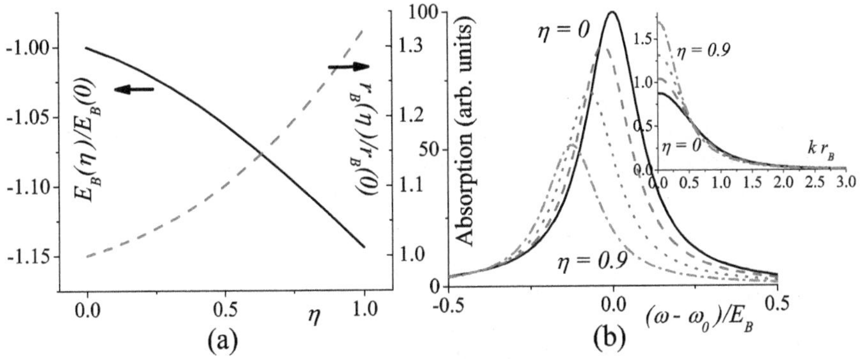

Fig. 4. Modification of bright excitons due to interaction with dark BE condensate. (a) Variation of the exciton binding energy (left scale) and the exciton radius (right scale) with the condensate density [see Eq. (51)]. (b) Absorption spectra for different values of η: solid, dashed, dotted and dash-dotted lines correspond to $\eta = 0, 0.3, 0.6, 0.9$, respectively. The inset shows the exciton wave function in the momentum representation.

The exchange by hole with condensate leads to renormalizations of the mass of the hole and its interaction with electron within the bright exciton. As a result the observed exciton states exhibit a blue shift for the binding energy and increasing radius (see Fig. 4a and inset in Fig. 4b) as functions of the condensate density. In Fig. 4 we have plotted the absorption spectrum for different densities taking into account the variation of the exciton oscillator strength, which is determined by $|\phi(\mathbf{x}=0)|^2 \propto 1/r_B^3(\eta)$.

3. Ab initio approach - time-dependent density functional theory

In this Section, we present an alternative, ab initio, approach to describe ultrafast processes, based on time-dependent density functional theory (TDDFT).(Runge & Gross, 1984) The main advantage of this approach compared to the Green's function many-body theory methods is its simplicity. Being an effective single-electron theory, TDDFT uses time-dependent single-electron density to describe the nonequilibrium response. The corresponding single-particle Kohn-Sham wave equation depends on one vector coordinate and one time variable, which gives a big advantage compared to the Green's function approaches, where in strongly nonequilibrium case one needs to take into account many-time Green's functions, and the truncation used to have a finite system of equations is often not well justified. Moreover, technically to solve this system of many-time Green's functions is a very complicated task since in general case there is no time translation invariance, and one cannot use Fourier transforms and needs to consider time variables on the complex Keldysh type time-contour (see for example (Kadanoff & Baym, 1962)). This analysis can consume a very large amount of time (see, e.g., Refs. (Freericks et al., 2006; Onida et al., 2002)).

Another difficulty in the Green's function case comes from the correct treating of the Coulomb interaction. In the case of pulses shorter than the Coulomb scattering time, one cannot take these effects into account within a phenomenological scattering time parameter, like in the Boltzmann equation approach. Thus, one needs to take into account the electron correlation effects more accurately in this case, which is a very complicated problem in the majority of cases. In TDDFT, provided one has the correct exchange-correlation (XC) potential, these effects are taken into account exactly in the framework of a simple single-particle Kohn-Sham equation with relatively simple XC potential responsible for the electron correlation effects. Even though the form of such a potential in the case of materials containing transition metal atoms or atoms with f-electrons in the valence band is not solved yet, in the case of familiar semiconductor and molecular systems standard local-density approximation (LDA) and generalized gradient approximation (GGA) are often proved to be good approximations, like in description of single-electron excitations in molecular systems.(Elliott et al., 2009)

One of the most important questions in TDDFT is its ability as an effective single-particle theory to describe multi-particle effects, including multiple-electron excitations and bound electron-hole states, like excitons and biexcitons. To be more specific, one needs to find the form of the corresponding XC potential to describe these effects. At the moment our knowledge about the structure of such potentials is rather limited. Though, recently significant progress in this direction has been made.

In the case of a weak and slow perturbation, the algorithm based on the many-body Bethe-Salpeter equation (BSE) to construct the XC potential able to describe excitonic effects was proposed in Refs. (Botti et al., 2004; Marini et al., 2003; Reining et al., 2002). It was shown also that in the linear response regime in frequency representation one can construct a pure TDDFT potential, the time-dependent optimized effective potential in exact-exchange approximation (XX-TDOEP), to describe excitonic effects in the optical absorption spectra.(Kim & Görling, 2002a;b) Despite the progress of the approaches mentioned above, these methods are rather complicated, which makes them difficult to apply in general nonlinear case of strong ultrafast perturbations. Moreover, the question whether they can describe higher order effects, like biexcitons or other nonlinear collective effects, including Bose-Einstein condensation of excitons remains open.

Recently, we proposed formally simple and rather general TDDFT approach based on the density-matrix representation, which is able to describe some of the phenomena mentioned above.(Turkowski et al., 2009; Turkowski & Ullrich, 2008; Turkowski et al., 2010) In this Section, we present the general formalism used in this approach. In particular, we present the system of the TDDFT equations for the density matrix elements, which is the TDDFT version of semiconductor Bloch equations, that includes nonlinear excitonic effects in all orders of magnitude. As we demonstrate, the solution of these equations in the case of excitons, and their generalization on the case of biexcitons, gives the corresponding energies of the bound states and wave functions, allows to analyze nonlinear effects like exciton-exciton interaction, which can be applied to the description of four-wave mixing experiments. Finally, we discuss the possibility to describe with the approach a highly nonlinear, collective effect of the Bose condensation of excitons.

3.1 General formalism

In TDDFT the time evolution of the system is studied by solving time-dependent Kohn-Sham equation:

$$i\frac{\partial}{\partial t}\Psi(\mathbf{r},t) = \hat{H}(\mathbf{r},t)\Psi(\mathbf{r},t), \tag{54}$$

where the time-dependent Hamiltonian

$$\hat{H}(t) = -\frac{\nabla^2}{2} + V_{nucl}(\mathbf{r}) + V_H[n](\mathbf{r},t) + V_{xc}[n](\mathbf{r},t) + V(\mathbf{r},t), \tag{55}$$

consists of the kinetic energy part (first term), the nuclear potential $V_{nucl}(\mathbf{r})$ for the electrons, the Hartree potential $V_H[n](\mathbf{r}) = \int d\mathbf{r}' n(\mathbf{r}')/|\mathbf{r} - \mathbf{r}'|$, the XC potential $V_{xc}[n](\mathbf{r},t)$ and the external field term V(r, t). Generally speaking, in the case of an external electric field $\mathbf{E}(\mathbf{r},t)$, this field has to be included into the Kohn-Sham Hamiltonian through the standard substitution $\nabla \to \nabla - (i/c)\mathbf{A}_{ext}(\mathbf{r},t)$, and the scalar potential term $\varphi_{ext}(\mathbf{r},t)$, where the external vector potential $\mathbf{A}_{ext}(\mathbf{r},t)$ and $\varphi_{ext}(\mathbf{r},t)$ define the electric field:

$$\mathbf{E}(\mathbf{r},t) = -\nabla\varphi_{ext}(\mathbf{r},t) - \frac{1}{c}\frac{\partial \mathbf{A}_{ext}(\mathbf{r},t)}{\partial t}. \tag{56}$$

In the case of extended systems one needs to use the vector potential in order to describe the periodic system. Moreover, one should use the current-TDDFT with the macroscopic current that leads to the periodicity of the system. However, in the case when the field frequency is larger than the level spacing, the scalar potential $V(\mathbf{r},t) = \varphi_{ext}(\mathbf{r},t) = -\mathbf{E}(t)\mathbf{r}$, which significantly simplifies the solution and which we shall use below, is a good approximation.(Schäfer & Wegener, 2002)

The XC potential takes into account all electron exchange and correlation effects beyond the Hartree approximation. The Hartree and the XC potentials depend on the time-dependent electron charge density, which can be expressed in terms of the wave-functions $\Psi^i_{\mathbf{k}}(\mathbf{r},t)$, solutions of Equation (54),

$$n(\mathbf{r},t) = 2\sum_{i,\mathbf{k}} |\Psi^i_{\mathbf{k}}(\mathbf{r},t)|^2 \theta(\varepsilon_F - \varepsilon^{v_i}_{\mathbf{k}}), \tag{57}$$

where $\mathbf{k}$ is the momentum and i is the band (with the energy $\varepsilon^{v_i}_{\mathbf{k}}$) and all other quantum number indices, and ε_F is the Fermi energy in the case of extended bulk system. The bandstructure and the Fermi energy can be obtained from the stationary solution of the Kohn-Sham problem (standard DFT theory(Kohn, 1999)). It is easy to generalize the corresponding problem on the case of finite systems. The simplest well-known approximation for the XC potential (the exchange only part) is the adiabatic LDA approximation: $V_{xLDA}(n(\mathbf{r})) = -(3/\pi)^{1/3} n^{1/3}(\mathbf{r})$. In the case of highly-correlated systems and strongly nonequilibrium processes more complicated forms for $V_{xc}[n](\mathbf{r},t)$ have to be used. In particular, in the case of the ultrafast processes the memory effects (dependence of the XC potential on the charge densities in all previous times) can play an important role.

In order to study the optical transitions in the system, it is convenient to express the time-dependent wave functions as linear combinations of the ground-state wave functions:

$$\Psi_{\mathbf{k}}^{v_i}(\mathbf{r},t) = \sum_j \left[c_{\mathbf{k}}^{v_i v_j}(t)\psi_{\mathbf{k}}^{v_j}(\mathbf{r}) + c_{\mathbf{k}}^{v_i c_j}(t)\psi_{\mathbf{k}}^{c_j}(\mathbf{r}) \right], \tag{58}$$

where $c_{\mathbf{k}}^{v_i l_j}(t)$ are momentum and time-dependent complex coefficients, and $l_i = v_i, c_i$ are the valence and conduction band indices. One can solve the problem by studying the time dependence of these coefficients, however from the physical point of view it is more convenient to analyze the solution for the density matrix: $\rho_{\mathbf{k};\mathbf{qp}}^{v_i;l_m \bar{l}_n}(t) = c_{\mathbf{kq}}^{v_i l_m}(t)\left[c_{\mathbf{kp}}^{v_i \bar{l}_n}(t)\right]^*$. The elements of this matrix correspond are related to the probability of the optical transitions between different bands and their occupation. The density matrix satisfies the Liouville-von Neumann equation:

$$i\frac{\partial}{\partial t}\rho_{\mathbf{k}}^{v_i;l_m l'_n}(t) = [H(t),\rho]_{\mathbf{k}}^{v_i;l_m l'_n} = \sum_{l''_j}\left[H_{\mathbf{kk}}^{l_m l''_j}(t)\rho_{\mathbf{k}}^{v_i;l''_j l'_n}(t) - \rho_{\mathbf{k}}^{v_i;l_m l''_j}(t) H_{\mathbf{k}}^{l''_j l'_n}(t)\right], \tag{59}$$

where

$$H_{\mathbf{kq}}^{l_m l'_n}(t) = \int_{cell} d\mathbf{r}\psi_{\mathbf{k}}^{l_m *}(\mathbf{r})H(t)\psi_{\mathbf{q}}^{l'_n}(\mathbf{r}) = \varepsilon_{\mathbf{k}}^{l_m}\delta_{l_m l'_n} + \mathbf{E}(t)\mathbf{d}_{\mathbf{kq}}^{l_m l'_n} + V_{H\mathbf{kq}}^{l_m l'_n}(t) + V_{xc\mathbf{kq}}^{l_m l'_n}(t), \tag{60}$$

$\mathbf{d}_{\mathbf{kq}}^{l_m l'_n} = \int_{cell} d\mathbf{r}\psi_{\mathbf{k}}^{l_m *}(\mathbf{r})\mathbf{r}\psi_{\mathbf{q}}^{l'_n}(\mathbf{r})$ are the dipole matrix elements, and the space integration is performed over the unit cell. In Eq. (60), $V_{H\mathbf{kq}}^{l_m l'_n}(t)$ and $V_{xc\mathbf{kq}}^{l_m l'_n}(t)$ are the matrix elements for the difference between the time-dependent and ground state (at $t \leq t_0$) potentials $V_H[n](\mathbf{r},t) - V_H[n](\mathbf{r},t_0)$ and $V_{XC}[n](\mathbf{r},t) - V_{XC}[n](\mathbf{r},t_0)$. The Liouville-von Neumann equation has to be solved together with the corresponding charge density equation

$$n(\mathbf{r},t) = 2\sum_{i,l_m,l'_n,\mathbf{k},\mathbf{q},\mathbf{p}} \rho_{\mathbf{k},\mathbf{q},\mathbf{p}}^{v_i;l_m l'_n}(t)\psi_{\mathbf{p}}^{l'_n *}(\mathbf{r})\psi_{\mathbf{q}}^{l_m}(\mathbf{r})\theta(\varepsilon_F - \varepsilon_{\mathbf{k}}^{v_i}). \tag{61}$$

As follows from the last equation, the density matrix defines the time-dependence of the electron charge density. It also allows to calculate many other physical quantities, including the dynamical polarization $\mathbf{D}(t) = \sum_{i,l_m,l'_n,\mathbf{k},\mathbf{q},\mathbf{p}} \rho_{\mathbf{k};\mathbf{qp}}^{v_i;l_m l'_n}(t)\mathbf{d}_{\mathbf{pq}}^{l'_n l_m}$.

3.2 Exciton states

To analyze the possibility of excitonic states in the optical absorption spectrum of the system, in principle it is enough to find the dynamical polarization from the solution of the Liouville-von Neumann equations (59) in the case of an external perturbation, like an external femtosecond pulse: $\mathbf{E}(t) = \mathbf{E}_0 e^{-t^2/\tau^2}$ ($\tau \sim 1 - 100$fs). The optical absorption spectrum $A(\omega)$ can be found from the expression for the polarization $P(\omega)$: $A(\omega) = -2\mathrm{Re}[P(\omega)/E(\omega)]$. In the last equation the total polarization is the sum of polarizations for all possible interband transitions:

$$P^{c_i v_j}(\omega) = i(4\pi/\sqrt{\epsilon_b})(\varepsilon_0^{c_i} - \varepsilon_0^{v_j})\Omega_{cell} \int \frac{d\mathbf{k}}{(2\pi)^3} \int dt e^{i\omega t} |d_{\mathbf{k}}^{v_j c_i}|^2 \rho_{\mathbf{k}}^{c_i v_j}(t) \tag{62}$$

where ϵ_b is the dielectric constant, $\varepsilon_0^{c_i} - \varepsilon_0^{v_j}$ is the corresponding bandgap and Ω_{cell} is the volume of the unit cell.

For simplicity, in this Subsection we shall concentrate on the two-band case with 2×2 density matrix with two independent matrix elements: the conduction band occupancy $\rho_{\mathbf{k}}^{cc}(t)$ and the polarization $\rho_{\mathbf{k}}^{cv}(t)$. The other two elements, the valence band occupancy $\rho_{\mathbf{k}}^{vv}(t)$ and the polarization de-excitation matrix element $\rho_{\mathbf{k}}^{vc}(t)$, can be found from the first two elements by using the conservation of particles equation $\rho_{\mathbf{k}}^{vv}(t) + \rho_{\mathbf{k}}^{cc}(t) = 1$ and the definition of the matrix elements, which gives $\rho_{\mathbf{k}}^{vc}(t) = \rho_{\mathbf{k}}^{cv*}(t)$. The independent TDDFT matrix equations have the following form:

$$\frac{\partial}{\partial t}\rho_{\mathbf{k}}^{vv}(t) = -2\mathrm{Im}\left[(E(t)d_{\mathbf{k}}^{cv} + V_{H\mathbf{k}}^{cv} + V_{xc\mathbf{k}}^{cv})\rho_{\mathbf{k}}^{vc}(t)\right], \tag{63}$$

$$\begin{aligned} \frac{\partial}{\partial t}\rho_{\mathbf{k}}^{cv}(t) = & -i[\varepsilon_{\mathbf{k}}^{c} - \varepsilon_{\mathbf{k}}^{v}]\rho_{\mathbf{k}}^{cv}(t) - i[\rho_{\mathbf{k}}^{vv}(t) - \rho_{\mathbf{k}}^{cc}(t)]E(t)d_{\mathbf{k}}^{cv} \\ & - i[\rho_{\mathbf{k}}^{vv}(t) - \rho_{\mathbf{k}}^{cc}(t)](V_{H\mathbf{k}}^{cv}(t) + V_{xc\mathbf{k}}^{cv}(t)) \\ & - i[V_{H\mathbf{k}}^{cc}(t) + V_{xc\mathbf{k}}^{cc}(t) - V_{H\mathbf{k}}^{vv}(t) - V_{xc\mathbf{k}}^{vv}(t)]\rho_{\mathbf{k}}^{cv}(t). \end{aligned} \tag{64}$$

They correspond to the many-body theory semiconductor Bloch equations,(Haug & Koch, 2004) but in the TDDFT case the correlation effects are taken into account exactly without making the Hartree-Fock truncation.

In order to get a better feeling of the correspondence between both theories, it is instructive to compare both systems of the equations. Applying the same expansion of the wave function in terms of the stationary wave functions and writing down the corresponding Liouville-von-Neumann equation for the density matrix in the case of the Hartree-Fock equation

$$\begin{aligned} i\frac{\partial \Psi_{\mathbf{k}}^{v}(\mathbf{r},t)}{\partial t} = & \left[-\frac{\nabla^2}{2} - \mathbf{E}(t)\mathbf{r} + \int d\mathbf{r}' \frac{\sum_{\mathbf{q}} \Psi_{\mathbf{q}}^{v*}(\mathbf{r}',t)\Psi_{\mathbf{q}}^{v}(\mathbf{r}',t)}{|\mathbf{r}-\mathbf{r}'|}\right] \Psi_{\mathbf{k}}^{v}(\mathbf{r},t) \\ & - \int d\mathbf{r}' \frac{\sum_{\mathbf{q}} \Psi_{\mathbf{q}}^{v*}(\mathbf{r}',t)\Psi_{\mathbf{q}}^{v}(\mathbf{r},t)}{|\mathbf{r}-\mathbf{r}'|} \Psi_{\mathbf{k}}^{v}(\mathbf{r}',t), \end{aligned} \tag{65}$$

one can obtain familiar set of the semiconductor Bloch equations (in linear expansion in the polarization, see below):

$$\frac{\partial}{\partial t}\rho_{\mathbf{k}}^{vv}(t) = -2\mathrm{Im}\left[\left(E(t)d_{\mathbf{k}}^{cv} + \int \frac{d\mathbf{q}}{(2\pi)^3} V(\mathbf{k}-\mathbf{q})\rho_{\mathbf{q}}^{cv}(t)\right)\rho_{\mathbf{k}}^{vc}(t)\right], \tag{66}$$

$$\begin{aligned} \frac{\partial}{\partial t}\rho_{\mathbf{k}}^{cv}(t) = & -i[\varepsilon_{\mathbf{k}}^{c} - \varepsilon_{\mathbf{k}}^{v} - \int \frac{d\mathbf{q}}{(2\pi)^3} V(\mathbf{k}-\mathbf{q})(\rho_{\mathbf{q}}^{cc}(t) - \rho_{\mathbf{q}}^{vv}(t))]\rho_{\mathbf{k}}^{cv}(t) \\ & - i[\rho_{\mathbf{k}}^{vv}(t) - \rho_{\mathbf{k}}^{cc}(t)]\left(E(t)d_{\mathbf{k}}^{cv} - \int \frac{d\mathbf{q}}{(2\pi)^3} V(\mathbf{k}-\mathbf{q})\rho_{\mathbf{q}}^{cv}(t)\right). \end{aligned} \tag{67}$$

Comparizon of the systems of equations (63), (64) and (66), (67) suggests that the electron-hole interaction term $\int \frac{d\mathbf{q}}{(2\pi)^3} V(\mathbf{k}-\mathbf{q})\rho_{\mathbf{q}}^{cv}(t)$ resposible for the Rydberg series of the bound states in the Bloch equation case is contained in the nonlinear functional of the polarization $\rho_{\mathbf{q}}^{cv}(t)$ of the matrix element $V_{xc\mathbf{k}}^{cv}(t)$ (through the polarization that enters in the charge density, Eq. (61)).

We have analyzed the optical absorption spectrum in the case of several XC potentials and found that in some cases the spectrum contains an excitonic peak below the conduction band edge.(Turkowski & Ullrich, 2008) In particular, we have found that the optical absorption spectrum demonstrates a pronounceable excitonic peak when the XC kernel contains the Coulomb singularity $1/q^2$, like in the case of the KLI and Slater potentials.(Krieger et al., 1992) This result is in agreement with Kim and Görling (Kim & Görling, 2002a;b) who showed that in the translational-invariant systems in order to have the excitonic peaks one needs to have such a singularity. On the other hand, it was found that the standard LDA and GGA potentials are "too weak" to produce the peaks.

In order to get a deeper understanding of the structure of the XC kernels necessary to produce the excitonic bound state one can analyze the linearized TDDFT equation for the polarization which corresponds to the Wannier equation $\left[-(\nabla^2/2m_r) - (1/\epsilon r)\right]\phi(\mathbf{r}) = E\phi(\mathbf{r})$ for the exciton eigenenergies and eigenfunctions (m_r is the reduced electron-hole effective mass and ϵ is the static dielectric constant of the material).(Wannier, 1937) The solution of the last equation demonstrates a Rydberg series of the excitonic binding energies qualitatively described by the Elliott formula (Haug & Koch, 2004).

The corresponding TDDFT Wannier equation can be obtained by linearizing equation (64):

$$\sum_{\mathbf{q}} \left[\omega_{\mathbf{q}}^{cv}\delta_{\mathbf{kq}} + F_{\mathbf{kq}}(\omega)\right]\rho_{\mathbf{q}}^{cv}(\omega) = \omega\rho_{\mathbf{k}}^{cv}(\omega) \tag{68}$$

with the effective electron-hole interaction

$$F_{\mathbf{kq}}(\omega) = \frac{2}{\Omega^2}\int_{\Omega} d^3r \int_{\Omega} d^3r' \, \psi_{c\mathbf{k}}^*(\mathbf{r})\psi_{v\mathbf{k}}(\mathbf{r}) f_{\rm XC}(\mathbf{r},\mathbf{r}',\omega)\psi_{v\mathbf{q}}^*(\mathbf{r}')\psi_{c\mathbf{q}}(\mathbf{r}') \tag{69}$$

(in the momentum representation). The corresponding real-space equation can be obtained after the Fourier transforms $\rho(\mathbf{R},\omega) = \sum_{\mathbf{k}} e^{-i\mathbf{kR}}\rho_{\mathbf{k}}^{cv}(\omega)$ and $V_{eh}(\mathbf{R},\mathbf{R}',\omega) = \sum_{\mathbf{k},\mathbf{q}} e^{-i\mathbf{kR}} F_{\mathbf{kq}}(\omega) e^{i\mathbf{qR}'}$, where $\mathbf{R}$ is a direct lattice vector. Since the excitonic wave function extends over many lattice sites, one can consider $\mathbf{R}$ as a continuous variable. In this case the TDDFT Wannier equation takes the following form

$$\left[-\frac{\hbar^2\nabla^2}{2m_r} + E_g^{\rm KS} - \omega\right]\rho_i(\mathbf{r}) + \int_{\substack{\text{all}\\\text{space}}} d^3r' V_{eh}(\mathbf{r},\mathbf{r}',\omega)\rho_i(\mathbf{r}') = 0\,, \tag{70}$$

($m_r^{-1} = m_c^{-1} + m_v^{-1}$ is the reduced mass and $E_g^{\rm KS}$ is the KS band gap). The solution of the last equation gives the exciton eigenfunctions $\rho(\mathbf{R},\omega)$ and eigenenergies, which are defined by a nonlocal, frequency-dependent electron-hole interaction $V_{eh}(\mathbf{r},\mathbf{r}',\omega)$. This interaction is defined by the XC kernel. The analysis of the solution in the case of the of LDA kernel, shows again that it is too weak to produce bound states. On the other hand, it was found that a phenomenological local kernel $f_{\rm XC}^{\rm contact}(\mathbf{r},\mathbf{r}') = -A\delta(\mathbf{r}-\mathbf{r}')$ (A is a positive constant), and a long-range kernel with the Coulomb singularity $\sim 1/q^2$ in the momentum space, $f_{\rm XC}^{\rm LRC}(\mathbf{r},\mathbf{r}') = -\alpha/4\pi|\mathbf{r}-\mathbf{r}'|$ (α is an adjustable parameter, which might be interpreted as an effective inverse screening), give correct lowest excitonic energy at proper choice of the parameters A and α. These values are of the same order of magnitude for one group of semiconductors - zincblende or wurtzite, but can be one or two orders of magnitude different for the semiconductors from different groups.(Turkowski et al., 2009) This suggests that these kernels, similarly to the LDA case, might be defined by system parameters, in particular the electron density and the volume of the unit cell. The generalization of this formalism, based on two-particle density matrix, was used to study the biexcitonic binding energies.(Turkowski et al., 2010)

3.3 Biexcitons

The above formalism can be generalized to the case of multiple excitations, in particular on the case of biexcitons, correlated double electronic excitations.(Turkowski et al., 2010) In principle, one can obtain double excitations and possibly coupled (biexcitonic) states within single-particle TDDFT in the case of non-adiabatic XC kernel. In this case, nonlinear Casida equation for the eigenenergies will have extra solutions in addition to single-particle excitations. In this Subsection, we analyze how one can obtain biexcitonic states within adiabatic approximation since this case corresponds to a transparent biexciton eigenproblem. In order to find such an approach, one may use the natural orbital (NO) representation for the stationary electron eigenfunctions.(Giesbertz et al., 2008; 2009; Pernal et al., 2007) In this case multi-particle excited states can be described by elements of one- and two-electron density matrices, defined as

$$\gamma(x_1, x_1', t) = N \int dx_2 \int dx_3 ... \int dx_N \Psi(x_1, x_2, ..., x_N, t) \Psi^*(x_1', x_2, ..., x_N, t), \tag{71}$$

$$\Gamma(x_1, x_2, x_1', x_2', t) = N(N-1) \int dx_3 ... \int dx_N \Psi(x_1, x_2, ..., x_N, t) \Psi^*(x_1', x_2', ..., x_N, t), \tag{72}$$

where Ψ is the N-particle wave function and $x_i = (\mathbf{r}_i, s_i)$ denotes the space coordinate and spin index.(Giesbertz et al., 2008; 2009; Pernal et al., 2007) In principle, all ground state properties can be obtained from the the single-particle matrix $\gamma(x_1, x_1')$, due to one-to-one correspondence between the matrix and the ground state many-body wave function Ψ (the density matrix functional theory generalization of the Hohenberg-Kohn theorem (Gilbert, 1975)). Though to study the excited states the two-electron density matrix is necessary. We shall concentrate on an effective two-electron theory described by the Hamiltonian

$$\hat{H}(\mathbf{r}_1, \mathbf{r}_2, t) = \hat{h}^{\text{ad}}(\mathbf{r}_1, t) + \hat{h}^{\text{ad}}(\mathbf{r}_2, t) + w[n_2](\mathbf{r}_1, \mathbf{r}_2, t), \tag{73}$$

where $\hat{h}^{\text{ad}}$ is the single-particle TDDFT Hamiltonian (55). In order to have biexcitonic states in the adiabatic approximation, one can introduce an effective two-particle interaction $w[n_2](\mathbf{r}_1, \mathbf{r}_2, t)$ which depends on two-particle density $n_2(\mathbf{r}_1, \mathbf{r}_2, t) = \Psi^*(\mathbf{r}_1, \mathbf{r}_2, t)\Psi(\mathbf{r}_1, \mathbf{r}_2, t)$. Similar to the excitonic case, one can expand the two-electron wave-function in terms of the NOs $\chi_k(\mathbf{r})$, which in the singlet case gives $\Psi(\mathbf{r}, \mathbf{r}', t) = \sum_{k,l} C_{kl}(t)\chi_k(\mathbf{r})\chi_l(\mathbf{r}')$, where $C_{kl}(t)$ is a symmetric matrix with respect to quantum number indices k abd l. In this case, it is easy to show that

$$\gamma(x_1, x_1', t) = \sum_{k,l} \gamma_{kl}(t)\chi_k(x_1)\chi_l^*(x_1'), \tag{74}$$

$$\Gamma(x_1, x_2, x_1', x_2', t) = \sum_{klmn} \Gamma_{klmn}(t)\chi_k(x_1)\chi_l(x_2)\chi_m^*(x_1')\chi_n^*(x_2'), \tag{75}$$

where $\gamma_{kl}(t) = 2\sum_m C_{km}(t)C^{*T}_{lm}(t)$ and $\Gamma_{klmn}(t) = 2C_{kl}(t)C^*_{mn}(t)$. Some of the elements of the last two matrices are proportional to the excitonic and the biexcitonic wave functions (with indices cv and ccvv in the notations of the previous Subsection, correspondingly). Matrix elements $C_{kl}(t)$ satisfy the following equation of motion:

$$i\frac{\partial C_{kl}(t)}{\partial t} = \sum_r (h_{kr}(t)C_{rl}(t) + C_{kr}(t)h_{rl}(t)) + \sum_{rs} w_{klrs}(t)C_{rs}(t) \tag{76}$$

with the initial condition $C_{kl}(t=0) \sim \delta_{kl}$. The matrix elements $h_{kr}(t)$ are defined above and

$$w_{klmn}(t) = \int d\mathbf{r}_1 \int d\mathbf{r}_2 \chi^*_k(\mathbf{r}_1)\chi^*_l(\mathbf{r}_2)w[n_2](\mathbf{r}_1, \mathbf{r}_2, t)\chi_m(\mathbf{r}_1)\chi_n(\mathbf{r}_2). \tag{77}$$

From equation (76) one can obtain the equations for $\gamma_{kl}(t)$ and $\Gamma_{klmn}(t)$:

$$i\frac{\partial \gamma_{kl}}{\partial t} = \sum_r (h_{kr}\gamma_{rl} - \gamma_{kr}h_{rl}) + \sum_{r,s,m} (\Gamma^*_{krsm}w^*_{msrl} - \Gamma_{krsm}w_{msrl}), \tag{78}$$

$$i\frac{\partial \Gamma_{klmn}}{\partial t} = \sum_r (h_{kr}\Gamma_{rlmn} + h_{rl}\Gamma_{krmn} - h_{rm}\Gamma_{klrn} - h_{rn}\Gamma_{klmr}) + \sum_{r,s} (w_{klrs}\Gamma_{rsmn} - w^*_{mnrs}\Gamma_{klrs}). \tag{79}$$

This is a closed system of equations and is the generalization of the single-electron problem problem from the previous Subsection (at $w = 0$ and $\Gamma = 0$) on the two-electron case. Namely, at $w = 0$ in the linear approximation for two bands one obtains the TDDFT-Wannier equation:

$$E^v_{n\mathbf{q}}\gamma^{cv}_{n\mathbf{k},\mathbf{q}} = \sum_{\mathbf{k}'} \left[\left(\varepsilon^c_{\mathbf{k}'+\mathbf{q}} - \varepsilon^v_{\mathbf{k}'} \right) \delta_{\mathbf{k}\mathbf{k}'} + F_{\mathbf{k}\mathbf{k}'} \right] \gamma^{cv}_{n\mathbf{k}',\mathbf{q}}, \tag{80}$$

where $\mathbf{k}$ is the electron momentum and $\mathbf{q}$ is the sum of the electron and hole momenta (we consider more general case of nonzero exciton momentum). The electron-hole interaction $F_{\mathbf{k}\mathbf{k}'}$ is defined in Eq. (69).

Similarly, in the case of two-bands one can obtain the equation for the biexciton function:

$$0 = \left[i\frac{\partial}{\partial t} - \varepsilon^c_{\mathbf{k}+\mathbf{q}} - \varepsilon^c_{\mathbf{k}'} + \varepsilon^v_{\mathbf{k}} + \varepsilon^v_{\mathbf{k}'+\mathbf{q}} \right] \Gamma^{ccvv}_{\mathbf{k}+\mathbf{q},\mathbf{k}',\mathbf{k},\mathbf{k}'+\mathbf{q}} - \sum_{\mathbf{k}'} G_{\mathbf{k}+\mathbf{q},\mathbf{k}';\bar{\mathbf{k}}+\bar{\mathbf{q}},\bar{\mathbf{k}}',\bar{\mathbf{k}},\bar{\mathbf{k}}'+\bar{\mathbf{q}}} \Gamma^{ccvv}_{\bar{\mathbf{k}}+\bar{\mathbf{q}},\bar{\mathbf{k}}',\bar{\mathbf{k}},\bar{\mathbf{k}}'+\bar{\mathbf{q}}}, \tag{81}$$

where $G_{\mathbf{k}+\mathbf{q},\mathbf{k}';\bar{\mathbf{k}}+\bar{\mathbf{q}},\bar{\mathbf{k}}',\bar{\mathbf{k}},\bar{\mathbf{k}}'+\bar{\mathbf{q}}}$ are the matrix elements of the one- and two-electron density kernels $g_1(\mathbf{r}, \mathbf{r}_1, \mathbf{r}_2) = \frac{\delta V_{xc}(\mathbf{r})}{\delta n(\mathbf{r}_1,\mathbf{r}_2)}$ and $g_2(\mathbf{r}_1, \mathbf{r}_2, \mathbf{r}_3, \mathbf{r}_4) = \frac{\delta w(\mathbf{r}_1,\mathbf{r}_2)}{\delta n(\mathbf{r}_3,\mathbf{r}_4)}$ with respect to the Kohn-Sham eigendunctions, similar to $F_{\mathbf{k}\mathbf{k}'}$ (we refer the reader to paper (Turkowski et al., 2010), where the explicit expression for the matrix elements is presented). Eq. (81) can be solved by expanding the biexcitonic function in terms of the complete set of the excitonic functions $\gamma^{cv}_{n,\mathbf{k},\mathbf{q}}$ with eigenenergies $E_{n,\mathbf{q}}$ (n is the number of the bound state). These quantities can be found from the solution of Eq. (80). The next step is to antisymmetrize the corresponding function with respect to interchange of holes and electrons. Then one can get the following expressions for the singlet (−) and triplet (+) biexcitonic functions(Schäfer & Wegener, 2002; Turkowski et al., 2010):

$$\tilde{\Gamma}^{cc'vv'\pm}_{\mathbf{k}+\mathbf{q},\mathbf{k}',\mathbf{k},\mathbf{k}'+\mathbf{q}} = \sum_{n,m} \left[\gamma^v_{n,\mathbf{k}+\mathbf{q},\mathbf{q}}\gamma^{v'}_{m,\mathbf{k}'+\mathbf{q},-\mathbf{q}}b^{\pm}_{nm,\mathbf{q}} \mp \gamma^v_{n,\mathbf{k}',\mathbf{k}'-\mathbf{k}}\gamma^{v'}_{m,\mathbf{k}+\mathbf{q},\mathbf{k}-\mathbf{k}'}b^{\pm}_{nm,\mathbf{k}'-\mathbf{k}} \right]. \tag{82}$$

The solution of the eigenproblem for $b^{\pm}_{nm,\mathbf{q}}$

$$\sum_{n',m',\mathbf{q}'} \left[(\omega - E_{n\mathbf{q}} - E_{m\mathbf{q}})\, \delta_{nn'}\delta_{mm'}\delta_{\mathbf{q}\mathbf{q}'} - H^{\pm}_{nm,n'm',\mathbf{q}\mathbf{q}'} \right] b^{\pm}_{n'm',\mathbf{q}'} = 0, \tag{83}$$

gives one the biexcitonic eigenvectors and eigenenergies. In the last equation $H^{\pm}_{nm,n'm',\mathbf{q}\mathbf{q}'}$ are functionals of the excitonic functions and interaction elements $F_{\mathbf{k}\mathbf{k}'}$ and $G_{\mathbf{k}\mathbf{k}'\mathbf{p}\mathbf{p}'}$ (see Ref. (Turkowski et al., 2010)).

We tested the formalism in the case of several semiconductors by using phenomenological local one-particle kernel $f^{\text{contact}}_{\text{XC}}(\mathbf{r},\mathbf{r}')$ and two-particle "contact biexciton" kernels $g^{\text{local}}_{1}(\mathbf{r},\mathbf{r}_1,\mathbf{r}_2) = -C_0 A_1 \delta(\mathbf{r}-\mathbf{r}_1)\delta(\mathbf{r}-\mathbf{r}_2)$ and $g^{\text{local}}_{2}(\mathbf{r},\mathbf{r}',\mathbf{r}_1,\mathbf{r}_2) = -A_2\delta(\mathbf{r}-\mathbf{r}')\delta(\mathbf{r}-\mathbf{r}_1)\delta(\mathbf{r}-\mathbf{r}_2)$. It was found that indeed the TDDFT can describe the biexcitonic states in the adiabatic approximation.

3.4 Nonlinear effects

It is straightforward to extend the formalism developed above to the nonlinear case, including dynamical exciton-exciton interaction and memory effects. These processes play an important role in the case of ultrafast processes, including four-wave mixing experiments. So far, these nonlinear effects were studied only in the framework of many-body effective models. In most cases, the problem was analyzed by solving a third-order polarization equation. Beyond the importance of developing the TDDFT approach to describe the ultrafast processes, there is another important reason for this. Namely, from the experimental data one can learn about the non-adiabatic structure of the XC kernels, since our knowledge on the non-adiabatic kernels is much more limited comparing to the static adiabatic case. Below, we analyze some possible types of the response of the system by taking into account the memory effects and by using the known asymptotic limits of the XC kernels at low and high frequencies, and compare qualitatively the TDDFT results with the corresponding phenomenological many-body solution.

From equation (64) one can obtain the system of equations for the first and the third order polarizations:

$$i\frac{\partial}{\partial t}P^{(1)}_{\mathbf{k}}(t) = [\varepsilon^{c}_{\mathbf{k}} - \varepsilon^{v}_{\mathbf{k}}]\, P^{(1)}_{\mathbf{k}}(t) + \sum_{\mathbf{q}} \int dt' \alpha_{\mathbf{k}\mathbf{q}}(t,t') P^{(1)}_{\mathbf{q}}(t') + \mathbf{d}^{cv}_{\mathbf{k}}\mathbf{E}(t), \tag{84}$$

$$i\frac{\partial}{\partial t}P^{(3)}_{\mathbf{k}} = [\varepsilon^{c}_{\mathbf{k}} - \varepsilon^{v}_{\mathbf{k}}]\, P^{(3)}_{\mathbf{k}} + \sum_{\mathbf{q}} \int dt' \bar{\alpha}_{\mathbf{k}\mathbf{q}}(t,t') P^{(1)}_{\mathbf{k}}(t) |P^{(1)}_{\mathbf{q}}(t')|^2$$
$$+ \sum_{\mathbf{q},\mathbf{p},\mathbf{p}'} \int dt' \int dt'' \int dt''' \beta_{\mathbf{k}\mathbf{q}\mathbf{p}\mathbf{p}'}(t,t',t'',t''') P^{*(1)}_{\mathbf{k}}(t') P^{(1)}_{\mathbf{p}}(t'') P^{(1)}_{\mathbf{p}'}(t'''), \tag{85}$$

where

$$\alpha_{\mathbf{k}\mathbf{q}}(t,t') = 2\int d\mathbf{r} \int d\mathbf{r}' \Psi^{c(0)*}_{\mathbf{k}}(\mathbf{r}) \Psi^{v(0)}_{\mathbf{k}}(\mathbf{r}) f_{\text{XC}}(\mathbf{r},\mathbf{r}',t,t') \Psi^{c(0)*}_{\mathbf{q}}(\mathbf{r}') \Psi^{v(0)}_{\mathbf{q}}(\mathbf{r}') \tag{86}$$

$$\bar{\alpha}_{\mathbf{k}\mathbf{q}}(t,t') = 2\int d\mathbf{r} \int d\mathbf{r}' \Psi^{c(0)*}_{\mathbf{k}}(\mathbf{r}) \Psi^{v(0)}_{\mathbf{k}}(\mathbf{r}) f_{\text{XC}}(\mathbf{r},\mathbf{r}',t,t') \Psi^{c(0)*}_{\mathbf{q}}(\mathbf{r}') \Psi^{v(0)}_{\mathbf{q}}(\mathbf{r}'), \tag{87}$$

and $\beta_{\mathbf{kqpp}'}(t,t',t'',t''')$ is a sum of matrix elements of f_{XC} and its first two derivatives with respect to the particle density, similar to Eqs. (86) and (87). In the case of four-wave mixing experiments the system of equations (84), (85) can be solved by solving first the linear equation, and then the nonlinear effects can be found by solving Eq. (85). To study nonlinear effects one can also analyze the approximate effective third order equation for the total polarization which corresponds to the system (84), (85):

$$i\frac{\partial P}{\partial t} = \left(\delta + \frac{\beta}{2}|P|^2\right)P - \frac{1}{2}\left(1 - \frac{|P|^2}{n_c}\right)\Omega - \frac{iP^*}{2}\int_{-\infty}^{t} F(t-t')P(t')^2 dt', \tag{88}$$

similar to the many-body equation analyzed in Refs. (Ostreich et al., 1995; 1998). In the last equation,

$$\begin{aligned}
\delta &= \varepsilon^c_{\mathbf{k}} - \varepsilon^v_{\mathbf{k}'}, \\
\beta &\simeq 2\left[\alpha^{cc(3)}_{H\mathbf{kk}} - \alpha^{vv(3)}_{H\mathbf{kk}} - 2\alpha^{cv(1)}_{H\mathbf{kk}} - 2\alpha^{cv(1)}_{\mathbf{kk}}\right](0,0) \\
&\quad + 2\left[\beta^{cc(2)}_{\mathbf{kkk}} - \beta^{vv(2)}_{\mathbf{kkk}}\right](0,0,0) + \gamma^{cv(2)}_{\mathbf{kkkk}}(0,0,0,0) \\
F(t,t') &\simeq \frac{1}{2}\int d\mathbf{r}d\mathbf{r}'dr''\Psi^{c(0)*}_{\mathbf{k}}(\mathbf{r})\Psi^{v(0)}_{\mathbf{k}}(\mathbf{r})f'_{xc}(\mathbf{r},t,\mathbf{r}',t,\mathbf{r}'',t')(|\Psi^{c(0)}_{\mathbf{k}}(\mathbf{r}')|^2 - |\Psi^{v(0)}_{\mathbf{k}}(\mathbf{r}')|^2) \\
&\quad \times \Psi^{c(0)*}_{\mathbf{k}}(\mathbf{r}'')\Psi^{v(0)}_{\mathbf{k}}(\mathbf{r}''),
\end{aligned} \tag{89}$$

Ω is the Rabi frequency and n_c is the maximum density corresponding to the Pauli blocking term (we neglect the momentum variable below).

As it follows from Eq. (89), the nonlinear time-dependent effects are defined by the memory function $F(t-t')$, which depends on the non-adiabatic part of the XC kernel. In the many-body approach, the memory function usually depends on the exciton-exciton correlation function, which is difficult to find, so in this case a phenomenological approach has to be used. For example, as it was proposed in Ref. (Ostreich & Sham, 1999), in the case of slowly varying polarization the memory term can be approximated by

$$-\frac{iP^*}{2}\int_{-\infty}^{t} F(t-t')P(t')^2 dt' \simeq -\frac{iP^*}{2}P(t)^2\int_0^t F(t-t')dt', \tag{90}$$

so the time-correlation effects are defined by the function $g(t) = \int_0^t F(t')dt'$. This function can be expressed in terms of the spectral density $\rho(\omega)$: $g(t) \sim \int_0^\infty d\omega\rho(\omega)\omega^{-1}e^{-i\omega t}$. In the low-frequency limit, which defines the long-time asymptotic behavior of the system, the spectral density can be approximated by a power low-function $\rho(\omega) \sim \omega^\alpha$. This function defines the dissipation processes in the system, i.e. the role of the environment (other excitons) on the behavior of given exciton. In the cases when α is smaller, equal or larger than 1, the dissipation is called "sub-ohmic", "ohmic" and "super-ohmic" (Caldeira & Leggett, 1983). Since the spectral function must decay at large frequencies, the general form of the spectral density was approximated by

$$\rho(\omega) = A\omega^\alpha e^{-\omega/\omega_F}, \tag{91}$$

where ω_F is the frequency scale and A is the normalization constant. Thus, the memory function can be approximated by(Ostreich & Sham, 1999)

$$F(t) = A\int_0^\infty d\omega\omega^\alpha e^{-\omega/\omega_F}e^{-i\omega t}. \tag{92}$$

In the case of an one-dimensional model for excitons, the authors of Ref. (Ostreich & Sham, 1999) found $\alpha = 1$.

One can in principle construct an f_{XC} such that the functions on the right hand sides of the Eqs. (89) and (92) are equal. In this case, the equation for polarization will coincide with the many-body equation, and the solution in both cases will be the same. However, as we show below due to some constraints on the frequency-dependent f_{XC}, the last equation should be corrected. Namely, it is known that the exact asymptotic of the XC kernel at large frequencies is $f_{XC} \sim a + b\omega^{-2}$ (van Leeuwen, 2001). In the case of low frequencies, the information about the exact behavior of f_{XC} is more limited. In particular, it is known that it can have poles in the case of finite system in the discrete part of the spectrum. To get an idea about possible frequency-dependence of the XC kernel for all ranges of frequencies one can consider the case of the homogeneous electron gas, when $f_{XC}(\omega \to 0) \to 0, f_{XC}(\omega \to \infty) \to \omega^{-3/2}$ (Marques & Gross, 2003). From these results one can suggest the following rather general form for the non-adiabatic part of the XC kernel:

$$f_{XC}(\omega) = A\frac{\omega^{\alpha}}{1 + (\omega/\omega_F)^{\alpha+\beta}}, \tag{93}$$

where α is of order of 1, and $\beta = 2$, though the case $\beta = 3/2$ is also worth of special attention. Below we solve Eq. (88) in the case of different values of α and $\beta = 1.5$ and 2. We approximate the XC kernel in the following way: $f_{XC}(\mathbf{r}, t, t') = f^{A}_{XC}[n(\mathbf{r}, t)]f_{XC}(t - t')$, where $f^{A}_{XC}[n(\mathbf{r}, t)]$ is the adiabatic part and $f_{XC}(t - t')$ the last term is defined in Eq. (93). Substitution of this expression into Eq. (89) leads to the following form of the memory function: $F(t) = A\int_0^{\infty} d\omega\omega^{\alpha}[1 + (\omega/\omega_F)^{\alpha+\beta}]^{-1}e^{-i\omega t}$, where A is the integral over the derivative of the adiabatic part with respect to the particle density multiplied by the static wave functions (see Eq. (89)).

We analyze qualitatively possible solutions of Eq. (88) by considering two characteristic cases: an approximate time-evolution of the excitonic density and collective excitations in the case of two different memory functions: the exponentially decaying kernel (91) and the TDDFT-type algebraically decaying kernel (93). The time-dependence of the excitonic density $n(t)$ can be obtained from the equation for polarization by using the ansatz $P_0 = \sqrt{n(t)}exp(i\phi)$, which gives $n(t) \simeq n(0)[1 + n(0)\mathrm{Re}\int_0^t g(t')dt']^{-1}$. One can show that the equilibration takes much longer time in a more realistic case of the TDDFT spectral function, comparing to the exponential one (see Fig. 5).

One can analyze the collective excitations in the excitonic system by separating the slow and fast components of polarization, $P = P_0 + P_1$, so the fast component satisfies the following approximate equation: $\partial P_1/\partial t = -|P_0|^2\left[i\beta P_1 + \int_0^t F(t - t')P_1(t')dt'\right]$. The eigenvalues of this equations can be found numerically from the corresponding equation in frequency representation:

$$\omega/n = \beta - F(0)\int_0^{\infty}\frac{\rho(\omega')d\omega'}{\omega' - \omega - i0} \tag{94}$$

(for details, in particular for the normalization of the function ρ and the spectral sum rule for F(t), see Ref. (Ostreich & Sham, 1999)). It is possible to show that the number of possible collective modes increases from 0 to 2 with the exciton density increasing (the zero-frequency

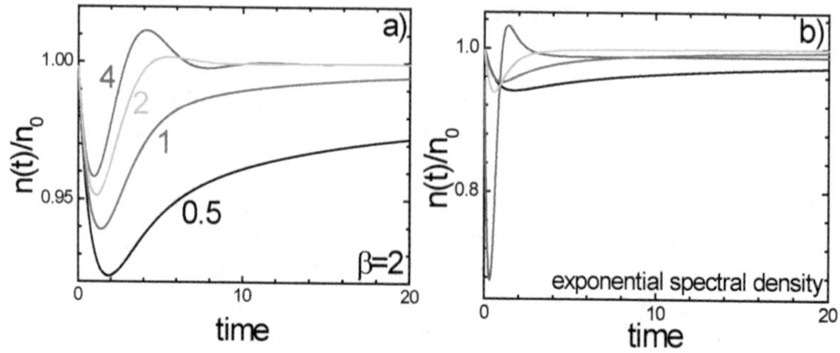

Fig. 5. The time-dependence of the exciton density in the case of $\beta = 2$" (a) and exponential spectral function (b). We have used the parameters for the 3D model of semiconductor from (Ostreich & Sham, 1999): $\beta = 52\omega_x/3$, $F(0) = 14\omega_x^2$, $\omega_x = 6.7meV$. The time is given in units of $1/\omega_x$ and $n(0) = 0.1$.

mode corresponds to the Goldstone mode). Since at small frequencies the change of the right hand side of Eq. (94) with frequency is faster in the case of power spectral function, the critical value for the density above which there are collective oscillations is lower in this case. Also, the corresponding energies for these oscillations are lower in the case of power spectral function.

Finally, similar to the many-body case, one can analyze the two-dimensional Fourier spectrum of the system by taking into account memory effects. It is possible to show that the presence of the memory function in Eq. (88) can not only result in a shift of the excitonic peak in the spectrum but also lead to coupled exciton-exciton states. The detailed results of these studies will be published in the nearest future.

To summarize, we have shown that our TDDFT approach for excitons, despite being at the early stage of the development, shows to be a very promising and powerful method that can be used in many applications, in particular in studies of ultrafast processes.

4. Acknowledgements

V.T. would like to thank Prof. C.A. Ullrich and Prof. T.S. Rahman for a collaboration which led to a part of the results presented in the paper. This work was supported in part by DoE grant no. DE-FG02-07ER15842 and grant no. DOE-DE-FG02-07ER46354 (V.T.). M.N.L. acknowledges support from NSF (Grant No. ECCS-0725514), DARPA/MTO (Grant No. HR0011-08-1-0059), NSF (Grant No. ECCS-0901784), AFOSR (Grant No. FA9550-09-1-0450), and NSF (Grant No. ECCS-1128597).

5. References

Botti, S., Sottile, F., Vast, N., Olevano, V. & Reining, L. (2004). Long-range contribution to the exchange-correlation kernel of time-dependent density functional theory, *Physical Review B* 69(15): 155112(1–14).

Butov, L., Gossard, A. & Chemla, D. (2002). Macroscopically ordered state in an exciton system, *Nature* 418: 751–754.

Butov, L., Lai, C., Ivanov, A., Gossard, A. & Chemla, D. (2002). Towards Bose-Einstein condensation of excitons in potential traps, *Nature* 417: 47–52.

Caldeira, A. O. & Leggett, A. J. (1983). Path integral approach to quantum brownian motion, *Physica A* 121(3): 587–616.

Combescot, M. & Leuenberger, M. (2009). General argument supporting Bose-Einstein condensate of dark excitons in single and double quantum wells, *Solid State Communications* 149: 567–571.

Elliott, P., Burke, K. & Furche, F. (2009). Excited states from time-dependent density functional theory, *in* K. B. Lipkowitz & T. R. Cundari (eds), *Reviews in Computational Chemistry*, Wiley, Hoboken Springer, New Jersey, pp. 91–165.

Erementchouk, M. & Leuenberger, M. (2010a). Entanglement of photons due to nonlinear optical response of quantum wells, *Physical Review B* 81: 195308–195308–9.

Erementchouk, M. & Leuenberger, M. (2010b). Nonlinear optics in semiconductor nanostructures, *in* K. Sattler (ed.), *Nanoelectronics and Nanophotonics*, CRC Press, Boca Raton, pp. 29–1–29–20.

Erementchouk, M., Leuenberger, M. & Li, X. (2011). 2d Fourier spectroscopy of disordered quantum wells, *physica status solidi (c)* 8: 1141–1144.

Freericks, J. K., Turkowski, V. M. & Zlatic, V. (2006). Nonequilibrium dynamical mean-field theory, *Physical Review Letters* 97(26): 266408(1–4).

Giesbertz, K. J. H., Baerends, E. J. & Gritsenko, O. (2008). Charge transfer, double and bond-breaking excitations with time-dependent density matrix functional theory, *Physical Review Letters* 101(3): 033004(1–4).

Giesbertz, K. J. H., Pernal, K., Gritsenko, O. V. & Baerends, E. J. (2009). Excitation energies with time-dependent density matrix functional theory: Singlet two-electron systems, *Journal of Chemical Physics* 130(11): 114104(1–16).

Gilbert, T. L. (1975). Hohenberg-kohn theorem for nonlocal external potentials, *Physical Review B* 12(6): 2111–2120.

Haug, H. & Koch, S. W. (2004). *Quantum theory of the optical and electronic properties of semiconductors*, 4th edn, World Scientific, Singapore.

Huang, K. (1987). *Statistical mechanics*, J. Wiley, New York.

Ivchenko, E. (2005). *Optical Spectroscopy of Semiconductor Nanostructures*, Alpha Science International, Harrow, UK.

Kadanoff, L. P. & Baym, G. (1962). *Quantum Statistical Mechanics*, Benjamin, New York.

Kim, Y. H. & Görling, A. (2002a). Exact Kohn-Sham exchange kernel for insulators and its long-wavelength behavior, *Physical Review B* 66(3): 035114(1–6).

Kim, Y. H. & Görling, A. (2002b). Excitonic optical spectrum of semiconductors obtained by time-dependent density-functional theory with the exact-exchange kernel, *Physical Review Letters* 89(9): 096402(1–4).

Kohn, W. (1999). Nobel lecture: Electronic structure of matter — wave functions and density functionals, *Review of Modern Physics* 71(5): 1253–1266.

Krieger, J. B., Li, Y. & Iafrate, G. J. (1992). Construction and application of an accurate local spin-polarized kohn-sham potential with integer discontinuity: Exchange-only theory, *Physical Review A* 45(1): 101–126.

Lifshitz, E. M. & Pitaevskii, L. P. (2002). *Statictical Physics, Part 2*, Pergamon Press.

Marini, A., Sole, R. D. & Rubio, A. (2003). Bound excitons in time-dependent density-functional theory: Optical and energy-loss spectra, *Physical Review Letters* 91(25): 256402(1–4).

Marques, M. A. L. & Gross, E. K. U. (2003). Time-dependent density-functional theory, *in* C. Fiolhais, F. Nogueira & M. A. L. Marques (eds), *A Primer in Density-Functional Theory, Lecture Notes in Physics, Vol. 620*, Springer, Berlin, pp. 144–184.

Moskalenko, S. A. & Snoke, D. W. (2000). *Bose-Einstein Condensation of Excitons and Biexcitons*, Cambridge University Press.

Onida, G., Reining, L. & Rubio, A. (2002). Electronic excitations: density-functional versus many-body greenâĂŹs-function approaches, *Review of Modern Physics* 74(2): 601–659.

Ostreich, T., Schonhammer, K. & Sham, L. J. (1995). Exciton-exciton correlation in the nonlinear optical regime, *Physical Review Letters* 74(23): 4698–4701.

Ostreich, T., Schonhammer, K. & Sham, L. J. (1998). Theory of exciton-exciton correlation in nonlinear optical response, *Physical Review B* 58(19): 12920–12936.

Ostreich, T. & Sham, L. J. (1999). Collective oscillations driven by correlation in the nonlinear optical regime, *Physical Review Letters* 83(17): 3510–3513.

Pernal, K., Gritsenko, O. & Baerends, E. J. (2007). Time-dependent density-matrix-functional theory, *Physical Review A* 75(1): 012506(1–8).

Reining, L., Olevano, V., Rubio, A. & Onida, G. (2002). Excitonic effects in solids described by time-dependent density-functional theory, *Physical Review Letters* 88(6): 066404(1–4).

Runge, E. & Gross, E. K. U. (1984). Density-functional theory for time-dependent systems, *Physical Review Letters* 52(12): 997–1000.

Schäfer, W. & Wegener, M. (2002). *Semiconductor Optics and Transport Phenomena*, Springer, Berlin.

Snoke, D., Denev, S., Liu, Y., Pfeiffer, L. & West, K. (2002). Long-range transport in excitonic dark states in coupled quantum wells, *Nature* 418: 754–757.

Turkowski, V., Leonardo, A. & Ullrich, C. A. (2009). Time-dependent density-functional approach for exciton binding energies, *Physical Review B* 79(23): 233201(1–4).

Turkowski, V. & Ullrich, C. A. (2008). Time-dependent density-functional theory for ultrafast interband excitations, *Physical Review B* 77(7): 075204(1–11).

Turkowski, V., Ullrich, C. A., Rahman, T. S. & Leuenberger, M. N. (2010). Time-dependent density-matrix functional theory for biexcitonic phenomena, *Physical Review B* 82(20): 205208(1–6).

van Leeuwen, R. (2001). Key concepts in time-dependent density-functional theory, *International Journal of Modern Physics B* 15(14): 1969–2023.

Wannier, G. H. (1937). The structure of electronic excitation levels in insulating crystals, *Physical Review* 52(3): 191–197.

5

Electroweak Interactions in a Chiral Effective Lagrangian for Nuclei

Brian D. Serot and Xilin Zhang
Department of Physics and Center for Exploration of Energy and Matter
Indiana University–Bloomington
USA

1. Introduction

The understanding of electroweak (EW) interactions in nuclei has played an important role in nuclear and particle physics. Previously, the electromagnetic (EM) interaction has provided valuable information about nuclear structure. On the other hand, weak interactions, which are intrinsically correlated with the EM interaction, can be complementary to the EM probe. Moreover, a good knowledge of (anti)neutrino–nucleus scattering cross sections is needed in other processes, including neutrino-oscillation experiments, neutrino astrophysics, and others.

To understand EW interactions in nuclei, we need to deal with the strong interaction that binds nucleons together. The fundamental theory of the strong interaction is quantum chromodynamics (QCD), which is a relativistic field theory with local gauge invariance, whose elementary constituents are colored quarks and gluons. In principle, QCD should provide a complete description of nuclear structure and dynamics. Unfortunately, QCD predictions at nuclear length scales with the precision of existing (and anticipated) experimental data are not available, and this state of affairs will probably persist for some time. Even if it becomes possible to use QCD to describe nuclei directly, this description is likely to be cumbersome and inefficient, since quarks cluster into hadrons at low energies.

How can we make progress towards understanding the EW interactions of nuclei? We will employ a framework based on Lorentz-covariant, effective quantum field theory and density functional theory. Effective field theory (EFT) embodies basic principles that are common to many areas of physics, such as the natural separation of length scales in the description of physical phenomena. In EFT, the long-range dynamics is included explicitly, while the short-range dynamics is parameterized generically; all of the dynamics is constrained by the symmetries of the interaction. When based on a local, Lorentz-invariant lagrangian (density), EFT is the most general way to parameterize observables consistent with the principles of quantum mechanics, special relativity, unitarity, gauge invariance, cluster decomposition, microscopic causality, and the required internal symmetries.

Covariant meson–baryon effective field theories of the nuclear many-body problem (often called quantum hadrodynamics or QHD) have been known for many years to provide a realistic description of the bulk properties of nuclear matter and heavy nuclei. [See Refs. (Furnstahl, 2003; Serot & Walecka, 1986; 1997; Serot, 2004), for example.] Some time ago, a QHD effective field theory (EFT) was proposed (Furnstahl et al., 1997) that includes all of the relevant symmetries of the underlying QCD. In particular, the spontaneously broken $SU(2)_L \otimes SU(2)_R$ chiral symmetry is realized nonlinearly. The motivation for this EFT and illustrations of some calculated results are discussed in Refs. (Furnstahl et al., 1997; Hu et al., 2007; Huertas, 2002; 2003; 2004; McIntire et al., 2007; McIntire, 2008; Serot, 2007; 2010), for example. This QHD EFT has also been applied to a discussion of the isovector axial-vector current in nuclei (Ananyan et al., 2002).

This QHD EFT has three desirable features: (1) It uses the same degrees of freedom to describe the currents and the strong-interaction dynamics; (2) It respects the same internal symmetries, both discrete and continuous, as the underlying QCD; and (3) Its parameters can be calibrated using strong-interaction phenomena, like π N scattering and the properties of finite nuclei (as opposed to EW interactions with nuclei).

In this work, we focus on the introduction of EW interactions in the QHD EFT, with the Delta (1232) resonance (Δ) included as manifest degrees of freedom. To realize the symmetries of QCD in QHD EFT, including both chiral symmetry $SU(2)_L \otimes SU(2)_R$ and discrete symmetries, we apply the background-field technique (Gasser & Leutwyler, 1984; Serot, 2007). Based on the EW synthesis in the Standard Model, a proper substitution of background fields in terms of EW gauge bosons in the lagrangian, as constrained by the EW interactions of quarks (Donoghue et al., 1992), leads to EW interactions of hadrons at low energy. This lagrangian has a linear realization of the $SU(2)_V$ isospin symmetry and a nonlinear realization of the spontaneously broken $SU(2)_L \otimes SU(2)_R$ (modulo $SU(2)_V$) chiral symmetry (when the pion mass is zero). It was shown in Ref. (Furnstahl et al., 1997) that by using Georgi's naive dimensional analysis (NDA) (Georgi, 1993) and the assumption of naturalness (namely, that all appropriately defined, dimensionless couplings are of order unity), it is possible to truncate the lagrangian at terms involving only a few powers of the meson fields and their derivatives, at least for systems at normal nuclear densities (Müller & Serot, 1996). It was also shown that a mean-field approximation to the lagrangian could be interpreted in terms of density functional theory (Kohn, 1999; Müller & Serot, 1996; Serot & Walecka, 1997), so that calibrating the parameters to observed bulk and single-particle nuclear properties (approximately) incorporates many-body effects that go beyond Dirac–Hartree theory. Explicit calculations of closed-shell nuclei provided such a calibration and verified the naturalness assumption. This approach therefore embodies the three desirable features needed for a description of electroweak interactions in the nuclear many-body problem.

Moreover, the technical issues involving spin-3/2 degrees of freedom in relativistic quantum field theory are also discussed here (Krebs et al., 2010; Pascalutsa, 2008). Following the construction of the lagrangian, we apply it to calculate certain matrix elements to illustrate the consequences of chiral symmetries in this theory, including the conservation of vector current (CVC) and the partial conservation of axial-vector current (PCAC). To explore the discrete symmetries, we talk about the manifestation of G parity in these current matrix elements.

This chapter is organized as follows: After a short introduction, we discuss chiral symmetry and discrete symmetries in QCD in the framework of background fields. The EW interactions of quarks are also presented, and this indicates the relation between the EW bosons and background fields. Then we consider the nonlinear realization of chiral symmetry and other symmetries in QHD EFT, as well as the EW interactions. Following that, we outline the lagrangian with the Δ included. Subtleties concerning the number of degrees of freedom and redundant interaction terms are discussed. Finally, some concrete calculations of matrix elements serve as examples and manifestations of symmetries in the theory. We also briefly touch on how this formalism can be used to study neutrino–nucleus scattering (Serot & Zhang, 2010; 2011a;b; Zhang, 2012).

2. QCD, symmetries, and electroweak synthesis

In this section, we talk about various symmetries in QCD including Lorentz-invariance, C, P, and T symmetries, and approximate $SU(2)_L \otimes SU(2)_R$ chiral symmetry (together with baryon number conservation). The last one is the major focus. Here, we consider only u and d quarks, and their antiquarks, while others are chiral singlets. Moreover, the EW interactions, realized in the electroweak synthesis of the Standard Model, are also discussed with limited scope.

2.1 Symmetries

To consider the symmetries, we apply the background-field technique (Gasser & Leutwyler, 1984). First we introduce background fields into the QCD lagrangian, including $\mathsf{v}^\mu \equiv \mathsf{v}^{i\mu}\tau_i/2$ (isovector vector), $\mathsf{v}^\mu_{(s)}$ (isoscalar vector), $\mathsf{a}^\mu \equiv \mathsf{a}^{i\mu}\tau_i/2$ (isovector axial-vector), $\mathsf{s} \equiv \mathsf{s}^i\tau_i/2$ (isovector scalar), and $\mathsf{p} \equiv \mathsf{p}^i\tau_i/2$ (isovector pseudoscalar), where $i = x, y, z$ or $+1, 0, -1$ (the convention about $i = \pm 1$, 0 will be shown in Sec. 3.1):

$$\begin{aligned}
\mathcal{L} &= \mathcal{L}_{QCD} + \bar{q}\gamma_\mu(\mathsf{v}^\mu + B\mathsf{v}^\mu_{(s)} + \gamma_5 \mathsf{a}^\mu)q - \bar{q}(\mathsf{s} - i\gamma_5 \mathsf{p})q \\
&= \mathcal{L}_{QCD} + \bar{q}_L\gamma_\mu(l^\mu + B\mathsf{v}^\mu_{(s)})q_L + \bar{q}_R\gamma_\mu(r^\mu + B\mathsf{v}^\mu_{(s)})q_R \\
&\quad - \bar{q}_L(\mathsf{s} - i\mathsf{p})q_R - \bar{q}_R(\mathsf{s} + i\mathsf{p})q_L \\
&\equiv \mathcal{L}_{QCD} + \mathcal{L}_{\text{ext}} \, .
\end{aligned} \tag{1}$$

Here, $r^\mu = \mathsf{v}^\mu + \mathsf{a}^\mu$, $l^\mu = \mathsf{v}^\mu - \mathsf{a}^\mu$, $q_L = \frac{1}{2}(1-\gamma_5)\,q$, $q_R = \frac{1}{2}(1+\gamma_5)\,q$, $q = (u\ ,d)^T$ and $B = 1/3$ is the baryon number. To preserve C, P, and T invariance of $\mathcal{L}$, the change of background fields under these discrete symmetry transformations are determined by the the properties of the currents coupled to them. The details are in Tabs. 1 and 2. Inside the tables, $\mathcal{P}^\mu_\nu = \mathrm{diag}(1,-1,-1,-1)_{\mu\nu}$ and $\mathcal{T}^\mu_\nu = \mathrm{diag}(-1,1,1,1)_{\mu\nu}$. Moreover, the Lorentz-invariance is manifest, considering the definition of these background fields.

	v^μ	$\mathsf{v}^\mu_{(s)}$	a^μ	s	p
C	$-\mathsf{v}^{T\mu}$	$-\mathsf{v}^\mu_{(s)}$	$\mathsf{a}^{T\mu}$	s^T	p^T
P	$\mathcal{P}^\mu_\nu \mathsf{v}^\nu$	$\mathcal{P}^\mu_\nu \mathsf{v}^\nu_{(s)}$	$-\mathcal{P}^\mu_\nu \mathsf{a}^\nu$	s	$-\mathsf{p}$
T	$-\mathcal{T}^\mu_\nu \mathsf{v}^\nu$	$-\mathcal{T}^\mu_\nu \mathsf{v}^\nu_{(s)}$	$-\mathcal{T}^\mu_\nu \mathsf{a}^\nu$	s	$-\mathsf{p}$

Table 1. Transformations of background fields under C, P, and T operations. The transformations of spacetime arguments are not shown here.

	r^μ	l^μ	$f_{R\mu\nu}$	$f_{L\mu\nu}$	$f_{s\mu\nu}$
C	$-l^{T\mu}$	$-r^{T\mu}$	$-f^T_{L\mu\nu}$	$-f^T_{R\mu\nu}$	$-f^T_{s\mu\nu}$
P	$\mathcal{P}^\mu_\nu l^\nu$	$\mathcal{P}^\mu_\nu r^\nu$	$\mathcal{P}^\lambda_\mu \mathcal{P}^\sigma_\nu f_{L\lambda\sigma}$	$\mathcal{P}^\lambda_\mu \mathcal{P}^\sigma_\nu f_{R\lambda\sigma}$	$\mathcal{P}^\lambda_\mu \mathcal{P}^\sigma_\nu f_{s\lambda\sigma}$
T	$-\mathcal{T}^\mu_\nu r^\nu$	$-\mathcal{T}^\mu_\nu l^\nu$	$-\mathcal{T}^\lambda_\mu \mathcal{T}^\sigma_\nu f_{R\lambda\sigma}$	$-\mathcal{T}^\lambda_\mu \mathcal{T}^\sigma_\nu f_{L\lambda\sigma}$	$-\mathcal{T}^\lambda_\mu \mathcal{T}^\sigma_\nu f_{s\lambda\sigma}$

Table 2. Continuation of Tab. 1.

To understand $SU(2)_L \otimes SU(2)_R \otimes U(1)_B$ symmetry, we can see that the $\mathcal{L}$ defined in Eq. (1) has this symmetry with the following *local* transformation rules:

$$q_{LA} \to \exp\left[-i\frac{\theta(x)}{3}\right]\left(\exp\left[-i\theta_{Li}(x)\,\frac{\tau^i}{2}\right]\right)_A^{\ B} q_{LB} \equiv \exp\left[-i\frac{\theta(x)}{3}\right](L)_A^{\ B} q_{LB}\,, \tag{2}$$

$$q_R \to \exp\left[-i\frac{\theta(x)}{3}\right]\exp\left[-i\theta_{Ri}(x)\,\frac{\tau^i}{2}\right] q_R \equiv \exp\left[-i\frac{\theta(x)}{3}\right] R q_R\,, \tag{3}$$

$$l^\mu \to L\, l^\mu L^\dagger + iL\,\partial^\mu L^\dagger\,, \tag{4}$$

$$r^\mu \to R\, r^\mu R^\dagger + iR\,\partial^\mu R^\dagger\,, \tag{5}$$

$$\mathrm{v}^\mu_{(s)} \to \mathrm{v}^\mu_{(s)} - \partial^\mu\theta\,, \tag{6}$$

$$\mathrm{s} + i\mathrm{p} \to R(\mathrm{s} + i\mathrm{p})L^\dagger\,. \tag{7}$$

We can also construct field strength tensors that transform homogeneously:

$$f_{L\mu\nu} \equiv \partial_\mu l_\nu - \partial_\nu l_\mu - i\left[l_\mu\,,\, l_\nu\right] \to L f_{L\mu\nu} L^\dagger\,, \tag{8}$$

$$f_{R\mu\nu} \equiv \partial_\mu r_\nu - \partial_\nu r_\mu - i\left[r_\mu\,,\, r_\nu\right] \to R f_{R\mu\nu} R^\dagger\,, \tag{9}$$

$$f_{s\mu\nu} \equiv \partial_\mu \mathrm{v}_{(s)\nu} - \partial_\nu \mathrm{v}_{(s)\mu} \to f_{s\mu\nu}\,. \tag{10}$$

2.2 Electroweak synthesis

Now we can discuss the electroweak synthesis ($SU_L(2) \otimes U_Y(1)$) of the Standard Model, which is mostly summarized in Tab. 3 (electric charge $Q = Y/2 + T^3_L$) (Donoghue et al., 1992; Itzykson & Zuber, 1980). We ignore the Higgs fluctuations and gauge boson self-interactions:

$$\begin{aligned}\mathcal{L}_I &= -\overline{q_L}\gamma^\mu\left(g\frac{\tau_i}{2}W^i_\mu + g'\frac{Y}{2}B_\mu\right)q_L - \overline{q_R}\gamma^\mu\left(g'\frac{Y}{2}B_\mu\right)q_R \\ &= -\overline{q_L}\gamma^\mu g\left(\frac{\tau_{+1}}{2}W^{+1}_\mu + \frac{\tau_{-1}}{2}W^{-1}_\mu\right)q_L - \overline{q_L}\gamma^\mu\left(g\frac{\tau_0}{2}W^0_\mu + g'\frac{Y}{2}B_\mu\right)q_L \\ &\quad - \overline{q_R}\gamma^\mu\left(g'\frac{Y}{2}B_\mu\right)q_R\,. \end{aligned} \tag{11}$$

	T_L	T_L^3	Q	Y	B
u_L	$\frac{1}{2}$	$\frac{1}{2}$	$\frac{2}{3}$	$\frac{1}{3}$	$\frac{1}{3}$
d_L	$\frac{1}{2}$	$-\frac{1}{2}$	$-\frac{1}{3}$	$\frac{1}{3}$	$\frac{1}{3}$
u_R	0	0	$\frac{2}{3}$	$\frac{4}{3}$	$\frac{1}{3}$
d_R	0	0	$-\frac{1}{3}$	$-\frac{2}{3}$	$\frac{1}{3}$

Table 3. Multiplets in electroweak synthesis.

Here g, g' and e are the $SU(2)_L$, $U(1)_Y$ and $U(1)_{EM}$ charges. To make sure that $U_{EM}(1)$ is preserved, we impose the following redefinition of excitations relative to the vacuum (θ_w is the weak mixing angle):

$$B^\mu = \cos\theta_w A^\mu - \sin\theta_w Z^\mu \,, \tag{12}$$

$$W^{0\mu} = \cos\theta_w Z^\mu + \sin\theta_w A^\mu \,, \tag{13}$$

$$g\sin\theta_w = g'\cos\theta_w \equiv e \,. \tag{14}$$

After substituting Eqs. (12) to (14) into Eq. (11), we have the right coupling for the EM interaction. Let's compare Eq. (11) with Eq. (1); we deduce the following (V_{ud} describes u and d mixing):

$$\begin{aligned} l_\mu &= -e\frac{\tau^0}{2}A_\mu + \frac{g}{\cos\theta_w}\sin^2\theta_w\frac{\tau^0}{2}Z_\mu \\ &- \frac{g}{\cos\theta_w}\frac{\tau^0}{2}Z_\mu - gV_{ud}\left(W_\mu^{+1}\frac{\tau_{+1}}{2} + W_\mu^{-1}\frac{\tau_{-1}}{2}\right), \end{aligned} \tag{15}$$

$$r_\mu = -e\frac{\tau^0}{2}A_\mu + \frac{g}{\cos\theta_w}\sin^2\theta_w\frac{\tau^0}{2}Z_\mu \,, \tag{16}$$

$$\mathsf{v}_{(s)\mu} = -e\frac{1}{2}A_\mu + \frac{g}{\cos\theta_w}\sin^2\theta_w\frac{1}{2}Z_\mu \,. \tag{17}$$

Furthermore,

$$\begin{aligned} f_{L\mu\nu} &= -e\frac{\tau^0}{2}A_{[\nu,\mu]} + \frac{g}{\cos\theta_w}\sin^2\theta_w\frac{\tau^0}{2}Z_{[\nu,\mu]} - \frac{g}{\cos\theta_w}\frac{\tau^0}{2}Z_{[\nu,\mu]} \\ &- gV_{ud}\frac{\tau^{+1}}{2}W_{+1[\nu,\mu]} - gV_{ud}\frac{\tau^{-1}}{2}W_{-1[\nu,\mu]} \\ &+ \text{interference terms including } (WZ), (WA), (WW), \text{ but no } (ZA)\,, \end{aligned} \tag{18}$$

$$f_{R\mu\nu} = -e\frac{\tau^0}{2}A_{[\nu,\mu]} + \frac{g}{\cos\theta_w}\sin^2\theta_w\frac{\tau^0}{2}Z_{[\nu,\mu]} \quad \text{(no interference terms)}\,, \tag{19}$$

$$f_{s\mu\nu} = -e\frac{1}{2}A_{[\nu,\mu]} + \frac{g}{\cos\theta_w}\sin^2\theta_w\frac{1}{2}Z_{[\nu,\mu]} \,. \tag{20}$$

Here $A_{[\nu,\mu]} \equiv \partial_\mu A_\nu - \partial_\nu A_\mu$ and so are the indices of other fields. If we define [see Eq. (1)]

$$\mathcal{L}_{\text{ext}} \equiv \mathsf{v}_{i\mu} V^{i\mu} - \mathsf{a}_{i\mu} A^{i\mu} + \mathsf{v}_{(s)\mu} J^{B\mu}$$

$$= J^L_{i\mu}\, l^{i\mu} + J^R_{i\mu}\, r^{i\mu} + \mathsf{v}_{(s)\mu} J^{B\mu} , \tag{21}$$

$$\mathcal{L}_I = -e J^{EM}_\mu A^\mu - \frac{g}{\cos\theta_w} J^{NC}_\mu Z^\mu - g V_{ud}\, J^L_{+1\mu} W^{+1\mu} - g V_{ud}\, J^L_{-1\mu} W^{-1\mu} , \tag{22}$$

and use Eqs. (15) to (17), we can discover

$$J^L_{i\mu} \equiv \frac{1}{2}\left(V_{i\mu} + A_{i\mu}\right) , \tag{23}$$

$$J^R_{i\mu} \equiv \frac{1}{2}\left(V_{i\mu} - A_{i\mu}\right) , \tag{24}$$

$$J^{EM}_\mu = V^0_\mu + \frac{1}{2} J^B_\mu , \tag{25}$$

$$J^{NC}_\mu = J^{L0}_\mu - \sin^2\theta_w\, J^{EM}_\mu . \tag{26}$$

Here J^B_μ is the baryon current, defined to be coupled to $\mathsf{v}^\mu_{(s)}$. These relations are consistent with the charge algebra $Q = T^0 + B/2$ (B is the baryon number). $V^{i\mu}$ and $A^{i\mu}$ are the isovector vector current and the isovector axial-vector current, respectively. J^{NC}_μ, $J^L_{\pm 1\mu}$ are the conventional neutral current (NC) and charged current (CC) up to normalization factors.

3. QHD EFT, symmetries, and electroweak interactions

Here we present parallel discussions about QHD EFT's symmetries and EW interactions. The QHD EFT, as an EFT of QCD at low energy, should respect all the symmetries of QCD. Moreover, the approximate global chiral symmetry $SU(2)_L \otimes SU(2)_R \otimes U(1)_B$ in two flavor QCD is spontaneously broken to $SU(2)_V \otimes U(1)_B$, and is also manifestly broken due to the small masses of the quarks. To implement such broken global symmetry in the phenomenological lagrangian using hadronic degrees of freedom, it was found that there exists a general nonlinear realization of such symmetry (Callan et al., 1969; Coleman et al., 1969; Weinberg, 1968). Here, we follow the procedure in Ref. (Gasser & Leutwyler, 1984). The discussion about the conventions is presented first.

3.1 Conventions

In this work, the metric $g_{\mu\nu} = \text{diag}(1,-1,-1,-1)_{\mu\nu}$, and for the Levi–Civita symbol $\epsilon^{\mu\nu\alpha\beta}$, the convention is $\epsilon^{0123} = 1$. Since we are going to talk about the Δ, which is the lowest isospin $I = 3/2$ nucleon resonance, we define the conventions for isospin indices. The following example, which shows the relation between two isospin representations for Δ, may help explain the convention:

$$\Delta^{*a} \equiv T^a_{\ iA} \Delta^{*iA} . \tag{27}$$

Here $a = \pm 3/2, \pm 1/2$, $i = \pm 1, 0$, and $A = \pm 1/2$. The upper components labeled as 'a', 'i', and 'A' furnish $\mathcal{D}^{(3/2)}$, $\mathcal{D}^{(1)}$, and $\mathcal{D}^{(1/2)}$ representations of the isospin $SU(2)$ group. (We work with spherical vector components for $I = 1$ isospin indices, which requires some care with signs.) We can immediately realize that $T^{a}_{\ iA} = \langle 1, \frac{1}{2}; i, A | \frac{3}{2}; a \rangle$, which are CG coefficients. It is well known that the complex conjugate representation of $SU(2)$ is equivalent to the representation itself, so we introduce a metric linking the two representations to raise or lower the indices a, i, and A. For example, $\Delta_a \equiv (\Delta^{*a})^* = T^{\dagger\, iA}_{a} \Delta_{iA}$, where $T^{\dagger\, iA}_{a} = \langle \frac{3}{2}; a | 1, \frac{1}{2}; i, A \rangle$, should also be able to be written as

$$\Delta_a = T^{iA}_{a} \Delta_{iA} \equiv T^{b}_{jB}\, \widetilde{\delta}_{ba}\, \widetilde{\delta}^{ji}\, \widetilde{\delta}^{BA}\, \Delta_{iA} \,. \tag{28}$$

Here, $\widetilde{\delta}$ denotes a metric for one of the three representations. It can be shown that in this convention, $T^{\dagger\, iA}_{a} = T^{iA}_{a}$. Details about the conventions are given in Appendix 7.A.

3.2 QHD's symmetry realizations

Now we proceed to discuss a low-energy lagrangian involving N^A, Δ^a, π^i, ρ^i_μ, and the chiral singlets V_μ and ϕ (Furnstahl et al., 1997; Serot & Walecka, 1997). Under the transformations shown in Eqs. (2) to (7), the symmetry is realized nonlinearly in terms of hadronic degrees of freedom (Gasser & Leutwyler, 1984):

$$U \equiv \exp\left[2i \frac{\pi_i(x)}{f_\pi} t^i\right] \to L U R^\dagger \,, \tag{29}$$

$$\xi \equiv \sqrt{U} = \exp\left[i \frac{\pi_i}{f_\pi} t^i\right] \to L \xi h^\dagger = h\, \xi R^\dagger \,, \tag{30}$$

$$\widetilde{v}_\mu \equiv \frac{-i}{2} [\xi^\dagger (\partial_\mu - i l_\mu) \xi + \xi (\partial_\mu - i r_\mu) \xi^\dagger] \equiv \widetilde{v}_{i\mu} t^i \to h\, \widetilde{v}_\mu h^\dagger - i h\, \partial_\mu h^\dagger \,, \tag{31}$$

$$\widetilde{a}_\mu \equiv \frac{-i}{2} [\xi^\dagger (\partial_\mu - i l_\mu) \xi - \xi (\partial_\mu - i r_\mu) \xi^\dagger] \equiv \widetilde{a}_{i\mu} t^i \to h\, \widetilde{a}_\mu h^\dagger \,, \tag{32}$$

$$\widetilde{\partial}_\mu U \equiv \partial_\mu U - i l_\mu U + i U r_\mu \to L\, \widetilde{\partial}_\mu U R^\dagger \,, \tag{33}$$

$$(\widetilde{\partial}_\mu \psi)_\alpha \equiv (\partial_\mu + i\, \widetilde{v}_\mu - i \mathrm{v}_{(s)\mu} B)^{\beta}_{\alpha}\, \psi_\beta \to \exp\left[-i\theta(x) B\right] h^{\beta}_{\alpha} (\widetilde{\partial}_\mu \psi)_\beta \,, \tag{34}$$

$$\widetilde{v}_{\mu\nu} \equiv -i[\widetilde{a}_\mu \,, \widetilde{a}_\nu] \to h\, \widetilde{v}_{\mu\nu} h^\dagger \,, \tag{35}$$

$$F^{(+)}_{\mu\nu} \equiv \xi^\dagger f_{L\mu\nu}\, \xi + \xi f_{R\mu\nu}\, \xi^\dagger \to h F^{(+)}_{\mu\nu} h^\dagger \,, \tag{36}$$

$$F^{(-)}_{\mu\nu} \equiv \xi^\dagger f_{L\mu\nu}\, \xi - \xi f_{R\mu\nu}\, \xi^\dagger \to h F^{(-)}_{\mu\nu} h^\dagger \,, \tag{37}$$

$$\widetilde{\partial}_\lambda F^{(\pm)}_{\mu\nu} \equiv \partial_\lambda F^{(\pm)}_{\mu\nu} + i[\widetilde{v}_\lambda \,, F^{(\pm)}_{\mu\nu}] \to h\, \widetilde{\partial}_\lambda F^{(\pm)}_{\mu\nu} h^\dagger \,. \tag{38}$$

In the preceding equations, t^i are the generators of reducible representations of $SU(2)$. Specifically, they could be generators of $\mathcal{D}^{(1/2)}_N \oplus \mathcal{D}^{(1)}_\rho \oplus \mathcal{D}^{(3/2)}_\Delta$, which operate on non-Goldstone isospin multiplets including the nucleon, ρ meson, and Δ. We generically label these fields by $\psi_\alpha = (N_A, \rho_i, \Delta_a)_\alpha$. Most of the time, the choice of t^i is clear from the context. B is the baryon number of the particle. The transformations of the isospin and chiral singlets V_μ and ϕ are trivial ($\phi \to \phi$, $V_\mu \to V_\mu$). h is generally a local $SU(2)_V$ matrix. We also make use of the dual field tensors, for example, $\overline{F}^{(\pm)\mu\nu} \equiv \epsilon^{\mu\nu\alpha\beta} F^{(\pm)}_{\alpha\beta}$, which have the same chiral transformations as the ordinary field tensors. Here we do not include the background fields s and p mentioned in Eq. (1), which are the source of manifest chiral-symmetry breaking in the Standard Model.

The C, P, and T transformation rules are summarized in Tabs. 4 and 5. A plus sign means normal, while a minus sign means abnormal, i.e., an extra minus sign exists in the transformation. The convention for *Dirac matrices* sandwiched by nucleon and/or Δ fields are

$$C\overline{N}\Gamma N C^{-1} = \begin{cases} -N^T \Gamma^T \overline{N}^T, & \text{normal}; \\ N^T \Gamma^T \overline{N}^T, & \text{abnormal}. \end{cases} \tag{39}$$

$$C(\overline{\Delta}\Gamma N + \overline{N}\Gamma\Delta)C^{-1} = \begin{cases} -\Delta^T \Gamma^T \overline{N}^T - N^T \Gamma^T \overline{\Delta}^T, & \text{normal}; \\ +\Delta^T \Gamma^T \overline{N}^T + N^T \Gamma^T \overline{\Delta}^T, & \text{abnormal}. \end{cases} \tag{40}$$

$$Ci(\overline{\Delta}\Gamma N - \overline{N}\Gamma\Delta)C^{-1} = \begin{cases} +i\Delta^T \Gamma^T \overline{N}^T - iN^T \Gamma^T \overline{\Delta}^T, & \text{normal}; \\ -i\Delta^T \Gamma^T \overline{N}^T + iN^T \Gamma^T \overline{\Delta}^T, & \text{abnormal}. \end{cases} \tag{41}$$

Here, in Eqs. (39), (40), and (41), the extra minus sign arises because the fermion fields anticommute. The factor of i in Eq. (41) is due to the requirement of Hermiticity of the lagrangian. To make the analysis easier for $\overline{\Delta}\Gamma N + C.C.$, we can just attribute a minus sign to an i under the C transformation. Whenever an i exists, the lagrangian takes the form $i(\overline{\Delta}\Gamma N - \overline{N}\Gamma\Delta)$. When no i exists, the lagrangian is like $\overline{\Delta}\Gamma N + \overline{N}\Gamma\Delta$.

For P and T transformations, the conventions are the same for N and Δ fields, except for an extra minus sign in the parity assignment for each Δ field (Weinberg, 1995a), so we list only the N case:

$$P\overline{N}\Gamma_\mu N P^{-1} = \begin{cases} \overline{N}\mathcal{P}_\mu^\nu \Gamma_\nu N, & \text{normal}; \\ -\overline{N}\mathcal{P}_\mu^\nu \Gamma_\nu N, & \text{abnormal}. \end{cases} \tag{42}$$

$$T\overline{N}\Gamma_\mu N T^{-1} = \begin{cases} \overline{N}\mathcal{T}_\mu^\nu \Gamma_\nu N, & \text{normal}; \\ -\overline{N}\mathcal{T}_\mu^\nu \Gamma_\nu N, & \text{abnormal}. \end{cases} \tag{43}$$

It is easy to generalize these results to $\Gamma_{\mu\nu}$, etc.

Now a few words about isospin structure are in order. Suppose an isovector object is denoted as $O_\mu \equiv O_{i\mu} t^i$, then the conventions are explained below:

$$CO_\mu C^{-1} = \begin{cases} O_\mu^T, & \text{normal}; \\ -O_\mu^T, & \text{abnormal}. \end{cases} \tag{44}$$

$$PO_\mu P^{-1} = \begin{cases} \mathcal{P}_\mu^\nu O_\nu, & \text{normal}; \\ -\mathcal{P}_\mu^\nu O_\nu, & \text{abnormal}. \end{cases} \tag{45}$$

$$TO_\mu T^{-1} = \begin{cases} \mathcal{T}_\mu^\nu O_\nu, & \text{normal}; \\ -\mathcal{T}_\mu^\nu O_\nu, & \text{abnormal}. \end{cases} \tag{46}$$

	γ^μ	$\sigma^{\mu\nu}$	1	$\gamma^\mu\gamma_5$	$i\gamma_5$	i	$i\overleftrightarrow{\partial}$	$\epsilon^{\mu\nu\alpha\beta}$
C	$-$	$-$	$+$	$+$	$+$	$-$	$-$	$+$
P	$+$	$+$	$+$	$-$	$-$	$+$	$+$	$-$
T	$-$	$-$	$+$	$-$	$-$	$-$	$-$	$-$

Table 4. Transformation properties of objects under C, P, and T. Here '$+$' means normal and '$-$' means abnormal.

	$\widetilde{a}_\mu$	$\widetilde{v}_\mu$	$\widetilde{v}_{\mu\nu}$	ρ_μ	$\rho_{\mu\nu}$	$\overline{\rho}_{\mu\nu}$	V_μ	$V_{\mu\nu}$	$\overline{V}_{\mu\nu}$	$F^{(\pm)}_{\mu\nu}$	$f_{s\mu\nu}$	$\overline{F}^{(\pm)}_{\mu\nu}$	$\overline{f}_{s\mu\nu}$
C	$+$	$-$	$-$	$-$	$-$	$-$	$-$	$-$	$-$	$\mp$	$-$	$\mp$	$-$
P	$-$	$+$	$+$	$+$	$+$	$-$	$+$	$+$	$-$	$\pm$	$+$	$\mp$	$-$
T	$-$	$-$	$-$	$-$	$-$	$+$	$-$	$-$	$+$	$-$	$-$	$+$	$+$

Table 5. Continuation of Tab. 4.

The same convention applies to the isovector (pseudo)tensors. For isovector (pseudo)scalars, the $\mathcal{P}$ and $\mathcal{T}$ should be changed to **1**. For the C transformation, O^T means transposing both isospin and Dirac matrices in the definition of O, if necessary.

3.3 QHD EFT lagrangian (without Δ) and electroweak interactions

Now we begin to discuss the QHD EFT lagrangian. Based on the symmetry transformation rules discussed above, we can construct the lagrangian as an invariant of these transformations by using the building blocks shown in Eqs. (29) to (38). In principle, there are an infinite number of possible interaction terms in this lagrangian. However, power counting (Furnstahl et al., 1997; Hu et al., 2007; McIntire et al., 2007) and Naive Dimensional Analysis (NDA) (Georgi & Manohar, 1984; Georgi, 1993) enable us to truncate this series of interactions to achieve a good approximation. Following the discussion in Ref. (Furnstahl et al., 1997), we associate with each interaction term a power-counting index:

$$\hat{v} \equiv d + \frac{n}{2} + b\,. \tag{47}$$

Here d is the number of derivatives (small momentum transfer) in the interaction, n is the number of fermion fields, and b is the number of heavy meson fields.

The QHD theory has been developed for some time. Details can be found in Refs. (Furnstahl et al., 1997; Serot & Walecka, 1997; Serot, 2007). *Here, we give a complete treatment of electroweak interactions in this theory.* (However, we do not discuss "seagull" terms of higher order in the couplings.) We begin with

$$\begin{aligned}
\mathcal{L}_{N(\hat{v}\leqslant 3)} &= \overline{N}(i\gamma^\mu[\widetilde{\partial}_\mu + ig_\rho\rho_\mu + ig_v V_\mu] + g_A\gamma^\mu\gamma_5\widetilde{a}_\mu - M + g_s\phi)N \\
&\quad - \frac{f_\rho g_\rho}{4M}\overline{N}\rho_{\mu\nu}\sigma^{\mu\nu}N - \frac{f_v g_v}{4M}\overline{N}V_{\mu\nu}\sigma^{\mu\nu}N - \frac{\kappa_\pi}{M}\overline{N}\,\widetilde{v}_{\mu\nu}\sigma^{\mu\nu}N \\
&\quad + \frac{4\beta_\pi}{M}\overline{N}N\,\mathrm{Tr}(\widetilde{a}_\mu\widetilde{a}^\mu) + \frac{i\kappa_1}{2M^2}\overline{N}\gamma_\mu\overleftrightarrow{\widetilde{\partial}}_\nu N\,\mathrm{Tr}(\widetilde{a}^\mu\widetilde{a}^\nu) \\
&\quad + \frac{1}{4M}\overline{N}\sigma^{\mu\nu}(2\lambda^{(0)}f_{s\mu\nu} + \lambda^{(1)}F^{(+)}_{\mu\nu})N\,,
\end{aligned} \tag{48}$$

where $\widetilde{\partial}_\mu$ is defined in Eq. (34), $\overleftrightarrow{\widetilde{\partial}}_\nu \equiv \widetilde{\partial}_\nu - (\overleftarrow{\partial}_\nu - i\widetilde{v}_\nu + iv_{(s)\nu})$, and the new field tensors are $V_{\mu\nu} \equiv \partial_\mu V_\nu - \partial_\nu V_\mu$ and

$$\rho_{\mu\nu} \equiv \partial_{[\mu}\rho_{\nu]} + i\overline{g}_\rho[\rho_\mu\,,\,\rho_\nu] + i([\widetilde{v}_\mu\,,\,\rho_\nu] - \mu \leftrightarrow \nu) \rightarrow h\,\rho_{\mu\nu}h^\dagger\,. \tag{49}$$

The superscripts $^{(0)}$ and $^{(1)}$ denote the isospin. In Appendix 7.B, details about the tilde objects (which are defined exactly above) are shown explicitly in terms of pion and background fields.

Next is a purely mesonic lagrangian:

$$\begin{aligned}
\mathcal{L}_{\text{meson}(\hat{\nu} \leqslant 4)} = {} & \frac{1}{2}\partial_\mu\phi\,\partial^\mu\phi + \frac{1}{4}f_\pi^2\,\mathrm{Tr}[\widetilde{\partial}_\mu U(\widetilde{\partial}^\mu U)^\dagger] + \frac{1}{4}f_\pi^2\,m_\pi^2\,\mathrm{Tr}(U + U^\dagger - 2) \\
& - \frac{1}{2}\,\mathrm{Tr}(\rho_{\mu\nu}\rho^{\mu\nu}) - \frac{1}{4}\,V^{\mu\nu}V_{\mu\nu} \\
& + \frac{1}{2}\left(1 + \eta_1\frac{g_s\phi}{M} + \frac{\eta_2}{2}\frac{g_s^2\phi^2}{M^2}\right) m_v^2\,V_\mu V^\mu + \frac{1}{4!}\,\zeta_0\,g_v^2(V_\mu V^\mu)^2 \\
& + \left(1 + \eta_\rho\frac{g_s\phi}{M}\right) m_\rho^2\,\mathrm{Tr}(\rho_\mu\rho^\mu) - \left(\frac{1}{2} + \frac{\kappa_3}{3!}\frac{g_s\phi}{M} + \frac{\kappa_4}{4!}\frac{g_s^2\phi^2}{M^2}\right) m_s^2\phi^2 \\
& + \frac{1}{2g_\gamma}\left(\mathrm{Tr}(F^{(+)\mu\nu}\rho_{\mu\nu}) + \frac{1}{3}\,f_s^{\mu\nu}\,V_{\mu\nu}\right)\,.
\end{aligned} \tag{50}$$

The $\nu = 3$ and $\nu = 4$ terms in $\mathcal{L}_{\text{meson}(\hat{\nu} \leqslant 4)}$ are important for describing the bulk properties of nuclear many-body systems (Furnstahl et al., 1995; 1996; 1997). The only manifest chiral-symmetry breaking is through the nonzero pion mass. It is well known that there are other $\hat{\nu} = 4$ terms involving pion-pion interactions. Since multiple pion interactions and chiral-symmetry-violating terms other than the pion mass term are not considered, this additional lagrangian is not shown here.

Finally, we have

$$\begin{aligned}
\mathcal{L}_{N,\pi(\hat{\nu}=4)} = {} & \frac{1}{2M^2}\,\overline{N}\gamma_\mu(2\beta^{(0)}\partial_\nu f_s^{\mu\nu} + \beta^{(1)}\widetilde{\partial}_\nu F^{(+)\mu\nu} + \beta_A^{(1)}\gamma_5\widetilde{\partial}_\nu F^{(-)\mu\nu})N \\
& - \omega_1\,\mathrm{Tr}(F_{\mu\nu}^{(+)}\,\widetilde{v}^{\mu\nu}) + \omega_2\,\mathrm{Tr}(\widetilde{a}_\mu\widetilde{\partial}_\nu F^{(-)\mu\nu}) + \omega_3\,\mathrm{Tr}\left(\widetilde{a}_\mu i\left[\widetilde{a}_\nu\,,\,F^{(+)\mu\nu}\right]\right) \\
& - g_{\rho\pi\pi}\,\frac{2f_\pi^2}{m_\rho^2}\,\mathrm{Tr}(\rho_{\mu\nu}\widetilde{v}^{\mu\nu}) \\
& + \frac{c_1}{M^2}\,\overline{N}\gamma^\mu N\,\mathrm{Tr}\left(\widetilde{a}^\nu\,\overline{F}_{\mu\nu}^{(+)}\right) + \frac{e_1}{M^2}\,\overline{N}\gamma^\mu\,\widetilde{a}^\nu N\,\overline{f}_{s\mu\nu} \\
& + \frac{c_{1\rho}g_\rho}{M^2}\,\overline{N}\gamma^\mu N\,\mathrm{Tr}\left(\widetilde{a}^\nu\,\overline{\rho}_{\mu\nu}\right) + \frac{e_{1v}g_v}{M^2}\,\overline{N}\gamma^\mu\,\widetilde{a}^\nu N\,\overline{V}_{\mu\nu}\,.
\end{aligned} \tag{51}$$

Note that $\mathcal{L}_{N,\pi(\hat{\nu}=4)}$ is *not* a complete list of all possible $\hat{\nu} = 4$ interaction terms. However, $\beta^{(0)}$ and $\beta^{(1)}$ are used in the form factors of the nucleon's vector current, $\omega_{1,2,3}$ contribute to the form factor of the pion's vector current, and $g_{\rho\pi\pi}$ is used in the form factors that incorporate vector meson dominance (VMD).[1] Special attention should be given to the $c_1, e_1, c_{1\rho}$, and e_{1v} couplings, since they are the only relevant $\hat{v} = 4$ terms for NC photon production (Serot & Zhang, 2011a;b).

The construction of these high-order terms, $\mathcal{L}_{N,\pi(\hat{v}=4)}$ for example, is carried out by exhaustion. Based on the various symmetry transformation rules, at a given order there are a finite number of interaction terms, although the number can be big. For example, the interaction terms involving two pions and only one nucleon at $\hat{v} = 4$ *without chiral symmetry breaking* are (Ellis & Tang, 1998)

$$\overline{N}\sigma^{\mu\nu} i\overset{\leftrightarrow}{\widetilde{\partial}^{\lambda}} N \,\mathrm{Tr}\left(\widetilde{\partial}_{\lambda}\widetilde{a}_{\mu}\widetilde{a}_{\nu}\right) \qquad \text{and other contractions of Lorentz indices}\,,$$

$$\overline{N}\gamma_{\mu} i \left[\widetilde{\partial}^{\mu}\widetilde{a}^{\nu}\,,\,\widetilde{a}_{\nu}\right] N \qquad \text{and other contractions of Lorentz indices}\,.$$

3.4 Introducing Δ resonances

The pathologies of relativistic field theory with spin-3/2 particles have been investigated in the canonical quantization framework for some time. There are two kinds of problems: one is the so-called Johnson–Sudarshan problem (Capri & Kobes, 1980; Hagen, 1971; Johnson & Sudarshan, 1961); the other one is the Velo–Zwanzinger problem (Capri & Kobes, 1980; Singh, 1973; Velo & Zwanziger, 1969). It was realized in (Kobayashi & Takahashi, 1987) that the two problems may both be related to the fact that the classical equation of motion, as the result of minimizing the action, fails to eliminate redundant spin components, because the invertibility condition of the constraint equation is not satisfied all the time. For example, in the Rarita–Schwinger formalism, the representation of the field is ψ^{μ}: $(\frac{1}{2}, \frac{1}{2}) \otimes \left((\frac{1}{2}, 0) \oplus (0, \frac{1}{2})\right) = (1, \frac{1}{2}) \oplus (\frac{1}{2}, 1) \oplus (\frac{1}{2}, 0) \oplus (0, \frac{1}{2})$ (Weinberg, 1995b). It can be shown that the spin-1/2 components are not dynamical in the free theory, which is generally not true after introducing interactions. Another issue is about the so-called *off-shell couplings*, which have the form $\gamma_{\mu}\psi^{\mu}, \partial_{\mu}\psi^{\mu}, \overline{\psi}^{\mu}\gamma_{\mu}$, and $\partial_{\mu}\overline{\psi}^{\mu}$ (still in the Rarita–Schwinger representation).

Recently, the problem has been investigated in a path-integral formalism in Ref. (Pascalutsa, 1998), where a gauge invariance is required for interactions. But this constraint conflicts with the manifest nonlinear chiral-symmetry realization in chiral EFT. Subsequently, in (Krebs et al., 2009; Pascalutsa, 2001), the authors realized that the commonly used non-invariant interactions are related to gauge-invariant interactions by field redefinitions, up to some contact interaction terms. Moreover, from the modern chiral EFT viewpoint, it has been concluded (Krebs et al., 2010; Tang & Ellis, 1996) that the off-shell couplings are redundant, since they lead to contributions to contact interactions without spin-3/2 degrees of freedom. Furthermore, it has been proved that off-shell couplings with ∂_{μ} changed to $\widetilde{\partial}_{\mu}$ are also redundant, which makes the manifest realization of chiral symmetry possible with a spin-3/2 particle.

However, the modern argument, which makes use of field redefinitions and gauge invariance for the EFT, looks abstract. The whole argument is that the field redefinitions, constructed to transform non-invariant terms to gauge-invariant terms, is applicable here, which requires us

[1] VMD in QHD EFT has been discussed in detail in Ref. (Serot, 2007). We will discuss VMD for the form factor of the transition current involving Δ and N.

to be far away from the singularities of these transformations, i.e., to stay at low-energy and in weak-field regions (Krebs et al., 2009). This leads us to give another interesting argument, based directly on this assumption. In the Hamiltonian formalism, these two issues are somewhat clarified,[2] however the quantization of the EFT and hence Lorentz-invariance are not straightforward. So we use the path-integral approach.

Let's focus on the spin-3/2 propagator in the Rarita–Schwinger representation. First, we can decompose the free propagator into different spin components:

$$\begin{aligned} S_F^{0\mu\nu}(p) &= \frac{-(\not{p}+m)}{p^2-m^2+i\epsilon}\left[g^{\mu\nu}-\frac{1}{3}\gamma^\mu\gamma^\nu+\frac{p^\mu\gamma^\nu-p^\nu\gamma^\mu}{3m}-\frac{2}{3m^2}p^\mu p^\nu\right] \\ &\equiv -\frac{1}{\not{p}-m+i\epsilon}P^{(\frac{3}{2})\mu\nu}-\frac{1}{\sqrt{3}m}P_{12}^{(\frac{1}{2})\mu\nu}-\frac{1}{\sqrt{3}m}P_{21}^{(\frac{1}{2})\mu\nu} \\ &\quad +\frac{2}{3m^2}(\not{p}+m)P_{22}^{(\frac{1}{2})\mu\nu}, \end{aligned} \tag{52}$$

$$P^{(\frac{3}{2})\mu\nu} = g^{\mu\nu}-\frac{1}{3}\gamma^\mu\gamma^\nu+\frac{1}{3p^2}\gamma^{[\mu}p^{\nu]}\not{p}-\frac{2}{3p^2}p^\mu p^\nu, \tag{53}$$

$$P_{11}^{(\frac{1}{2})\mu\nu} = \frac{1}{3}\gamma^\mu\gamma^\nu-\frac{1}{3p^2}\gamma^{[\mu}p^{\nu]}\not{p}-\frac{1}{3p^2}p^\mu p^\nu, \tag{54}$$

$$P_{12}^{(\frac{1}{2})\mu\nu} = \frac{1}{\sqrt{3}p^2}(-p^\mu p^\nu+\gamma^\mu p^\nu\not{p}), \tag{55}$$

$$P_{21}^{(\frac{1}{2})\mu\nu} = \frac{1}{\sqrt{3}p^2}(p^\mu p^\nu-\gamma^\nu p^\mu\not{p}), \tag{56}$$

$$P_{22}^{(\frac{1}{2})\mu\nu} = \frac{1}{p^2}p^\mu p^\nu. \tag{57}$$

By using the identities shown in Eqs. (114) to (119) in Appendix 7.C, we can immediately write down

$$\begin{aligned} S_F^0(p) &= P^{(\frac{3}{2})}\frac{-1}{\not{p}-m+i\epsilon}P^{(\frac{3}{2})} \\ &\quad +P^{(\frac{3}{2}\perp)}\left[-\frac{1}{\sqrt{3}m}P_{12}^{(\frac{1}{2})}-\frac{1}{\sqrt{3}m}P_{21}^{(\frac{1}{2})}+P_{22}^{(\frac{1}{2})}\frac{2}{3m^2}(\not{p}+m)P_{22}^{(\frac{1}{2})}\right]P^{(\frac{3}{2}\perp)} \\ &\equiv S_F^{0(\frac{3}{2})}+S_F^{0(\frac{3}{2}\perp)}. \end{aligned} \tag{58}$$

---

[2] In the perturbative calculation of EFT, the time-ordered free propagator defined in the Hamiltonian formalism for a spin-3/2 particle always satisfies the constraint on the degrees of freedom. (Assume we have a well defined Hamiltonian for the EFT.) For finite sums of the series of diagrams involving this propagator, the constraint is always satisfied. Moreover, those off-shell terms when either contracted to external legs or to the internal propagator of spin-3/2 degrees of freedom, give zero value. We may conclude that they are redundant. However, it is not clear whether the two conclusions hold for infinite sums. Moreover, as we know, the time-ordered propagator is not covariant, and leads to the difficulty of understanding Lorentz-invariance.

In principle, the decomposition shown in Eq. (58) should be obvious in the beginning, because Lorentz-invariance is preserved. However the key is that only the spin-3/2 component has pole structure, while the spin-1/2 components resemble contact vertices.

Furthermore, given certain interaction terms, we can carry out the calculation of the self-energy insertion, as done in Ref. (Ellis & Tang, 1998), for example. Based on the same argument as given above, the self-energy for renormalization should also be decomposed into a diagonal form for the spin. The details are as follows. The self-energy of the Δ can be defined as $\Sigma_{\mu\nu} = \Sigma^{\Delta} g_{\mu\nu} + \delta\Sigma_{\mu\nu}$. We see immediately that $\delta\Sigma_{\mu\nu}$'s indices can only have a structure like the products of $(\gamma_\mu, p_\mu) \times$ (Dirac matrices) $\times (\gamma_\nu, p_\nu)$. Then we find

$$\Sigma = \Sigma^{\Delta} g + \delta\Sigma \tag{59}$$

$$= P^{(\frac{3}{2})}\Sigma^{\Delta} g P^{(\frac{3}{2})} + P^{(\frac{3}{2}\perp)}\Sigma P^{(\frac{3}{2}\perp)}$$
$$+ P^{(\frac{3}{2})}(\Sigma^{\Delta} g + \delta\Sigma) P^{(\frac{3}{2}\perp)} + P^{(\frac{3}{2}\perp)}(\Sigma^{\Delta} g + \delta\Sigma) P^{(\frac{3}{2})} . \tag{60}$$

So, we can conclude that $\Sigma = P^{(3/2)}\Sigma^{\Delta} P^{(3/2)} + P^{(3/2\perp)}\Sigma P^{(3/2\perp)} \equiv \Sigma^{(3/2)} + \Sigma^{(3/2\perp)}$. In the proof, we make use of $\left[P^{(3/2)}\, ,\, \Sigma^{\Delta}\right] = 0$, $\left[P^{(3/2\perp)}\, ,\, \Sigma^{\Delta}\right] = 0$, because the only possible spin structures of Σ^{Δ} are $\mathbf{1}$, $\not{p}$ and γ_5 (parity violation), which commute with the two projection operators. Then $P^{(3/2)}\Sigma^{\Delta} g P^{(3/2\perp)} = 0$ and $P^{(3/2)\perp}\Sigma^{\Delta} g P^{(3/2)} = 0$. Also we make use of Eqs. (115) and (116), so we get $P^{(3/2)}\delta\Sigma P^{(3/2\perp)} = 0$ and $P^{(3/2\perp)}\delta\Sigma P^{(3/2)} = 0$.

Based on previous discussions, we can have the following renormalization of the spin-3/2 propagator:

$$S_F = (S_F^{0(\frac{3}{2})} + S_F^{0(\frac{3}{2}\perp)}) + (S_F^{0(\frac{3}{2})} + S_F^{0(\frac{3}{2}\perp)})(\Sigma^{(\frac{3}{2})} + \Sigma^{(\frac{3}{2}\perp)})(S_F^{0(\frac{3}{2})} + S_F^{0(\frac{3}{2}\perp)}) + \dots$$
$$= S_F^{0(\frac{3}{2})} + S_F^{0(\frac{3}{2})}\Sigma^{(\frac{3}{2})}S_F^{0(\frac{3}{2})} + \dots \tag{61}$$
$$+ S_F^{0(\frac{3}{2}\perp)} + S_F^{0(\frac{3}{2}\perp)}\Sigma^{(\frac{3}{2}\perp)}S_F^{0(\frac{3}{2}\perp)} + \dots . \tag{62}$$

So the renormalized propagator is decomposed into two different components: $S_F \equiv S_F^{(3/2)} + S_F^{(3/2\perp)}$. The resonant contribution is $S_F^{(3/2)} = S_F^{0(3/2)} + S_F^{0(3/2)}\Sigma^{(3/2)}S_F^{(3/2)}$. The background contribution is $S_F^{(3/2\perp)} = S_F^{0(3/2\perp)} + S_F^{0(3/2\perp)}\Sigma^{(3/2\perp)}S_F^{(3/2\perp)}$. The renormalization shifts the pole position of the resonant part. For the nonresonant part, as long as power counting is valid, i.e., $O(\Sigma/m) \ll 1$, we are away from any unphysical pole in the renormalized nonresonant part; $[1 - O(\Sigma/m)]^{-1}$ never diverges. This also suggests that we will not see the unphysical pole in the renormalized propagator, when working in the low-energy perturbative region. Meanwhile, the argument helps to clarify the redundancy of the off-shell couplings. We have seen that the self-energy due to these couplings does not contribute in the renormalization of $S_F^{(3/2)}$. But it indeed changes the nonresonant part. However, the effect is power expandable. So essentially it is the same as higher-order contact terms without the Δ. This justifies the redundancy of these couplings. To ignore them in a way which does not break chiral symmetry on a term-by-term basis, we can always associate the ∂^μ with π fields so that it becomes $\widetilde{\partial}^\mu$. *This indicates that those couplings having $\widetilde{\partial}^\mu$ or γ^μ contracted with Δ_μ can be ignored without breaking manifest chiral symmetry.*

A few words on the singularity of $1/p^2$ are in order here. [See Eqs. (52) to (57).] The whole calculation is only valid in the low-energy limit, and in this limit we should not find any diagrams with Δ's that are far "off shell". Take pion scattering for example; we assume the pion energy to be small, and hence p^2 is always roughly equal to the incoming nucleon's invariant mass. So the singularity in $1/p^2$ should not be a problem in the low-energy theory from a very general perspective.

3.5 QHD with Δ

Consider first $\mathcal{L}_\Delta$ ($\hat{\nu} \leqslant 3$), which is essentially a copy of the corresponding lagrangian for the nucleon as shown in Eq. (48):

$$\begin{aligned}
\mathcal{L}_\Delta = {} & \frac{-i}{2}\overline{\Delta}_\mu^{\,a}\{\sigma^{\mu\nu}, (i\,\widetilde{\not\partial} - h_\rho\,\not\rho - h_v\,\not V - m + h_s\phi)\}_a^{\,b}\,\Delta_{b\nu} + \widetilde{h}_A\overline{\Delta}_\mu^{\,a}\,\widetilde{\not a}_a^{\,b}\gamma_5\Delta_b^\mu \\
& - \frac{\widetilde{f}_\rho h_\rho}{4m}\,\overline{\Delta}_\lambda\,\rho_{\mu\nu}\sigma^{\mu\nu}\Delta^\lambda - \frac{\widetilde{f}_v h_v}{4m}\,\overline{\Delta}_\lambda V_{\mu\nu}\sigma^{\mu\nu}\Delta^\lambda \\
& - \frac{\widetilde{\kappa}_\pi}{m}\,\overline{\Delta}_\lambda\widetilde{v}_{\mu\nu}\sigma^{\mu\nu}\Delta^\lambda + \frac{4\widetilde{\beta}_\pi}{m}\,\overline{\Delta}_\lambda\Delta^\lambda\,\mathrm{Tr}(\widetilde{a}^\mu\,\widetilde{a}_\mu)\,. \qquad (63)
\end{aligned}$$

Here the sub- and superscripts $a, b = (\pm 3/2, \pm 1/2)$, and the isospin conventions and T matrix have been discussed in Sec. 3.1.

To produce the $N \leftrightarrow \Delta$ transition currents, we construct the following lagrangians ($\hat{\nu} \leqslant 4$):

$$\mathcal{L}_{\Delta,N,\pi} = h_A\overline{\Delta}^{a\mu}\,T_a^{\dagger\,iA}\,\widetilde{a}_{i\mu}N_A + \text{C.C.}\,, \qquad (64)$$

$$\begin{aligned}
\mathcal{L}_{\Delta,N,\text{background}} = {} & \frac{ic_{1\Delta}}{M}\,\overline{\Delta}_\mu^{\,a}\gamma_\nu\gamma_5\,T_a^{\dagger\,iA}F_i^{(+)\mu\nu}N_A + \frac{ic_{3\Delta}}{M^2}\,\overline{\Delta}_\mu^{\,a}\,i\gamma_5\,T_a^{\dagger\,iA}(\widetilde{\partial}_\nu F^{(+)\mu\nu})_i N_A \\
& + \frac{c_{6\Delta}}{M^2}\,\overline{\Delta}_\lambda^{\,a}\sigma_{\mu\nu}T_a^{\dagger\,iA}(\widetilde{\partial}^\lambda\overline{F}^{(+)\mu\nu})_i N_A \\
& - \frac{d_{2\Delta}}{M^2}\,\overline{\Delta}_\mu^{\,a}\,T_a^{\dagger\,iA}(\widetilde{\partial}_\nu F^{(-)\mu\nu})_i N_A - \frac{id_{4\Delta}}{M}\,\overline{\Delta}_\mu^{\,a}\gamma_\nu\,T_a^{\dagger\,iA}F_i^{(-)\mu\nu}N_A \\
& - \frac{id_{7\Delta}}{M^2}\,\overline{\Delta}_\lambda^{\,a}\sigma_{\mu\nu}T_a^{\dagger\,iA}(\widetilde{\partial}^\lambda F^{(-)\mu\nu})_i N_A + \text{C.C.}\,, \qquad (65)
\end{aligned}$$

$$\begin{aligned}
\mathcal{L}_{\Delta,N,\rho} = {} & \frac{ic_{1\Delta\rho}}{M}\,\overline{\Delta}_\mu^{\,a}\,\gamma_\nu\gamma_5\,T_a^{\dagger\,iA}\rho_i^{\mu\nu}N_A + \frac{ic_{3\Delta\rho}}{M^2}\,\overline{\Delta}_\mu^{\,a}\,i\gamma_5\,T_a^{\dagger\,iA}(\widetilde{\partial}_\nu\rho^{\mu\nu})_i N_A \\
& + \frac{c_{6\Delta\rho}}{M^2}\,\overline{\Delta}_\lambda^{\,a}\sigma_{\mu\nu}\,T_a^{\dagger\,iA}(\widetilde{\partial}^\lambda\,\overline{\rho}^{\mu\nu})_i N_A + \text{C.C.}\,. \qquad (66)
\end{aligned}$$

It can be checked that the interaction terms respect all of the required symmetries. Terms omitted from these lagrangians are either redundant or are not relevant to the transition interaction involving N and Δ (at tree level). The construction of terms is by means of exhausting all the possibilities. Here we give an example:

$$\overline{\Delta}_\mu\gamma_\nu N\epsilon^{\mu\nu\alpha\beta}F_{\alpha\beta}^{(+)} = 2i\overline{\Delta}_\mu\gamma_\nu\gamma_5 NF^{(+)\mu\nu} + iF_{\alpha\beta}^{(+)}\overline{\Delta}_\mu\gamma_5(\gamma^\mu\gamma^\alpha\gamma^\beta - g^{\alpha\beta}\gamma^\mu)N\,. \qquad (67)$$

The preceding identity indicates that $\overline{\Delta}^{a}_{\mu}\gamma_{\nu}T^{\dagger\, iA}_{a}\overline{F}^{(+)\mu\nu}_{i}N_{A}$ differs from the $c_{1\Delta}$ coupling in Eq. (65) by off-shell terms, which can be ignored.

Moreover, the terms in the lagrangian in Eq. (66) and the $1/g_{\gamma}$ coupling in Eq. (50) are necessary for the realization of transition form factors using VMD. First, we make the following definitions:

$$\langle \Delta, a, p_{\Delta}|V^{i\mu}(A^{i\mu})|N, A, p_{N}\rangle \equiv T^{\dagger\, iA}_{a}\, \overline{u}_{\Delta\alpha}(p_{\Delta})\, \Gamma^{\alpha\mu}_{V(A)}(q)\, u_{N}(p_{N})\, . \tag{68}$$

Based on the lagrangians given previously, formulas shown in Appendix 7.B, and the definitions of currents in Eq. (21), we find (note that $\sigma_{\mu\nu}\epsilon^{\mu\nu\alpha\beta} \propto i\sigma^{\alpha\beta}\gamma_{5}$)

$$\begin{aligned}\Gamma^{\alpha\mu}_{V} &= \frac{2c_{1\Delta}(q^2)}{M}\,(q^{\alpha}\gamma^{\mu} - \not{q}g^{\alpha\mu})\gamma_{5} + \frac{2c_{3\Delta}(q^2)}{M^2}\,(q^{\alpha}q^{\mu} - g^{\alpha\mu}q^{2})\gamma_{5} \\ &\quad - \frac{8c_{6\Delta}(q^2)}{M^2}\, q^{\alpha}\sigma^{\mu\nu}iq_{\nu}\gamma_{5}\, , \\ c_{i\Delta}(q^2) &\equiv c_{i\Delta} + \frac{c_{i\Delta\rho}}{2g_{\gamma}}\frac{q^2}{q^2 - m_{\rho}^{2}} \qquad i = 1, 3, 6,\end{aligned} \tag{69}$$

$$\begin{aligned}\Gamma^{\alpha\mu}_{A} &= -h_{A}\left(g^{\alpha\mu} - \frac{q^{\alpha}q^{\mu}}{q^{2} - m_{\pi}^{2}}\right) + \frac{2d_{2\Delta}}{M^2}\,(q^{\alpha}q^{\mu} - g^{\alpha\mu}q^{2}) - \frac{2d_{4\Delta}}{M}\,(q^{\alpha}\gamma^{\mu} - g^{\alpha\mu}\,\not{q}) \\ &\quad - \frac{4d_{7\Delta}}{M^2}\, q^{\alpha}\sigma^{\mu\nu}iq_{\nu}\, ,\end{aligned} \tag{70}$$

where h_{A} is from Eq. (64). Quite similar to the $c_{i\Delta}(q^2)$, we can introduce axial-vector meson $[a_{1}(1260)]$ exchange into the axial transition current, which leads to a structure for the $d_{i\Delta}(q^2)$ that is similar to the vector transition current form factors. There is one subtlety associated with the realization of $h_{A}(q^2)$: with our lagrangian, we have the pion-pole contribution associated only with the h_{A} coupling, and all the higher-order terms contained in $\delta h_{A}(q^2) \equiv h_{A}(q^2) - h_{A}$ conserve the axial transition current. With the limited information about manifest chiral-symmetry breaking, we ignore this subtlety and still use the form similar to the $c_{1\Delta}(q^2)$ to parameterize $h_{A}(q^2)$. The axial-vector meson couplings $h_{\Delta a_{1}}$ and $d_{i\Delta a_{1}}$ are the combinations of $g_{a_{1}}$ (a_{1} and isovector axial-vector external field coupling strength) and the coupling strength of the $\overline{\Delta}a_{1}N$ interaction. $m_{a_{1}}$ is the 'mass' of the meson. So we have

$$h_{A}(q^2) \equiv h_{A} + h_{\Delta a_{1}}\,\frac{q^2}{q^2 - m_{a_{1}}^{2}}\, , \tag{71}$$

$$d_{i\Delta}(q^2) \equiv d_{i\Delta} + d_{i\Delta a_{1}}\,\frac{q^2}{q^2 - m_{a_{1}}^{2}} \qquad i = 2, 4, 7. \tag{72}$$

To determine the coefficients in the transition form factors shown in Eqs. (69), (71), and (72), we need to compare ours with the conventional ones used in the literature. In

Refs. (Graczyk et al., 2009; Hernández et al., 2007) for example, the definition is

$$\langle \Delta, \frac{1}{2}| j^{\mu}_{cc+} |N, -\frac{1}{2}\rangle \equiv \overline{u}_{\alpha}(p_{\Delta}) \left\{ \left[\frac{C_3^V}{M} (g^{\alpha\mu} \not{q} - q^{\alpha}\gamma^{\mu}) + \frac{C_4^V}{M^2} (q \cdot p_{\Delta}\, g^{\alpha\mu} - q^{\alpha} p^{\mu}_{\Delta}) \right.\right.$$
$$\left. + \frac{C_5^V}{M^2} (q \cdot p_N\, g^{\alpha\mu} - q^{\alpha} p^{\mu}_{N}) \right] \gamma_5$$
$$+ \left[\frac{C_3^A}{M} (g^{\alpha\mu} \not{q} - q^{\alpha}\gamma^{\mu}) + \frac{C_4^A}{M^2} (q \cdot p_{\Delta}\, g^{\alpha\mu} - q^{\alpha} p^{\mu}_{\Delta}) \right.$$
$$\left.\left. + C_5^A g^{\alpha\mu} + \frac{C_6^A}{M^2} q^{\mu} q^{\alpha} \right] \right\} u(p_N) . \tag{73}$$

The basis given above is known to be complete. The determination of the couplings through comparing our results with the conventional ones has been given in Ref. (Serot & Zhang, 2010). There we find that our meson dominance form factors are accurate up to $Q^2 \approx 0.3\,\mathrm{GeV}^2$. Moreover, CVC and PCAC can be easily checked for the transition currents. The details can be found in Ref. (Serot & Zhang, 2010).

4. Application

In this section, we briefly discuss the weak production of pions from nucleons. We focus only on two properties of the Feynman diagrams in this problem, including the G parity and the current's Hermiticity. Then we talk about the production from nuclei, in which Δ dynamics is the key component (for both the interaction mechanism and the final state interaction of the pion). This points out the importance of understanding the strong interaction, associated with nuclear structure and Δ dynamics, in the study of the electroweak response of nuclei. So it is necessary to have a framework that includes the two and also provides for efficient calculations. The details of these subjects are presented in Refs. (Serot & Zhang, 2010) and (Serot & Zhang, 2011a;b).

4.1 Weak production of pions from free nucleons

The relevant Feynman diagrams are shown in Fig. 1 for weak production of pions due to (anti)neutrino scattering off free nucleons. The 'C' in the figure stands for various currents including the vector current, axial current, and baryon current, of which both CC and NC are composed according to Sec. 2.2. The details about these diagrams can be found in (Serot & Zhang, 2010). Here we begin with G parity. We use $\langle N, B, \pi, j | J^{\mu} | N, A\rangle$ to represent the contribution of diagrams, where 'A' and 'B' denote isospin-1/2 projections. From G parity, we have

$$G A^{i\mu} G^{-1} = -A^{i\mu} ,$$
$$G V^{i\mu}(J^{\mu}_{B}) G^{-1} = V^{i\mu}(J^{\mu}_{B}) .$$

By applying this to the current's matrix elements, we get

$$\langle N, B, \pi, j | A^{i\mu} | N, A\rangle = \langle \overline{N}, B, \pi, j | A^{i\mu} | \overline{N}, A\rangle , \tag{74}$$

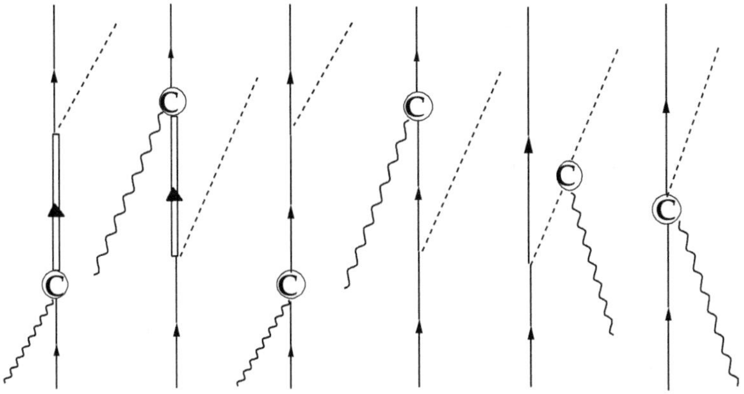

Fig. 1. Feynman diagrams for pion production. Here, **C** stands for various types of currents including vector, axial-vector, and baryon currents. Some diagrams may be zero for some specific type of current. For example, diagrams (a) and (b) will not contribute for the (isoscalar) baryon current. Diagram (e) will be zero for the axial-vector current. The pion-pole contributions to the axial current in diagrams (a) (b) (c) (d) and (f) are included in the vertex functions of the currents.

$$\langle N,B,\pi,j|V^{i\mu}(J_B^{\mu})|N,A\rangle = -\langle \overline{N},B,\pi,j|V^{i\mu}(J_B^{\mu})|\overline{N},A\rangle \,. \tag{75}$$

Eqs. (74) and (75) give a relation between a current's matrix element involving nucleon states and a matrix element involving antinucleon states. Because of the isospin symmetry, we can define

$$\begin{aligned}
&\langle N,B,p_f;\pi,j,k_\pi|\; A^{i\mu}\; |N,A,p_i\rangle \\
&\quad\equiv \delta^i_j\delta^A_B\, \overline{u}(p_f)\Gamma^{\mu}_{sym}(p_f,k_\pi;p_i,q)u(p_i) \\
&\qquad + i\epsilon^i{}_{jk}\left(\frac{\tau^k}{2}\right)^A_B \overline{u}(p_f)\Gamma^{\mu}_{asym}(p_f,k_\pi;p_i,q)u(p_i)\,.
\end{aligned} \tag{76}$$

Vector currents can be decomposed in the same way. From crossing symmetry, we can see

$$\begin{aligned}
&\langle \overline{N},B,p_f;\pi,j,k_\pi|\; A^{i\mu}\; |\overline{N},A,p_i\rangle \\
&= -\delta^i_j\delta^A_B\, \overline{v}(p_i)\Gamma^{\mu}_{sym}(-p_i,k_\pi;-p_f,q)v(p_f) \\
&\quad - i\epsilon^i{}_{jk}\left(-\frac{\tau^{k\,T}}{2}\right)^B_A \overline{v}(p_i)\Gamma^{\mu}_{asym}(-p_i,k_\pi;-p_f,q)v(p_f) \\
&= \delta^i_j\delta^A_B\, \overline{u}(p_f)\left(-C\Gamma^{T\mu}_{sym}(-p_i,k_\pi;-p_f,q)C\right)u(p_i) \\
&\quad - i\epsilon^i{}_{jk}\left(\frac{\tau^k}{2}\right)^A_B \overline{u}(p_f)\left(-C\Gamma^{T\mu}_{asym}(-p_i,k_\pi;-p_f,q)C\right)u(p_i)\,.
\end{aligned} \tag{77}$$

In Eq. (77), the $-\frac{1}{2}\tau^{k^T}$ appears because antiparticles furnish the complex conjugate representation. (It is equivalent to the original representation.) $\mathcal{C}$ is the charge conjugation matrix applied to a Dirac spinor, i.e., $\psi^C(x) = \mathcal{C}(\overline{\psi}(x))^T$. By comparing Eq. (77) with Eq. (74), we have the following constraint on the axial current's matrix element:

$$-\mathcal{C}\Gamma^{T\mu}_{(a)sym}(-p_i, k_\pi; -p_f, q)\mathcal{C} = \underset{(-)}{+}\Gamma^{\mu}_{(a)sym}(p_f, k_\pi; p_i, q) . \tag{78}$$

Similarly, we have the following constraint on vector current's matrix element:

$$-\mathcal{C}\Gamma^{T\mu}_{(a)sym}(-p_i, k_\pi; -p_f, q)\mathcal{C} = \underset{(+)}{-}\Gamma^{\mu}_{(a)sym}(p_f, k_\pi; p_i, q) . \tag{79}$$

For the baryon current $\langle N, B', \pi, j|J^{\mu}_{B}|N, A\rangle \equiv (\frac{1}{2}\tau_j)^{A}_{B'}\,\overline{u}(p_f)\Gamma^{\mu}_{B}(p_f, k_\pi; p_i, q)u(p_i)$, G parity indicates

$$-\mathcal{C}\Gamma^{T\mu}_{B}(-p_i, k_\pi; -p_f, q)\mathcal{C} = -\Gamma^{\mu}_{B}(p_f, k_\pi; p_i, q) . \tag{80}$$

Now we can see how adding a crossed diagram involving the Δ is necessary to satisfy G parity. For example, let's talk about the vector current's matrix element. If we define it for diagrams (a) and (b) in Fig. 1 as follows:

$$\langle V^{i\mu}\rangle_a \equiv T^{a}_{Bj}T^{\dagger\, iA}_{a}\overline{u}_f\Gamma^{\mu}_{dir}(p_f, k_\pi; p_i, q)u_i , \tag{81}$$

$$\langle V^{i\mu}\rangle_b \equiv T^{ai}_{B}T^{\dagger\, A}_{ja}\overline{u}_f\Gamma^{\mu}_{cross}(p_f, k_\pi; p_i, q)u_i . \tag{82}$$

Then by using Eq. (98), we get [here we include only diagram (a) and (b) contributions]

$$\Gamma^{\mu}_{(a)sym} = \frac{2}{3}\left(\Gamma^{\mu}_{dir} \underset{(-)}{+} \Gamma^{\mu}_{cross}\right) . \tag{83}$$

By calculating the diagrams, it is straightforward to prove that

$$-\mathcal{C}\Gamma^{T\mu}_{cross}(-p_i, k_\pi; -p_f, q)\mathcal{C} = -\Gamma^{\mu}_{dir}(p_f, k_\pi; p_i, q) . \tag{84}$$

This equation justifies the G parity of the vector current's matrix elements. Other currents' matrix elements can be justified in a similar way.

Now we discuss the Hermiticity of the current. Let's consider $\langle N, \pi\, out|J^{\mu}|N, in\rangle$:

$$\begin{aligned}\langle N, p_f, \pi, k_\pi, out|J^{\mu}|N, p_i, in\rangle^* &= \langle N, p_i, in|J^{\dagger\mu}|N, p_f, \pi, k_\pi, out\rangle \\ &\neq \langle N, p_i, out|J^{\dagger\mu}|N, p_f, \pi, k_\pi, in\rangle .\end{aligned} \tag{85}$$

But how do we generally understand $\langle i, in|O|f, out\rangle$? Naively, we would have the following:

$$\begin{aligned}\langle i, in|O|f, out\rangle &= \langle i|U(-\infty, 0)U(0, t)o(t)U(t, 0)U(0, +\infty)|f\rangle \\ &= \langle i|\overline{T}o(t)\exp[i\int dt H_I(t)]|f\rangle .\end{aligned} \tag{86}$$

Here O and $o(t)$ are the operators in the Heisenberg and interaction pictures. $\overline{T}$ is another type of time ordering: $\overline{T}H_I(t_1)H_I(t_2) = \theta(t_2 - t_1)H_I(t_1)H_I(t_2) + \theta(t_1 - t_2)H_I(t_2)H_I(t_1)$. It is easy to realize that in momentum space, if we mirror the pole of the T defined Green's function, and apply $(-)$ to the overall Green's function, we get the $\overline{T}$ defined Green's function. Second, each interaction vertex in the $\langle i, in|O|f, out\rangle$ calculation differs from that of $\langle i, out|O|f, in\rangle$ by a $(-)$ sign. Third, since now all the poles are in the first and third quadrants in the complex momentum plane, the corresponding loop integration differs from the normal loop integration by a $(-)$ sign! So, without a rigorous proof, we have that after calculating $\langle i, out|O|f, in\rangle$, if we mirror all the poles relative to the real axis for the propagator and apply a phase $(-)^{(V-V_o)+I+L} = (-)^{V_o-1}$ to it, then we get the corresponding $\langle i, in|O|f, out\rangle$. Here V, V_o, I, and L are the number of vertices in the graph, vertices in the operator O, internal lines, and loops. For the current operator J^μ, $V_o = 1$ and hence the phase is $(+)$.

Now let's proceed to see the consequence of the Hermiticity of $J^\mu(x=0)$, i.e., $J^{i\mu\dagger} = J^\mu_i$:

$$
\begin{aligned}
&\langle N, B, p_f, \pi, j, k_\pi, out|\, J^{i\mu}\, |N, A, p_i, in\rangle^* \\
&\qquad = \langle N, A, p_i, in|J^\mu_i|N, B, p_f, \pi, j, k_\pi, out\rangle \\
&\qquad = \langle N, A, p_i, out|J^\mu_i|N, B, p_f, \pi, j, k_\pi, in\rangle|_{pm} \\
&\qquad = \delta_{ii'}\delta^{jj'}\langle N, A, p_i, \pi, j', -k_\pi, out|J^{i'\mu}|N, B, p_f, in\rangle|_{pm} \,.
\end{aligned} \tag{87}
$$

Here $|_{pm}$ indicates poles are mirrored with respect to the real axis. In the following, we decompose the general current matrix element into symmetric and antisymmetric parts, as we did in in Eq. (76):

$$
\begin{aligned}
&\langle N, B, p_f, \pi, j, k_\pi, out|J^{i\mu}|N, A, p_i, in\rangle^* \\
&= \delta^j_i \delta^B_A\, \overline{u}(p_i)\overline{\Gamma}^\mu_{sym}(p_f, k_\pi; p_i, q)u(p_f) - i\epsilon_i^{\ jk}\left(\frac{\tau_k}{2}\right)_A^{\ B} \overline{u}(p_i)\overline{\Gamma}^\mu_{asym}(p_f, k_\pi; p_i, q)u(p_f) \,.
\end{aligned} \tag{88}
$$

Here, $\overline{\Gamma} = \gamma^0\Gamma^\dagger\gamma^0$. Meanwhile, Eq. (87) can be rewritten as

$$
\begin{aligned}
&\langle N, A, p_i, \pi, j', -k_\pi, out|\, J^{i'\mu}\, |N, B, p_f, in\rangle|_{pm}\, \delta_{ii'}\delta^{jj'} \\
&\qquad = \delta_{ii'}\delta^{jj'}\Bigg[\delta^{i'}_{j'}\delta^B_A\, \overline{u}(p_i)\Gamma^\mu_{sym}(p_i, -k_\pi; p_f, -q)u(p_f) \\
&\qquad\quad + i\epsilon^{i'}_{\ j'k}\left(\frac{\tau^k}{2}\right)_A^{\ B} \overline{u}(p_i)\Gamma^\mu_{asym}(p_i, -k_\pi; p_f, -q)u(p_f)\Bigg]_{pm} \\
&\qquad = \delta^j_i\delta^B_A\, \overline{u}(p_i)\Gamma^\mu_{sym}(p_i, -k_\pi; p_f, -q)u(p_f)|_{pm} \\
&\qquad\quad + i\epsilon_i^{\ jk}\left(\frac{\tau_k}{2}\right)_A^{\ B} \overline{u}(p_i)\Gamma^\mu_{asym}(p_i, -k_\pi; p_f, -q)u(p_f)|_{pm} \,.
\end{aligned} \tag{89}
$$

If we compare Eq. (88) with Eq. (89), we see the Hermiticity constraint is

$$
\gamma^0[\Gamma^\mu_{(a)sym}(p_f, k_\pi; p_i, q)]^\dagger\gamma^0 = \underset{(-)}{+}\Gamma^\mu_{(a)sym}(p_i, -k_\pi; p_f, -q)|_{pm} \,. \tag{90}
$$

Now let's focus on the constraint on diagrams (a) and (b) in Fig. 1. We can check by calculating diagrams:

$$\overline{\Gamma}^{\mu}_{dir}(p_f, k_\pi; p_i, q) = \Gamma^{\mu}_{cross}(p_i, -k_\pi; p_f, -q)|_{pm} \,. \tag{91}$$

We can choose kinematics where no poles and cuts arise, i.e., there is no phase shift, and then test the constraint without $|_{pm}$. The preceding observation, with Eq. (83) taken into account, leads to the satisfaction of the constraint in Eq. (90). The Hermiticity of the baryon current can be studied in a similar way, and hence is not shown explicitly here. Moreover, it is interesting to see that the higher-order contact terms satisfy the requirements due to G parity and Hermiticity on a term-by-term basis.

4.2 Weak production of pions from nuclei, Δ dynamics

With the development of neutrino-oscillation experiments, precise knowledge about the neutrino (antineutrino)-nuclei scattering cross sections is needed for the understanding of the experiments' background. Take MiniBooNE (Aguilar-Arevalo et al., 2009; 2010), for example; the median energy of the neutrino (antineutrino) beam is around 0.6 (0.5) GeV, and the high-energy tail extends up to 2 GeV. In this regime, the Δ is the most important resonance for the interaction mechanism, except in the *very* low-energy region. Therefore, to understand pion production, we need to study Δ dynamics in the nucleus. This subject has been extensively discussed in the nonrelativistic framework (Hirata et al., 1976; Horikawa et al., 1980; Oset & Salcedo, 1987), and it has also been initiated in the relativistic framework in (Herbert et al., 1992; Wehrberger et al., 1989; Wehrberger & Wittman, 1990; Wehrberger, 1993). It is shown that the Δ width increases in the normal nuclear medium, since new decay channels are opened, like $\Delta N \to NN$, for example. The real part of the Δ's self-energy has also been studied. From the lagrangian in Eq. (63), we can see that the two parameters h_s and h_v in the lagrangian are important.[3] However, the information in (Boguta, 1982; Kosov et al., 1998; Wehrberger et al., 1989; Wehrberger, 1993) is still limited. In (Serot & Zhang, 2011a;b), we have realized that these Δ-meson couplings are responsible for the Δ's spin-orbit coupling in the nucleus, and based on this we provide some information about the couplings from this new perspective.

Meanwhile, the Δ dynamics is also strongly correlated with the pion dynamics in the nuclear medium, and hence is important for understanding the pion's final state interactions, especially in the energy regime of these neutrino-oscillation experiments.

5. Summary

In this work, we have studied EW interactions in QHD EFT. First, we discuss the EW interactions at the quark level. Then we include EW interactions in QHD EFT by using the background-field technique. The completed QHD EFT has a nonlinear realization of $SU(2)_L \otimes SU(2)_R \otimes U(1)_B$ (chiral symmetry and baryon number conservation), as well as realizations of other symmetries including Lorentz-invariance, C, P, and T. Meanwhile, as we know, chiral symmetry is manifestly broken due to the nonzero quark masses; the P and C symmetries are also broken because of weak interactions. All these breaking patterns are parameterized in a general way in the EFT. Moreover, we have included the

[3] h_ρ should not play an important role in normal nuclei with small asymmetry.

Δ resonance as manifest degrees of freedom in our QHD EFT. This enables us to discuss physics at the kinematics where the resonance becomes important. As a result, the effective theory uses hadronic degrees of freedom, satisfies the constraints due to QCD (symmetries and their breaking pattern), and is calibrated to strong-interaction phenomena. (The EW interaction of *individual hadrons*, like the transition currents discussed in this work, need to be parameterized.) So this effective field theory satisfies the three listed points laid out in the Introduction.

The technical issues that arise when introducing the Δ in the EFT need to be emphasized here. It has been proven that the general EFT with conventional interactions has no redundant degrees of freedom (Krebs et al., 2009). (Unphysical degrees of freedom have been considered in the canonical quantization scheme as the reason for pathologies in field theory with high-spin fields.) However, the proof rests on the work of (Pascalutsa, 1998), which claims that gauge invariance could eliminate the redundant degrees of freedom. Here, we have provided another perturbative argument about this issue, which indicates that as long as we work in the low-energy and weak-field limit, the unphysical degrees of freedom do not show up. This condition is satisfied in the EFT. Throughout the argument, we do not need to make use of the gauge-invariance requirement. And in this way, we can easily see the redundancy of *off-shell* interactions, which has also been rigorously addressed in (Krebs et al., 2010). Moreover, the argument can be easily generalized to other high-spin fields.

To appreciate the importance of the symmetries realized in QHD EFT, we have discussed the currents' matrix elements in pion production from nucleons. The calculation and results are detailed in (Serot & Zhang, 2010). Here, we first briefly mention the consequence of chiral symmetry (and its breaking), i.e., CVC and PCAC. These two principles provide important constraints on the EW interactions at the hadronic level. The G parity is then studied for pion production. This provides another constraint on the analytical structure of matrix elements. Meanwhile, it also points out the importance of including cross diagrams involving the Δ. When combining the Δ's contribution in the s and u channels, the full result respects G parity. Moreover, the constraint due to the Hermiticity of current operators is explored. It is important to notice that other contact terms respect all these constraints. So, it is necessary to have a theoretical framework that satisfies these constraints. The QHD EFT, with symmetries included, clearly provides such a framework.

However, the calibration of a model on the hadronic level does not guarantee its success at the nuclear level. To study EW interactions in nuclei, we clearly have to understand how the nucleons are bound together to form nuclei. QHD has been applied extensively to this kind of problem (Serot & Walecka, 1986; 1997), and the recently developed chiral QHD EFT has also been tested in the nuclear many-body problem (Furnstahl et al., 1997). The mean-field approximation is understood in terms of density functional theory (Kohn, 1999), and hence the theory calibrated to nuclear properties includes many-body correlations beyond the Hartree approximation. Moreover, the power counting of diagrams in terms of $O(k/M)$ (k can be the Fermi momentum, mean-field strength, or other dimensional quantities) in the many-body calculations has also been studied in this framework with the justification that fitted parameters are *natural* (Hu et al., 2007; McIntire et al., 2007). This enables us to discuss the EW interactions order-by-order in the nuclear many-body system using QHD EFT.

As mentioned before, we have initiated the study of weak production of pions due to neutrino and antineutrino scattering off nuclei in this framework (Serot & Zhang, 2010; 2011a;b).

Moreover, we also studied the production of photons, in which the conservation of the EM current is clearly crucial. The discussion of power counting has been presented in these references. Furthermore, we should also anticipate the importance of Δ dynamics modified in nuclei. It has been studied in the nonrelativistic framework, but just started in the relativistic framework. The study indicates that the Δ decay width increases at normal nuclear density because the reduced pion-decay phase space is more than compensated by the opening of other decay channels. But a detailed discussion on this is still needed. The real part of the Δ self-energy is still unclear. As we pointed out, the h_s and h_v couplings in Eq. (63) play important roles, but there are still limited constraints on them. (Some constraints have been gained from an equation of state perspective, and others come from electron scattering.) As we realized in (Serot & Zhang, 2011a;b), the phenomenologically fitted spin-orbit coupling of the Δ in the nucleus may shed some light on this issue. Clearly, more efforts are needed to study Δ dynamics, which in the meantime is closely related to pion dynamics in the nuclear many-body system.

6. Acknowledgements

This work was supported in part by the US Department of Energy under Contract No. DE–FG02–87ER40365.

7. Appendix

A. Isospin indices, T matrices

Suppose $\vec{t}$ are the generators of some (ir)reducible representation of $SU(2)$; then it is easy to prove that ($\underline{\widetilde{\delta}} \equiv -\underline{e}^{-i\pi t^y}$)

$$(-\vec{t}^{\,T})^{\,i}{}_{j} = \widetilde{\delta}^{ik}\,\vec{t}_{k}{}^{l}\,\widetilde{\delta}_{lj} \equiv \vec{t}^{\,i}{}_{j} \quad ,\text{i.e.}, \quad -\underline{\vec{t}}^{\,T} = \underline{\widetilde{\delta}}\,\underline{\vec{t}}\,\underline{\widetilde{\delta}}^{-1} . \tag{92}$$

Here the superscript T denotes transpose. This equation justifies the use of $\widetilde{\delta}$ as a metric linking the representation and the equivalent complex-conjugate representation. One easily finds for $\mathcal{D}^{(3/2)}$, $\mathcal{D}^{(1)}$, and $\mathcal{D}^{(1/2)}$,

$$\widetilde{\delta}^{ab} = \begin{pmatrix} 0 & 0 & 0 & 1 \\ 0 & 0 & -1 & 0 \\ 0 & 1 & 0 & 0 \\ -1 & 0 & 0 & 0 \end{pmatrix} , \qquad \widetilde{\delta}_{ab} = \begin{pmatrix} 0 & 0 & 0 & -1 \\ 0 & 0 & 1 & 0 \\ 0 & -1 & 0 & 0 \\ 1 & 0 & 0 & 0 \end{pmatrix} , \tag{93}$$

$$\widetilde{\delta}^{ij} = \begin{pmatrix} 0 & 0 & -1 \\ 0 & 1 & 0 \\ -1 & 0 & 0 \end{pmatrix} , \qquad \widetilde{\delta}_{ij} = \begin{pmatrix} 0 & 0 & -1 \\ 0 & 1 & 0 \\ -1 & 0 & 0 \end{pmatrix} , \tag{94}$$

$$\widetilde{\delta}^{AB} = \begin{pmatrix} 0 & 1 \\ -1 & 0 \end{pmatrix} , \qquad \widetilde{\delta}_{AB} = \begin{pmatrix} 0 & -1 \\ 1 & 0 \end{pmatrix} . \tag{95}$$

We now turn to the T matrices. As discussed in Sec. 3.1,

$$\begin{aligned} T_a^{\dagger\, iA} &= \langle \frac{3}{2}; a | 1, \frac{1}{2}; i, A \rangle \,, \\ T^a_{\ iA} &= \langle 1, \frac{1}{2}; i, A | \frac{3}{2}; a \rangle \,. \end{aligned} \tag{96}$$

It is easy to prove the following relations (here τ^i is a Pauli matrix):

$$\tau^i \tau_j = \widetilde{\delta}^i_j + i \widetilde{\epsilon}^{\ i}_{jk} \tau^k \,, \tag{97}$$

$$\left(P^j_i\right)^{\ B}_{A} \equiv T^a_{\ iA}\, T_a^{\dagger\, jB} = \widetilde{\delta}^j_i \widetilde{\delta}^B_A - \frac{1}{3}(\tau_i \tau^j)^{\ B}_{A} \,, \tag{98}$$

$$T_a^{\dagger\, iA}\, T^b_{\ iA} = \widetilde{\delta}^b_a \,. \tag{99}$$

Here P^j_i is a projection operator that projects $\mathcal{H}^{(1/2)} \otimes \mathcal{H}^{(1)}$ onto $\mathcal{H}^{(3/2)}$.

A few words about $\widetilde{\epsilon}^{\ i}_{jk}$ are in order here. We have the following transformations of pion fields:

$$\pi^i = \pi^I u^i_I \qquad \text{here, } i = +1, 0, -1\,; \quad I = x, y, z \,,$$

$$\left(\pi^{+1},\, \pi^0,\, \pi^{-1}\right) = \left(\pi^x,\, \pi^y,\, \pi^z\right) \begin{pmatrix} \frac{-1}{\sqrt{2}} & 0 & \frac{1}{\sqrt{2}} \\ \frac{-i}{\sqrt{2}} & 0 & \frac{-i}{\sqrt{2}} \\ 0 & 1 & 0 \end{pmatrix} . \tag{100}$$

Under such transformations,

$$\begin{aligned} \widetilde{\epsilon}^{\,ijk} &\equiv u^i_I\, u^j_J\, u^k_K\, \epsilon^{IJK} = \det(\underline{u^i_I}) \epsilon^{ijk} = -i\, \epsilon^{ijk} \\ \Longrightarrow \quad \widetilde{\epsilon}^{\,ijk} &= \begin{cases} -i, & \text{if } ijk = +1, 0, -1\,; \\ -i\, \delta_{\mathcal{P}}, & \text{if } ijk = \mathcal{P}(+1, 0, -1)\,. \end{cases} \end{aligned} \tag{101}$$

Here $\delta_{\mathcal{P}}$ is the phase related with the $\mathcal{P}$ permutation. It is $+$ $(-)$ with an even (odd) number of permutations. To simplify the notation, we will ignore the *tilde* on $\widetilde{\delta}$ and $\widetilde{\epsilon}$ in other places.

B. Expansion of tilde objects

Here we show some details about $\widetilde{v}_\mu$, $\widetilde{a}_\mu$, $F^{(\pm)}_{\mu\nu}$ and others, which are needed for understanding electroweak interactions in QHD EFT. The pion-decay constant is $f_\pi \approx 93$ MeV.

$$\mathrm{Tr}(\frac{\tau^i}{2}[U\,, \partial^\mu U^\dagger]) \approx 2i\epsilon^{ijk} \frac{\pi_j}{f_\pi} \frac{\partial_\mu \pi_k}{f_\pi} \,, \tag{102}$$

$$\mathrm{Tr}(\frac{\tau^i}{2}\{U\,, \partial^\mu U^\dagger\}) \approx -2i\, \frac{\partial_\mu \pi^i}{f_\pi} \,, \tag{103}$$

$$\xi^{\dagger}\frac{\tau^i}{2}\xi+\xi\frac{\tau^i}{2}\xi^{\dagger}\approx\tau^i\,, \tag{104}$$

$$\xi^{\dagger}\frac{\tau^i}{2}\xi-\xi\frac{\tau^i}{2}\xi^{\dagger}\approx-\epsilon^{ijk}\,\frac{\pi_j}{f_\pi}\,\tau_k\,, \tag{105}$$

$$\widetilde{v}_\mu\approx\frac{1}{2f_\pi^2}\,\epsilon^{ijk}\pi_j\partial_\mu\pi_k\frac{\tau_i}{2}-\mathrm{v}_{i\mu}\frac{\tau^i}{2}-\epsilon^{ijk}\,\frac{\pi_j}{f_\pi}\,\frac{\tau_k}{2}\,\mathrm{a}_{i\mu}\,, \tag{106}$$

$$\widetilde{a}_\mu\approx\frac{1}{f_\pi}\,\partial_\mu\pi^i\,\frac{\tau_i}{2}+\mathrm{a}_{i\mu}\frac{\tau^i}{2}+\epsilon^{ijk}\,\frac{\pi_j}{f_\pi}\,\frac{\tau_k}{2}\mathrm{v}_{i\mu}\,, \tag{107}$$

$$\begin{aligned}\widetilde{v}_{\mu\nu}\approx{}&\frac{1}{f_\pi^2}\,\epsilon^{ijk}\partial_\mu\pi_j\partial_\nu\pi_k\,\frac{\tau_i}{2}\\&-\left(i\left[\frac{1}{f_\pi}\,\partial_\mu\pi^i\,\frac{\tau_i}{2}\,,\,\mathrm{a}_\nu+\epsilon^{ijk}\,\frac{\pi_j}{f_\pi}\,\frac{\tau_k}{2}\,\mathrm{v}_{i\nu}\right]-(\mu\leftrightarrow\nu)\right)\\&+\text{background interference terms,}\end{aligned} \tag{108}$$

$$\rho_{\mu\nu}=\partial_{[\mu}\rho_{\nu]}+i\bar{g}_\rho[\rho_\mu\,,\,\rho_\nu]+i([\widetilde{v}_\mu\,,\,\rho_\nu]-\mu\leftrightarrow\nu)\,, \tag{109}$$

$$f_{L\mu\nu}+f_{R\mu\nu}=2\partial_{[\mu}\mathrm{v}_{\nu]}-2i[\mathrm{v}_\mu\,,\,\mathrm{v}_\nu]-2i[\mathrm{a}_\mu\,,\,\mathrm{a}_\nu]\,, \tag{110}$$

$$f_{L\mu\nu}-f_{R\mu\nu}=-2\partial_{[\mu}\mathrm{a}_{\nu]}+2i[\mathrm{v}_\mu\,,\,\mathrm{a}_\nu]+2i[\mathrm{a}_\mu\,,\,\mathrm{v}_\nu]\,, \tag{111}$$

$$F^{(+)}_{\mu\nu}\approx2\partial_{[\mu}\mathrm{v}_{\nu]}+2\epsilon^{ijk}\,\frac{\pi_j}{f_\pi}\,\frac{\tau_k}{2}\,\partial_{[\mu}\mathrm{a}_{i\nu]}+\text{background interference,} \tag{112}$$

$$F^{(-)}_{\mu\nu}\approx-2\partial_{[\mu}\mathrm{a}_{\nu]}-2\epsilon^{ijk}\,\frac{\pi_j}{f_\pi}\,\frac{\tau_k}{2}\,\partial_{[\mu}\mathrm{v}_{i\nu]}+\text{background interference.} \tag{113}$$

C. Properties of projection operators in the spin-3/2 propagator

We have properties about these spin projectors:

$$(P^{(I)}_{ij})^{\mu\nu}(P^{(J)}_{kl})_{\nu\lambda}=\delta_{IJ}\delta_{jk}(P^{(I)}_{il})^{\mu}_{\ \lambda}\,, \tag{114}$$

$$\gamma^\mu P^{(\frac{3}{2})}_{\mu\nu}=P^{(\frac{3}{2})}_{\mu\nu}\gamma^\nu=0\,, \tag{115}$$

$$p^\mu P^{(\frac{3}{2})}_{\mu\nu}=P^{(\frac{3}{2})}_{\mu\nu}p^\nu=0\,. \tag{116}$$

Based on the above identities, we can prove that

$$P^{(\frac{3}{2})}+P^{(\frac{1}{2})}_{11}+P^{(\frac{1}{2})}_{22}=\mathbf{1}\,, \tag{117}$$

$$P_{11}^{(\frac{1}{2})} + P_{22}^{(\frac{1}{2})} \equiv P^{(\frac{3}{2}\perp)}, \tag{118}$$

$$\left[P^{(\frac{3}{2})}, \not{p}\right] = \left[P_{11}^{(\frac{1}{2})}, \not{p}\right] = \left[P_{22}^{(\frac{1}{2})}, \not{p}\right] = 0. \tag{119}$$

8. References

Aguilar-Arevalo, A. A. *et al.*, (MiniBooNE Collaboration) (2009). Unexplained Excess of Electronlike Events from a 1-GeV Neutrino Beam. *Phys. Rev. Lett.*, Vol. 102, No. 10, Mar. 2009, 101802-1–101802-5, ISSN: 0031-9007.

Aguilar-Arevalo, A. A. *et al.*, (MiniBooNE Collaboration) (2010). Event Excess in the MiniBooNE Search for $\bar{\nu}_\mu \rightarrow \bar{\nu}_e$ Oscillations. *Phys. Rev. Lett.*, Vol. 105, No. 18, Oct. 2010, 181801-1–181801-5, ISSN: 0031-9007.

Ananyan, S. M.; Serot, B. D. & Walecka, J. D. (2002). Axial-Vector Current in Nuclear Many-Body Physics. *Phys. Rev. C*, Vol. 66, No. 5, Nov. 2002, 055502-1–055502-18, ISSN: 0003-4916.

Boguta, J. (1982). Baryonic Degrees of Freedom Leading to Density Isomers. *Phys. Lett. B*, Vol. 109, No. 4, Feb. 1982, 251–254, ISSN: 0370-2693.

Callan, Jr., C. G.; Coleman, S.; Wess, J. & Zumino, B. (1969). Structure of Phenomenological Lagrangians II. *Phys. Rev.*, Vol. 177, No. 5, Jan. 1969, 2247–2250.

Capri, A. Z. & Kobes, R. L. (1980). Further Problems in Spin-3/2 Field Theories. *Phys. Rev. D*, Vol. 22, No. 8, Oct. 1980, 1967–1978, ISSN:1550-7998.

Coleman, S.; Wess, J. & Zumino, B. (1969). Structure of Phenomenological Lagrangians I. *Phys. Rev.*, Vol. 177, No. 5, Jan. 1969, 2239-2247.

Donoghue, J. F.; Golowich, E. & Holstein, B. (1992). *Dynamics of the Standard Model*, ch. 2, Cambridge Press, ISBN: 0-521-36288-1, New York.

Ellis, P. J. & Tang, H.-B. (1998). Pion-Nucleon Scattering in a New Approach to Chiral Perturbation Theory. *Phys. Rev. C*, Vol. 57, No. 6, Jun. 1998, 3356–3375, ISSN: 0003-4916.

Furnstahl, R. J.; Tang, H.-B. & Serot, B. D. (1995). Vacuum Contributions in a Chiral Effective Lagrangian For Nuclei. *Phys. Rev. C*, Vol. 52, No. 3, Sep. 1995, 1368–1379, ISSN: 0003-4916.

Furnstahl, R. J.; Tang, H.-B. & Serot, B. D. (1996). Analysis of Chiral Mean-Field Models for Nuclei. *Nucl. Phys. A*, Vol. 598, No. 4, Mar. 1996, 539–582, ISSN: 0375-9474.

Furnstahl, R. J.; Serot, B. D. & Tang, H.-B. (1997). A Chiral Effective Lagrangian for Nuclei. *Nucl. Phys. A*, Vol. 615, No. 4, Apr. 1997, 441–482, ISSN: 0375-9474. [Erratum: *ibid.*, A640 (1998) 505.]

Furnstahl, R. J. (2003). Recent Developments in the Nuclear Many-Body Problem. *Int. J. Mod. Phys. B*, Vol. 17, No. 28, Nov. 2003, 5111–5126, ISSN: 0217-9792.

Gasser, J. & Leutwyler, H. (1984). Chiral Perturbation Theory to One Loop. *Ann. Phys.*, Vol. 158, No. 1, Nov. 1984, 142–210, ISSN: 0003-4916.

Georgi, H. & Manohar, A. (1984). Chiral Quarks and the Non-Relativistic Quark Model. *Nucl. Phys. B*, Vol. 234, No. 1, Mar. 1984, 189–212, ISSN: 0550-3213.

Georgi, H. (1993). Generalized Dimensional Analysis. *Phys. Lett. B*, Vol. 298, No. 1-2, Jan. 1993, 187–189, ISSN: 0370-2693.

Graczyk, K. M.; Kiełczewska, D.; Przewłocki, P. & Sobczyk, J. T. (2009). C_5^A Axial Form Factor from Bubble Chamber Experiments. *Phys. Rev. D*, Vol. 80, No. 9, Nov. 2009, 093001-1–093001-14, ISSN:1550-7998.

Hagen, C. R. (1971). New Inconsistencies in the Quantization of Spin-3/2 Fields. *Phys. Rev. D*, Vol. 4, No. 8, Oct. 1971, 2204–2208, ISSN:1550-7998.

Herbert, T.; Wehrberger, K. & Beck, F. (1992). The Pion Propagator in the Walecka Model with the Delta Baryon. *Nucl. Phys. A*, Vol. 541, No. 4, May 1992, 699–713, ISSN: 0375-9474.

Hernández, E.; Nieves, J. & Valverde, M. (2007). Weak Pion Production off the Nucleon. *Phys. Rev. D*, Vol. 76, No. 3, Aug. 2007, 033005-1–033005-22, ISSN:1550-7998.

Hirata, M.; Koch, J. H.; Lenz, F. & Moniz, E. J. (1979). Isobar-Hole Doorway States and Pion ^{16}O Scattering. *Ann. of Phys.*, Vol. 120, No. 1, Jul. 1979, 205–248, ISSN: 0003-4916.

Horikawa, Y.; Thies, M. & Lenz, F. (1980). The Delta-Nucleus Spin-Orbit Interaction in Pion-Nucleus Scattering. *Nucl. Phys. A*, Vol. 345, No. 2, Aug. 1980, 386–408, ISSN: 0375-9474.

Hu, Y.; McIntire, J. & Serot, B. D. (2007). Two-Loop Corrections for Nuclear Matter in a Chiral Effective Field Theory. *Nucl. Phys. A*, Vol. 794, No. 3-4, Oct. 2007, 187–209, ISSN: 0375-9474.

Huertas, M. A. (2002). Effective Lagrangian Approach to Structure of Selected Nuclei Far From Stability. *Phys. Rev. C*, Vol. 66, No. 2, Aug. 2002, 024318-1–024318-11. [Erratum: *ibid.*, 67 (2003) 019901.]

Huertas, M. A. (2003). Effective Field Theory Approach to Electroweak Transitions of Nuclei Far From stability. *Acta Phys. Polon. B*, Vol. 34, No. 8, Aug. 2003, 4269–4292.

Huertas, M. A. (2004). Extension of Effective Lagrangian Approach to Structure of Selected Nuclei Far from Stability. *Acta Phys. Polon. B*, Vol. 35, No. 2, Feb. 2004, 837–858.

Itzykson, C. & Zuber, J.-B. (1980). *Quantum Field Theory*, ch. 12, McGraw–Hill, ISBN: 0486445682, New York.

Johnson, K. & Sudarshan, E. C. G. (1961). Inconsistency of the Local Field Theory of Charged Spin-3/2 Particles. *Ann. of Phys.*, Vol. 13, No. 1, Apr. 1961, 126–145, ISSN: 0003-4916.

Kobayashi, M. & Takahashi, Y. (1987). The Rarita-Schwinger Paradoxes. *J. Phys, A: Math. Gen.*, Vol. 20, No. 18, Dec. 1987, 6581–6589, ISSN: 0305-4470.

Kohn, W. (1999). Nobel Lecture: Electronic Structure of Matter—Wave Functions and Density Functionals. *Rev. Mod. Phys.*, Vol. 71, No. 5, Oct. 1999, 1253–1266, ISSN: 0034-6861.

Kosov, D. S.; Fuchs, C.; Martemyanov, B. V. & Faessler, A. (1998). Constraints on the Relativistic Mean Field of Delta-Isobar in Nuclear Matter. *Phys. Lett. B*, Vol. 421, No. 1-4, Mar. 1998, 37–40, ISSN: 0370-2693.

Krebs, H.; Epelbaum, E. & Meißner, Ulf-G. (2009). On-Shell Consistency of the Rarita–Schwinger Field Formulation. *Phys. Rev. C*, Vol. 80, No. 2, Aug. 2009, 028201-1–028201-4, ISSN: 0003-4916.

Krebs, H.; Epelbaum, E. & Meißner, Ulf-G. (2010). Redundancy of the Off-Shell Parameters in Chiral Effective Field Theory with Explicit Spin-3/2 Degrees of Freedom. *Phys. Lett. B*, Vol. 683, No. 2-3, Jan. 2010, 222–228, ISSN: 0370-2693.

McIntire, J.; Hu, Y. & Serot, B. D. (2007). Loop Corrections and Naturalness in a Chiral Effective Field Theory. *Nucl. Phys. A*, Vol. 794, No. 3-4, Oct. 2007, 166–186, ISSN: 0375-9474.

McIntire, J. (2008). Three-Loop Corrections in a Covariant Effective Field Theory. *Ann. of Phys.*, Vol. 323, No. 6, Jun. 2008, 1460–1477, ISSN: 0003-4916.

Müller, H. & Serot, B. D. (1996). Relativistic Mean-Field Theory and the High-Density Nuclear Equation of State. *Nucl. Phys. A*, Vol. 606, No. 3-4, Sep. 1996, 508–537, ISSN: 0375-9474.

Oset, E. & Salcedo, L.L. (1987). Delta Self-Energy in Nuclear Matter. *Nucl. Phys. A*, Vol. 468, No. 3-4, Jul. 1987, 631–652, ISSN: 0375-9474.

Pascalutsa, V. (1998). Quantization of an Interacting Spin-3/2 Field and the Delta Isobar. *Phys. Rev. D*, Vol. 58, No. 9, Sep. 1998, 096002-1–096002-9, ISSN:1550-7998.

Pascalutsa, V. (2001). Correspondence of Consistent and Inconsistent Spin-3/2 Couplings via the Equivalence Theorem. *Phys. Lett. B*, Vol. 503, No. 1-2, Mar. 2001, 85–90, ISSN: 0370-2693.

Pascalutsa, V. (2008). The Delta(1232) Resonance in Chiral Effective Field Theory. *Prog. Part. Nucl. Phys.*, Vol. 61, No. 1, Jul. 2008, 27–33, ISSN: 0146-6410.

Serot, B. D. & Walecka, J. D. (1986). The Relativistic Nuclear Many-Body Problem. In: *Advances In Nuclear Physics*, Vol. 16, J. W. Negele and Erich Vogt, eds., pp. 1–327, Plenum Press, ISBN: 0-306-41997-1, New York.

Serot, B. D. & Walecka, J. D. (1997). Recent Progress in Quantum Hadrodynamics. *Int. J. Mod. Phys. E*, Vol. 6, No. 4, Dec. 1997, 515–631, ISSN: 0218-3013.

Serot, B. D. (2004). Covariant Effective Field Theory for Nuclear Structure and Nuclear Currents. In: *Lecture Notes in Physics*, Vol. 641, G. A. Lalazissis, P. Ring, and D. Vretenar, eds., pp. 31–63, Springer, ISSN: 0075-8450, Berlin, Heidelberg.

Serot, B. D. (2007). Electromagnetic Interactions in a Chiral Effective Lagrangian for Nuclei. *Ann. of Phys.*, Vol. 322, No. 12, Dec. 2007, 2811–2830, ISSN: 0003-4916.

Serot, B. D. (2010). Field-Theoretic Parametrization of Low-Energy Nucleon Form Factors. *Phys. Rev. C*, Vol. 81, No. 3, Mar. 2010, 034305-1–034305-6, ISSN: 0556-2813.

Serot, B. D. & Zhang, X. (2010). Weak Pion and Photon Production off Nucleons in a Chiral Effective Field Theory. arXiv:1011.5913 [nucl-th].

Serot, B. D. & Zhang, X. (2011)a. Incoherent Weak Pion and Photon Production from Nuclei in a Chiral Effective Field Theory.

Serot, B. D. & Zhang, X. (2011)b. Coherent Weak Pion and Photon Production from Nuclei in a Chiral Effective Field Theory.

Singh, L. P. S. (1973). Noncausal Propagation of Classical Rarita–Schwinger Waves. *Phys. Rev. D*, Vol. 7, No. 4, Feb. 1973, 1256–1258, ISSN:1550-7998.

Tang, H.-B. & Ellis, P. J. (1996). Redundance of Delta-Isobar Parameters in Effective Field Theories. *Phys. Lett. B*, Vol. 387, No. 1, Oct. 1996, 9–13, ISSN: 0370-2693.

Velo, G. & Zwanziger, D. (1969). Propagation and Quantization of Rarita–Schwinger Waves in an External Electromagnetic Potential. *Phys. Rev.*, Vol. 186, No. 5, Oct. 1969, 1337–1341.

Wehrberger, K.; Bedau, C. & Beck, F. (1989). Electromagnetic Excitation of the Delta-Baryon in Quantum Hadrodynamics. *Nucl. Phys. A*, Vol. 504, No. 4, Nov. 1989, 797–817, ISSN: 0375-9474.

Wehrberger, K. & Wittman, R. (1990). Pauli Blocking for Effective Deltas in Nuclear Matter. *Nucl. Phys. A*, Vol. 513, No. 3-4, Jul. 1990, 603–620, ISSN: 0375-9474.

Wehrberger, K. (1993). Electromagnetic Response Functions in Quantum Hadrodynamics. *Phys. Rep.*, Vol. 225, No. 5-6, Apr. 1993, 273–362, ISSN: 0370-1573.

Weinberg, S. (1968). Nonlinear Realizations of Chiral Symmetry. *Phys. Rev.*, Vol. 166, No. 5, Feb. 1968, 1568–1577.

Weinberg, S. (1995)a. *The Quantum Theory of Fields, vol. I: Foundations*, Sec. 5.3, Cambridge, ISBN:0-521-55001-7, New York.

Weinberg, S. (1995)b. *The Quantum Theory of Fields, vol. I: Foundations*, Sec. 5.6, Cambridge, ISBN:0-521-55001-7, New York.

Zhang, X. (2012). *Electroweak Response of Nuclei in a Chiral Effective Field Theory*. Ph.D. thesis, Indiana University, 2012.

6

Ginzburg–Landau Theory of Phase Transitions in Compactified Spaces

C.A. Linhares[1], A.P.C. Malbouisson[2] and I. Roditi[2]
[1]*Instituto de Física, Universidade do Estado do Rio de Janeiro, Rio de Janeiro*
[2]*Centro Brasileiro de Pesquisas Físicas, Rio de Janeiro*
Brazil

1. Introduction

In this chapter, we review some aspects of physical systems described by quantum fields defined on spaces with compactified dimensions. For a D-dimensional space, this means we are considering a space which has a topology of the type $\Gamma_D^d = (S^1)^d \times \mathbf{R}^{D-d}$, with d $(\leq D)$ being the number of compactified dimensions, each with the topology of a circle. This is the type of topology that emerges, for instance, in quantum field theory at finite temperature: the Matsubara formalism imposes that the time direction is compactified in a circle with length $\beta = 1/T$, where T is the temperature; its topology is then $\Gamma_4^1 = S^1 \times \mathbf{R}^3$ in the notation introduced above. Another important example involves spacetimes of dimensions D larger than four, with the "extra" or "hidden" dimensions being compactified and assumed to be very small, as in Kaluza–Klein and string theories. In any case, the topology Γ_D^d mentioned above corresponds to a generalized Matsubara formalism, in which imaginary-time and spatial coordinates may be simultaneously compactified.

In the last few decades, this generalized Matsubara formalism has been employed in many instances of condensed-matter and particle physics. Some of them are: (1) the Casimir effect, studied in various geometries, topologies, fields, and physical boundary conditions [Bordag et al. (2001); Milonni (1993); Mostepanenko & Trunov (1997)], in a diversity of subjects ranging from nanodevices to cosmological models [Bordag et al. (2001); Boyer (2003); Levin & Micha (1993); Milonni (1993); Mostepanenko & Trunov (1997); Seife (1997)]; (2) the confinement/deconfinement phase transition of hadronic matter, in the Gross–Neveu and Nambu–Jona-Lasinio models as effective theories for quantum chromodynamics [Abreu et al. (2009); Khanna et al. (2010); Malbouisson et al. (2002)]; (3) quantum electrodynamics with one extra compactified dimension, which leads to estimates of the size of extra dimensions compatible with present-day experimental data [Ccapa Tira et al. (2010)]; (4) the study of superconductors in the form of films, wires and grains [Abreu et al. (2003; 2005); Khanna et al. (2009); Linhares et al. (2006; 2007); Malbouisson (2002); Malbouisson et al. (2009)], in which the Ginzburg–Landau model for phase transitions is defined on a three-dimensional Euclidean space with one, two or three dimensions compactified.

When studying the compactification of spatial coordinates, however, it is argued in Khanna et al. (2009) from topological considerations, that we may have a quite different interpretation of the generalized Matsubara prescription: it provides a general and practical way to account for systems confined in limited regions of space at finite temperatures. Distinctly, we shall be concerned here with stationary field theories and employ the generalized Matsubara prescription to study bounded systems by implementing the compactification of spatial coordinates; no imaginary-time compactification will be done, temperature will be introduced through the mass parameter in the Ginzburg-Landau Hamiltonian. We will consider a topology of the type $\Gamma_D^d = \mathbf{R}^{D-d} \times (S^1)_1 \times (S^1)_2 \times \cdots \times (S^1)_d$, where $(S^1)_1, \ldots, (S^1)_d$ refer to the compactification of d *spatial* dimensions.

In the following, we shall concentrate on Euclidean scalar field theories defined on such spaces, with the Matsubara formalism applied to spatial coordinates. Our aim is to describe the influence of compactification on physical phenomena as phase transitions in which, for instance, the critical temperature depends on the parameters of compactification, that is, on the "size" of the system. This means that, for instance, superconductors inside spatially bound spaces such as films, wires and grains may have a critical temperature distinct from the same material in the bulk form.

In this chapter, the way in which the critical temperature for a second-order phase transition is affected by the presence of confining boundaries is investigated on general grounds. We consider that the system is a portion of material of some size, the behavior of which in the critical region is derived from a quantum field theory calculation of the dependence of the physical mass parameter on its size. We focus in particular on the mathematical aspects of the formalism, which furnish the tools to study boundary effects on the phase transition. We consider the D-dimensional Ginzburg–Landau model compactified in d ($\leq D$) of the spatial dimensions. The Ginzburg–Landau Hamiltonian, considering only the term $\lambda\varphi^4$, is known to lead to second-order transitions. In its version with N-components, in the large-N limit, we are able to take into account nonperturbatively corrections to the coupling constant. In this case, we shall obtain expressions for the transition temperature in the general situation. Particularizing for $D = 3$ and $d = 1$, $d = 2$ and $d = 3$, we have the critical temperature $T_c(L)$ for the system in the form of a film of thickness L, an infinitely long wire having a square cross-section L^2, and for a cubic grain of volume L^3, respectively. We show that $T_c(L)$ decreases as the size L is diminished and a minimal size for the suppression of the second-order transition is obtained.

We also consider the model which, besides the quartic scalar field self-interaction, a sextic one is present. The model with both interactions taken together leads to a renormalizable quantum field theory in three dimensions and it may describe *first-order* phase transitions. We consider this formalism in a general framework, taking the Euclidean D-dimensional $-\lambda |\varphi|^4 + \eta |\varphi|^6$ ($\lambda, \eta > 0$) model with $d = 1, 2, 3$ compactified dimensions. It is known that such potential *ensures* that the system undergoes a *first-order* transition. We obtain formulas for the dependence of the transition temperature on the parameters delimiting the spatial region within which the system is confined. Surely, there are other potentials which may be considered, for instance, the Halperin–Lubensky–Ma potential [Halperin et al. (1974)], which also engender first-order transitions in superconducting materials by effect of integration over the gauge field and takes the form $-\alpha\varphi^3 + \beta\varphi^4$.

We start from the effective potential, which is related to the physical mass and coupling constant through renormalization conditions. These conditions, however, reduce considerably the number of relevant contributing Feynman diagrams, if one wishes to be restricted to 1- or 2-loop approximations. For second-order transitions, we need to consider only the tadpole diagram to correct the mass and the 1-loop four-point function to correct the coupling constant. For first-order transitions, we will not, for simplicity, make corrections to the coupling constant. In this case, just two diagrams need to be considered: a tadpole graph with the φ^4 coupling (one loop) and a "shoestring" graph with the φ^6 coupling (two loops). No diagram with both couplings needs to be considered. The size dependence appears from the treatment of the loop integrals. The dimensions of finite extent are treated in momentum space using the formalism of Khanna et al. (2009).

It is worth noticing that for superconducting films with thickness L, a qualitative agreement of our theoretical L-dependent critical temperature is found with experiments. This occurs in particular for thin films (in the case of first-order transitions) and for a wide range of values of L for second-order transitions [Linhares et al. (2006)]. Moreover, available experimental data for superconducting wires are compatible with our theoretical prediction of the first-order critical temperature as a function of the transverse cross section of the wire.

Finally, we discuss the infrared behavior and the fixed-point structure for the N-component $\lambda\varphi^4$ model in the large-N limit, with a compactified subspace. We study the cases in which the system has no external influence and in which the system is submitted to the action of an external magnetic field. In both situations, with or without a magnetic field, we get the result that the existence of an infrared stable fixed-point depends only on the space dimension; it does not depend on the number of compactified dimensions.

2. Critical behavior of the compactified $\lambda\varphi^4$ model

We start by considering the complex scalar field model described by the Ginzburg–Landau Hamiltonian density in a Euclidean D-dimensional space, in the absence of any geometrical constraints, given by (in natural units, $\hbar = c = k_B = 1$)

$$\mathcal{H} = \frac{1}{2}\left|\partial_\mu\varphi\right|\left|\partial^\mu\varphi\right| + \frac{1}{2}m_0^2\left|\varphi\right|^2 + \frac{\lambda}{4}\left|\varphi\right|^4, \tag{1}$$

where $\lambda > 0$ is the physical coupling constant. As usual, near criticality, the bare mass is taken as $m_0^2 = \alpha(T - T_0)$, with $\alpha > 0$ and T_0 being a parameter with the dimension of temperature, which is interpreted as the bulk transition temperature.

Let us now take the system in D dimensions confined to a region of space delimited by $d \leq D$ pairs of parallel planes. Each plane of a pair j is at a distance L_j from the other member of the pair, $j = 1, 2, \ldots, d$, and is orthogonal to all other planes belonging to distinct pairs $\{i\}$, $i \neq j$. This may be pictured as a parallelepipedal box embedded in the D-dimensional space, whose parallel faces are separated by distances $L_1, L_2, \ldots, L_d$. To simplify matters, we shall take all $L_i = L$. Let us define Cartesian coordinates $r = (x_1, x_2, \ldots, x_d, z)$, where z is a $(D-d)$-dimensional vector, with corresponding momentum $k = (k_1, k_2, \ldots, k_d, q)$, q being a $(D-d)$-dimensional vector in momentum space. The generating functional of Schwinger functions is written in the form

$$Z = \int \mathcal{D}\varphi\,\mathcal{D}\varphi^* \exp\left(-\int_0^{L_1} dx_1 \cdots \int_0^{L_d} dx_d \int d^{D-d}z\,\mathcal{H}(|\varphi|, |\nabla\varphi|)\right), \tag{2}$$

with the field $\varphi(x_1, ..., x_d, z)$ satisfying the condition of confinement inside the box, $\varphi(x_i \leq 0, z) = \varphi(x_i \geq 0, z) = \text{const}$. Then, following the procedure developed in Khanna et al.

(2009), we introduce a generalized Matsubara prescription, in which the Feynman rules are modified through the replacements

$$\int \frac{dk_i}{2\pi} \to \frac{1}{L} \sum_{n_i=-\infty}^{+\infty} ; \qquad k_i \to \frac{2n_i\pi}{L} \ , \ i = 1, 2..., d. \tag{3}$$

Notice that compactification can be implemented in different ways, as for instance by imposing specific conditions on the fields at spatial boundaries. We here choose periodic boundary conditions.

In principle, the effective potential for systems with spontaneous symmetry breaking is obtained, following the Coleman–Weinberg analysis [Coleman & Weinberg (1973)], as an expansion in the number of loops in Feynman diagrams. Accordingly, to the free propagator and to the tree diagrams, radiative corrections are added, with increasing number of loops. Thus, at the 1-loop approximation, we get the infinite series of 1-loop diagrams with all numbers of insertions of the φ^4 vertex (two external legs in each vertex).

At the 1-loop approximation, the contribution of loops with only $|\varphi|^4$ vertices to the effective potential in unbounded space is

$$U_1(\varphi_0) = \sum_{s=1}^{\infty} \frac{(-1)^{s+1}}{2s} \left[3\lambda|\varphi_0|^2\right]^s \int \frac{1}{(2\pi)^D} \frac{d^D k}{(k^2+m^2)^s}, \tag{4}$$

where m is the *physical* mass and the parameter s counts the number of vertices on the loop.

In the following, to deal with dimensionless quantities in the regularization procedures, we introduce parameters $c^2 = m^2/4\pi^2$, $L^2 = a^{-1}$, $g = 3\lambda/8\pi^2$, where φ_0 is the normalized vacuum expectation value of the field (the classical field). In terms of these parameters and performing the Matsubara replacements (3), the one-loop contribution to the effective potential can be written in the form

$$U_1(\phi_0, a) = a^{d/2} \sum_{s=1}^{\infty} \frac{(-1)^{s+1}}{2s} g^s |\varphi_0|^{2s}$$

$$\times \sum_{n_1,\ldots,n_d=-\infty}^{+\infty} \int \frac{d^{D-d}q}{\left[a\left(n_1^2+\cdots+n_d^2\right)+c^2+q^2\right]^s}. \tag{5}$$

It is easily seen that only the $s = 1$ term contributes to the renormalization condition

$$\left.\frac{\partial^2 U(\varphi_0)}{\partial \varphi_0^2}\right|_{\varphi_0=0} = m^2. \tag{6}$$

It corresponds to the tadpole diagram. The integral over the $D - d$ noncompactified momentum variables is performed using a well-known dimensional regularization formula [Zinn-Justin (2002)] so that, for $s = 1$, we obtain

$$U_1(\phi_0, a) = \frac{1}{2} a^{d/2} \pi^{(D-d)/2} \Gamma\left(1 - \frac{D-d}{2}\right) g|\varphi_0|^2 Z_d^{c^2}\left(\frac{2-D+d}{2}; a\right), \tag{7}$$

where $Z_d^{c^2}(\frac{2-D+d}{2};a)$ is one of the Epstein–Hurwitz zeta functions, defined by

$$Z_d^{c^2}(\nu;a_1,...,a_d) = \sum_{n_1,...,n_d=-\infty}^{+\infty} (a_1 n_1^2 + \cdots + a_d n_d^2 + c^2)^{-\nu}, \tag{8}$$

valid for $\mathrm{Re}(\nu) > 1$.

Next, we can use the generalization to several dimensions of the mode-sum regularization prescription described in Elizalde (1995). It results that the multidimensional Epstein–Hurwitz function has an analytic extension to the whole ν complex plane, which may be written as

$$\begin{aligned} Z_d^{c^2}(\nu;L) = \frac{2^{\nu-\frac{d}{2}+1}\pi^{2\nu-\frac{d}{2}}L^{d/2}}{\Gamma\left(1-\frac{D-d}{2}\right)\Gamma(\nu)} \Bigg[& 2^{\nu-\frac{d}{2}-1}m^{d-2\nu}\Gamma\left(\nu-\frac{d}{2}\right) \\ & +2d\sum_{n=1}^{\infty}\left(\frac{m}{Ln_i}\right)^{\frac{d}{2}-\nu} K_{\nu-\frac{d}{2}}(mLn) + \cdots \\ & +2^d \sum_{n_1,...,n_d=1}^{\infty}\left(\frac{m}{L\sqrt{n_1^2+\cdots+n_d^2}}\right)^{\frac{d}{2}-\nu} \\ & \times K_{\nu-\frac{d}{2}}\left(mL\sqrt{n_1^2+\cdots+n_d^2}\right)\Bigg], \end{aligned} \tag{9}$$

where the $K_\nu(z)$ are modified Bessel functions of the second kind. Taking $\nu = (2-D+d)/2$ in Eq. (9), we obtain from Eq. (7) the effective potential in D dimensions with a compactified d-dimensional subspace:

$$\begin{aligned} U_1(\varphi_0,L) = \frac{3\lambda|\varphi_0|^2}{(2\pi)^{D/2}} \Bigg[& 2^{-D/2-1}m^{D-2}\Gamma\left(\frac{2-D}{2}\right) \\ & +d\sum_{n=1}^{\infty}\left(\frac{m}{Ln}\right)^{D/2-1} K_{D/2-1}(mLn) + \cdots \\ & +2^{d-1} \sum_{n_1,...,n_d=1}^{\infty}\left(\frac{m}{L\sqrt{n_1^2+\cdots+n_d^2}}\right)^{D/2-1} \\ & \times K_{D/2-1}\left(mL\sqrt{n_1^2+\cdots+n_d^2}\right)\Bigg], \end{aligned} \tag{10}$$

where we have returned to the original variables, λ and L.

Notice that in Eq. (10) there is a term proportional to $\Gamma\left(\frac{2-D}{2}\right)$, which is divergent for even dimensions $D \geq 2$, and should be subtracted in order to obtain finite physical parameters. For odd D, the above gamma function is finite, but we also subtract this term (corresponding to a finite renormalization), for the sake of uniformity. We get

$$U_{1,R}(\varphi_0, L) = \frac{3\lambda|\varphi_0|^2}{(2\pi)^{D/2}}\left[d\sum_{n=1}^{\infty}\left(\frac{m}{Ln}\right)^{D/2-1}K_{D/2-1}(mLn)+\cdots\right.$$
$$+2^{d-1}\sum_{n_1,\ldots,n_d=1}^{\infty}\left(\frac{m}{L\sqrt{n_1^2+\cdots+n_d^2}}\right)^{D/2-1}$$
$$\left.\times K_{D/2-1}\left(mL\sqrt{n_1^2+\cdots+n_d^2}\right)\right]. \tag{11}$$

Then the physical mass is obtained from Eq. (6), using Eq. (11) and also taking into account the contribution at the tree level; it satisfies a generalized Dyson–Schwinger equation depending on the finite extension L of the confining box:

$$m^2(L) = m_0^2 + \frac{6\lambda}{(2\pi)^{D/2}}\left[d\sum_{n=1}^{\infty}\left(\frac{m}{Ln}\right)^{D/2-1}K_{D/2-1}(mLn)+\cdots\right.$$
$$+2^{d-1}\sum_{n_1,\ldots,n_d=1}^{\infty}\left(\frac{m}{L\sqrt{n_1^2+\cdots+n_d^2}}\right)^{D/2-1}$$
$$\left.\times K_{D/2-1}\left(mL\sqrt{n_1^2+\cdots+n_d^2}\right)\right]. \tag{12}$$

It is not envisageable to solve the above equation analytically for the mass. However, if we limit ourselves to the neighborhood of criticality, then we can put $m^2(L) \approx 0$, and we may also use an asymptotic formula for a Bessel function with a small argument, $K_\nu(z) \approx \frac{1}{2}\Gamma(\nu)(2/z)^\nu$ $(z \sim 0)$. In this way, the coefficients and arguments of the Bessel functions cancel out and we rewrite (12) as

$$m^2(L) \approx m_0^2 + \frac{3\lambda}{\pi^{D/2}}\Gamma\left(\frac{D}{2}-1\right)\left[\frac{d}{2}E_1\left(\frac{D}{2}-1;L\right)\right.$$
$$\left.+d(d-1)E_2\left(\frac{D}{2}-1;L\right)+\cdots+2^{d-2}E_d\left(\frac{D}{2}-1;L\right)\right], \tag{13}$$

where the $E_p(\nu;L)$ are generalized Epstein–Hurwitz zeta functions defined by Kirsten (1994)

$$E_p(\nu;L) = L^\nu\sum_{n_1=1}^{\infty}\cdots\sum_{n_p=1}^{\infty}\left(n_1^2+\cdots+n_p^2\right)^{-\nu}, \tag{14}$$

[for details, see Malbouisson et al. (2002)]. Notice that, for $p=1$, E_p reduces to the Riemann zeta function $\zeta(z) = \sum_{n=1}^{\infty} n^{-z}$.

Having developed the general case of a d-dimensional compactified subspace, we consider an illustrative example. We choose $d=1$, the compactification of just one dimension, along the

x_1-axis, say, meaning that we are considering that the system is confined between two planes, separated by a distance L (film of thickness L). Then, Eq. (13) simplifies to

$$m^2(L) \approx m_0^2(L) + \frac{3\lambda}{2\pi^{D/2}L^{D-2}}\Gamma\left(\frac{D}{2}-1\right)\zeta(D-2), \tag{15}$$

where $\zeta(z)$ is the Riemann zeta function. This equation is well defined for $D > 3$, but not for $D = 3$, due to the pole of the zeta function. However, we can assign it a meaning for the significative dimension $D = 3$ by adopting a regularization procedure: we use the well-known formula

$$\lim_{z\to 1} \zeta(z) = \frac{1}{z-1} + \gamma, \tag{16}$$

where $\gamma \approx 0.5772$ is the Euler–Mascheroni constant, for $\zeta(D-2)$ in Eq. (15) and afterwards we suppress the pole term at $D = 3$ ($z = 1$). Then, remembering that $m_0^2 = \alpha(T - T_0)$, we get the L-dependent critical temperature,

$$\begin{aligned} T_c^{\text{film}}(L) &= T_0 - C_1\frac{\lambda}{\alpha L}, \\ \text{with } C_1 &= \frac{3\gamma}{2\pi}. \end{aligned} \tag{17}$$

We see that, for $L < (3\gamma/2\pi)(\lambda/\alpha T_0)$, the critical temperature becomes negative, meaning that the transition does not occur.

With analogous steps, we can take the cases of $d = 2$ and $d = 3$, in which the system is confined within an infinite wire of rectangular cross section $L^2 \equiv A$ and a grain of volume $L^3 \equiv V$, respectively. In those cases, it is not necessary to renormalized the bare mass, as we have done for a film, as the divergences coming from the zeta and gamma functions completely cancel out algebraically. One obtains [Abreu et al. (2005)]

$$\begin{aligned} T_c^{\text{wire}}(A) &= T_0 - C_2\frac{\lambda}{\alpha A^{1/2}}, \\ T_c^{\text{grain}}(V) &= T_0 - C_3\frac{\lambda}{\alpha V^{1/3}}, \end{aligned} \tag{18}$$

where C_2 and C_3 are numerical constants. We note that, in all cases, it is found that the boundary-dependent critical temperature decreases linearly with the inverse of the linear dimension L, $T_c(L) = T_0 - C_d\lambda/\alpha L$, where α and λ are the Ginzburg–Landau parameters, T_0 is the bulk transition temperature and C_d is a constant depending on the number of compactified dimensions. This is in accordance with arguments raised from finite-size scaling [Zinn-Justin (2002)].

Such behavior suggests the existence of a minimal size of the system, below which the transition is suppressed. It seems to be in qualitative agreement with experimental results which indicate a minimal thickness of a film for the disapearance of superconductivity [Abreu et al. (2004); Kodama et al. (1983)]; also, the behavior of nanowires and nanograins have been studied [Shanenko et al. (2006); Zgirski et al. (2005)], searching for a limit on its size for the material while retaining its superconducting character.

3. First-order phase transitions

In the previous section, we have studied the Ginzburg–Landau Hamiltonian density, solely containing the interaction term $\lambda\varphi^4$, with $\lambda > 0$, which describes second-order phase transitions. Here we pass to consider the Ginzburg–Landau model in a Euclidean D-dimensional space, including both φ^4 and φ^6 interactions, in the absence of external fields; its Hamiltonian is given by (again, in natural units, $\hbar = c = k_B = 1$)

$$\mathcal{H} = \frac{1}{2}\left|\partial_\mu \varphi\right| \left|\partial^\mu \varphi\right| + \frac{1}{2} m_0^2 \left|\varphi\right|^2 - \frac{\lambda}{4}\left|\varphi\right|^4 + \frac{\eta}{6}\left|\varphi\right|^6, \tag{19}$$

where $\lambda > 0$ and $\eta > 0$ are the physical quartic and sextic coupling constants. Near criticality, the bare mass is given by $m_0^2 = \alpha(T/T_0 - 1)$, with $\alpha > 0$ and T_0 being a parameter with the dimension of temperature. A potential of this type, with the minus sign in the quartic term, ensures that the system undergoes a first-order transition. Recall that the critical temperature for a first-order transition described by the Hamiltonian above is higher than T_0. This will be explicitly stated in Eq. (25) below. Our purpose will be to develop the general case of compactifying a d-dimensional subspace, in order to compare results for films, wires and grains with the second-order ones given above.

We thus consider the system in D dimensions confined to a region of space delimited by $d \leq D$ pairs of parallel planes, as was done in the previous section, and introduce a generalized Matsubara prescription as in Eq. (3), with periodic boundary conditions. We again start from establishing the effective potential, related to the physical mass through a renormalization condition, Eq. (6). This condition, however, reduces considerably the number of relevant Feynman diagrams contributing to the mass, if we restrict ourselves to first-order terms in both coupling constants: in fact, just two diagrams need to be considered in this approximation, a tadpole graph with the φ^4 coupling (1 loop) and a "shoestring" graph with the φ^6 coupling (2 loops).

Within our approximation, we do not take into account the renormalization conditions for the interaction coupling constants, i.e., they are considered as already renormalized when they are written in the Hamiltonian (the same was assumed in the previous section).

At the 1-loop approximation, the contribution of loops with only $|\varphi|^4$ vertices to the effective potential is obtained directly from the previous section, Eq. (5). As before, we see that only the $s = 1$ term contributes to the renormalization condition in Eq. (6). It corresponds to the tadpole diagram. It is then also clear that all $|\varphi_0|^6$-vertex and mixed $|\varphi_0|^4$- and $|\varphi_0|^6$-vertex insertions on the 1-loop diagrams do not contribute when one computes the second derivative of similar expressions with respect to the field at zero field: only diagrams with two external legs should survive. This is impossible for a $|\varphi_0|^6$-vertex insertion at the 1-loop approximation. Therefore, the first contribution from the $|\varphi_0|^6$ coupling must come from a higher-order term in the loop expansion. Two-loop diagrams with two external legs and only $|\varphi_0|^4$ vertices are of second order in its coupling constant, and we neglect them, as well as all possible diagrams with vertices of mixed type. However, the 2-loop shoestring diagram, with only one $|\varphi_0|^6$ vertex and two external legs is a first-order (in η) contribution to the effective potential, according to our approximation.

The tadpole contribution to the effective potential was treated in the previous section, through dimensional and Epstein–Hurwitz zeta-function regularizations and subtraction of a polar term, resulting in the expression $U_{1,R}$ of Eq. (11), in terms of modified Bessel functions. Now, proceeding analogously for the 2-loop shoestring diagram contribution, we arrive at

$$U_{2,R}(\varphi_0, L_1, \ldots, L_d) = \frac{\eta|\varphi_0|^2}{4(2\pi)^D}\left[d\sum_{n=1}^{\infty}\left(\frac{m}{Ln}\right)^{D/2-1}K_{D/2-1}(mLn) + \cdots\right.$$
$$+2^{d-1}\sum_{n_1,\ldots,n_d=1}^{\infty}\left(\frac{m}{L\sqrt{n_1^2+\cdots+n_d^2}}\right)^{D/2-1}$$
$$\left.\times K_{D/2-1}\left(mL\sqrt{n_1^2+\cdots+n_d^2}\right)\right]^2. \tag{20}$$

Then the physical mass $m^2(L)$ with both contributions is obtained from Eq. (6), using Eqs. (11), (20) and also taking into account the contribution at the tree level; it satisfies a generalized Dyson–Schwinger equation depending on the extensions L of each dimension of the confining box, as in Eq. (12). We should remember that the tadpole part has a change of sign with respect to (12), reflecting the sign of λ in the Hamiltonian (19).

A first-order transition occurs when all the three minima of the potential

$$U(\varphi_0) = \frac{1}{2}m^2(L)|\varphi_0|^2 - \frac{\lambda}{4}|\varphi_0|^4 + \frac{\eta}{6}|\varphi_0|^6, \tag{21}$$

where $m(L)$ is the renormalized mass defined above, are simultaneously on the line $U(\varphi_0) = 0$. This gives the condition

$$m^2(L) = \frac{3\lambda^2}{16\eta}. \tag{22}$$

For $D = 3$, the Bessel functions have an explicit form, $K_{1/2}(z) = \sqrt{\pi}e^{-z}/\sqrt{2z}$, which is to be replaced in the expression for the renormalized mass. Performing the resulting sums, and remembering that $m_0^2 = \alpha(T/T_0 - 1)$, we get

$$m^2(L) = \alpha\left(\frac{T}{T_0} - 1\right) + \frac{3\lambda}{4\pi}\left[\frac{d}{L}\ln\left(1 - e^{-m(L)L}\right) + \cdots\right.$$
$$\left.+2^{d-1}\sum_{n_1,\ldots,n_d=1}^{\infty}\frac{e^{-m(L)L\sqrt{n_1^2+\cdots+n_d^2}}}{\sqrt{n_1^2+\cdots+n_d^2}}\right]$$
$$+\frac{\eta\pi}{8\,(2\pi)^3}\left[\frac{d}{L}\ln\left(1 - e^{-m(L)L}\right) + \cdots\right.$$
$$\left.+2^{d-1}\sum_{n_1,\ldots,n_d=1}^{\infty}\frac{e^{-m(L)L\sqrt{n_1^2+\cdots+n_d^2}}}{L\sqrt{n_1^2+\cdots+n_d^2}}\right]^2. \tag{23}$$

Then, introducing the value of the mass, Eq. (22), in Eq. (23), one obtains the critical temperature

$$T_c(L) = T_c\left\{1 - \left(1 + \frac{3\lambda^2}{16\eta\alpha}\right)^{-1}\left\{\frac{3\lambda}{4\pi\alpha}\left[\frac{d}{L}\ln\left(1 - e^{-L\sqrt{\frac{3\lambda^2}{16\eta}}}\right) + \cdots \right.\right.\right.$$
$$\left. + 2^{d-1}\sum_{n_1,\ldots,n_d=1}^{\infty}\frac{e^{-L\sqrt{\frac{3\lambda^2}{16\eta}}\sqrt{n_1^2+\cdots+n_d^2}}}{L\sqrt{n_1^2+\cdots+n_d^2}}\right]$$
$$- \frac{\eta}{64\pi^2\alpha}\left[\frac{d}{L}\ln\left(1 - e^{-L\sqrt{\frac{3\lambda^2}{16\eta}}}\right) + \cdots \right.$$
$$\left.\left.\left. + 2^{d-1}\sum_{n_1,\ldots,n_d=1}^{\infty}\frac{e^{-L\sqrt{\frac{3\lambda^2}{16\eta}}\sqrt{n_1^2+\cdots+n_d^2}}}{L\sqrt{n_1^2+\cdots+n_d^2}}\right]^2\right\}\right\}, \tag{24}$$

where

$$T_c = T_0\left(1 + \frac{3\lambda}{16\eta\alpha}\right) \tag{25}$$

is the bulk ($L \to \infty$) critical temperature for the first-order phase transition.

Specific formulas for particular values of d are now given. If we choose $d = 1$, this corresponds physically to a film of superconducting material, and we have that the transition occurs at the critical temperature $T_c^{\text{film}}(L)$ given by

$$T_c^{\text{film}}(L) = T_c\left\{1 - \left(1 + \frac{3\lambda^2}{16\eta\alpha}\right)^{-1}\left[\frac{3\lambda}{4\pi\alpha L}\ln\left(1 - e^{-L\sqrt{\frac{3\lambda^2}{16\eta}}}\right)\right.\right.$$
$$\left.\left. - \frac{\eta}{64\pi^2\alpha L^2}\left(\ln(1 - e^{-L\sqrt{\frac{3\lambda^2}{16\eta}}})\right)^2\right]\right\}. \tag{26}$$

In the case of a wire, $d = 2$, the critical temperature is written in terms of L as

$$T_c^{\text{wire}}(L) = T_c\left\{1 - \left(1 + \frac{3\lambda^2}{16\eta\alpha}\right)^{-1}\right.$$
$$\times\left[\frac{3\lambda}{2\pi\alpha L}\left[\ln\left(1 - e^{-L\sqrt{\frac{3\lambda^2}{16\eta}}}\right) + \ln\left(1 - e^{-L\sqrt{\frac{3\lambda^2}{16\eta}}}\right)\right.\right.$$
$$\left. + 2\sum_{n_1,n_2=1}^{\infty}\frac{e^{-L\sqrt{\frac{3\lambda^2}{16\eta}}\sqrt{n_1^2+n_2^2}}}{\sqrt{n_1^2+n_2^2}}\right]$$
$$- \frac{\eta}{32\pi^2\alpha L^2}\left(\ln\left(1 - e^{-L\sqrt{\frac{3\lambda^2}{16\eta}}}\right) + \ln\left(1 - e^{-L\sqrt{\frac{3\lambda^2}{16\eta}}}\right)\right.$$
$$\left.\left.\left. + 2\sum_{n_1,n_2=1}^{\infty}\frac{e^{-L\sqrt{\frac{3\lambda^2}{16\eta}}\sqrt{n_1^2+n_2^2}}}{\sqrt{n_1^2+n_2^2}}\right)^2\right]\right\}. \tag{27}$$

Finally, if we compactify all three dimensions ($d = 3$), which leaves us with a system in the form of a cubic "grain" of some material, the dependence of the critical temperature on its linear dimension L is given by

$$\begin{aligned} T_c^{\text{grain}}(L) = T_c \Bigg\{ 1 - \left(1 + \frac{3\lambda^2}{16\eta\alpha}\right)^{-1} \Bigg\{ \frac{3\lambda}{2\pi\alpha L} \\ \times \Bigg[3\ln(1 - e^{-L\sqrt{\frac{3\lambda^2}{16\eta}}}) + \cdots \\ + 4 \sum_{n_1,\ldots,n_3=1}^{\infty} \frac{e^{-L\sqrt{\frac{3\lambda^2}{16\eta}}\sqrt{n_1^2+n_2^2+n_3^2}}}{\sqrt{n_1^2+n_2^2+n_3^2}} \Bigg] \\ - \frac{\eta}{32\pi^2\alpha L^2} \Bigg[3\ln(1 - e^{-L\sqrt{\frac{3\lambda^2}{16\eta}}}) + \cdots \\ + 4 \sum_{n_1,\ldots,n_3=1}^{\infty} \frac{e^{-L\sqrt{\frac{3\lambda^2}{16\eta}}\sqrt{n_1^2+n_2^2+n_3^2}}}{\sqrt{n_1^2+n_2^2+n_3^2}} \Bigg]^2 \Bigg\} \Bigg\}. \end{aligned} \tag{28}$$

Comparing Eqs. (26)-(28) with the general behavior of the critical temperature obtained in the previous section, we see that in all cases (film, wire or grain), there is a sharp contrast between the simple inverse linear behavior of $T_c(L)$ for second-order transitions and the rather involved dependence on L of the critical temperature for first-order transitions.

In Linhares et al. (2006; 2007), we have shown that our general formalism could be not of a purely academic interest, but that it could be used to describe some experimentally observable situations. Experimental data on the critical temperature obtained from superconducting films and wires can be compared with our theoretical expressions. In Linhares et al. (2006), the coupling constants λ and η have been determined as functions of the microscopic parameters of the material, which was done generalizing Gorkov's [Kleinert (1989)] microscopic derivation done for the $\lambda\varphi^4$ model, in order to include the additional interaction term $\eta\varphi^6$ in the free energy. See Linhares et al. (2006; 2007) for details.

As described in Linhares et al. (2006), the transition temperature as a function of the thickness for a film grows from zero at a nonnull minimal allowed film thickness above the bulk transition temperature T_c as the thickness is enlarged, reaching a maximum and afterwards starting to decrease, going asymptotically to T_c as $L \to \infty$. Our theoretical curve is in qualitatively good agreement with measurements, especially for thin films [Strongin et al. (1970)]. This is illustrated in Figure 1. This behavior can be contrasted with the one shown by the critical temperature for a second-order transition. As one can see in Figure 2, in this case, the critical temperature increases monotonically from zero, again corresponding to a finite minimal film thickness, going asymptotically to the bulk transition temperature as $L \to \infty$ [Abreu et al. (2004)]. Such behavior has been experimentally found for a variety of transition-metal materials [Kodama et al. (1983); Minhaj et al. (1994); Pogrebnyakov et al. (2003); Raffy et al. (1983)]. Since in this section a first-order transition is explicitly assumed, it is tempting to infer that the transition described in the experiments of Strongin et al. (1970) is first order. In other words, one could say that an experimentally observed behavior of the critical temperature as a function of the film thickness may serve as a possible criterion

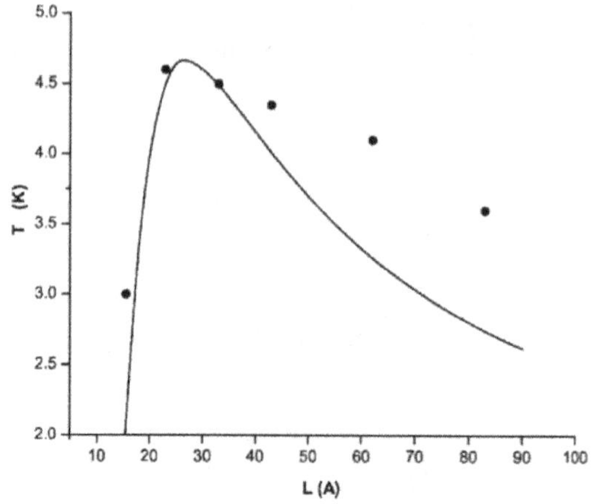

Fig. 1. Critical temperature $T_c^{\text{film}}(K)$ as a function of the thickness $L(\text{Å})$, with data from Strongin et al. (1970) for a superconducting film made from aluminum.

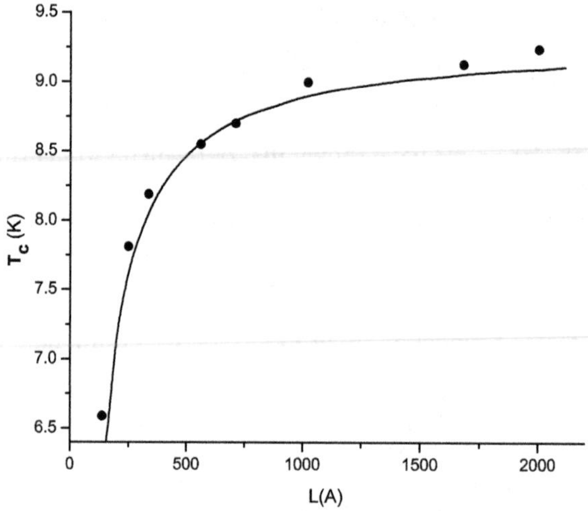

Fig. 2. Critical temperature $T_c^{\text{film}}(K)$ as a function of the thickness $L(\text{Å})$ for a second-order transition, with data from Kodama et al. (1983) for a superconducting film made from niobium.

to decide about the order of the superconductivity transition: a monotonically increasing critical temperature as L grows would indicate that the system undergoes a second-order transition, whereas if the critical temperature presents a maximum for a value of L larger than the minimal allowed one, this would be signaling the occurrence of a first-order transition. If we consider a sample of superconducting material in the form of an infinitely long wire with a cross section L^2, the same arguments and rescaling procedures used for films apply. In this case, the theoretical curve $T_c^{\rm wire}$ vs. L, together with Al data from Shanenko et al. (2006); Zgirski et al. (2005) agree quite well, for not extremely thin wires. One may conclude that the phase transition of these superconducting aluminum wires is first order, just as for aluminum films. The interested reader will find details in Linhares et al. (2006; 2007).

4. Coupling-constant corrections for second-order transitions

We have so far discussed the critical properties of confined superconducting matter under the assumption that the coupling constants, as they appear in the Hamiltonian, are the physical ones. It is however expected that the compactification of spatial dimensions as we have described also has an influence on the coupling constants and consequently on the behavior of the transition temperature with respect to the size of the compactified space. To undertake such study, we shall consider the four-point function at zero external momenta, which is the basic object for our definition of the renormalized coupling constant. We shall analyze it in the O(N)-symmetric version of the D-dimensional Ginzburg–Landau model, described by the Hamiltonian density

$$\mathcal{H} = \partial_\mu \varphi_a \partial^\mu \varphi_a + m_0^2(T)\varphi_a\varphi_a + \frac{\lambda}{N}\left(\varphi_a\varphi_a\right)^2, \tag{29}$$

and take the large-N limit. In Eq. (29), λ is the coupling constant and $m_0^2(T) = \alpha(T - T_0)$ is the bare mass, as before. The compactification procedure is the same as that implemented in section 2 and we look for the 1-loop contribution from φ^4 vertices for the effective potential after compactification of d dimensions. We may use directly Eq. (10), taking care that the convention for the coupling constant has changed: $\lambda/4 \to \lambda$. The mass is obtained from the normalization condition (6) and the coupling constant from

$$\left.\frac{\partial^4}{\partial\varphi_0^2}U(\varphi_0)\right|_{\varphi_0=0} = \frac{\lambda}{N}, \tag{30}$$

where U is the sum of the tree-level and 1-loop contributions to the effective potential.

The coupling constant is defined in terms of the 4-point function for zero external momenta, which, at leading order in $1/N$, is given by the sum of all chains of 1-loop diagrams, which has the formal expression

$$\Gamma_D^{(4)}(p=0,m,L) = \frac{\lambda/N}{1+\lambda\Pi(m,L)}, \tag{31}$$

where $\Pi(m,L) \equiv \Pi(p = 0, m, L)$ corresponds to the one-loop four-point diagram, after compactification. Next, we use the renormalization condition (30), from which we deduce formally that the one-loop four-point function $\Pi(m,L)$ is obtained from the coefficient of the fourth power of the field ($s = 2$) in Eq. (10). A divergent (for even dimensions) term is subtracted to give the finite one-loop four-point function $\Pi_R(m,L)$, which corresponds to (11). Such subtraction is performed even in the case of odd dimensions, where no pole singularity occurs (finite renormalization). From the properties of Bessel functions, we see that $\Pi_R(m,L) \to 0$ as $L \to \infty$, whereas it diverges when $L \to 0$. We conclude that the renormalized one-loop four-point function is positive for all values of D and L.

Let us define the the L-dependent renormalized coupling constant $\lambda_R(m,L)$, at leading order in $1/N$, as

$$N\Gamma^{(4)}_{D,R}(p=0,m,L) \equiv \lambda_R(m,L) = \frac{\lambda}{1+\lambda\Pi_R(m,L)}. \tag{32}$$

In the absence of constraints, the $L\to\infty$ limit of $\Gamma^{(4)}_{D,R}(p=0,m,L)$ defines the corresponding renormalized coupling constant $\lambda_R(m)$. We get simply that $\lambda_R(m)=\lambda$. This means that a renormalization scheme has been chosen so that the constant λ appearing in the Hamiltonian corresponds to the renormalized coupling constant in the absence of boundaries.

The physical mass is obtained at 1-loop from (12), with $\lambda/4\to\lambda$, and (6), after also changing $\lambda\to\lambda_R(m,L)$, given by (32). One should remember, however, that $\lambda_R(m,L)$ is itself a function of $m=m(T,L)$. Therefore, $m(T,L)$ is given by a complicated set of coupled equations. Just like in the situation in section 2, without the corrections in λ, it has no analytical solution in general. Nevertheless, as before, if we limit ourselves to the neighborhood of criticality, $m^2(T,L)\approx 0$, the behavior of the system can be studied by using the approximation $K_\nu(z)\approx \frac{1}{2}\Gamma(\nu)(2/z)^\nu$, for $z\sim 0$. The same kind of simplifications occurs and we regain Eq. (13), with $\lambda\to\lambda_R(D,L)$ given by

$$\lambda_R(D,L)\approx \lambda\Big\{1+\lambda C(D)L^{4-D}\left[d\zeta(D-4)+2d(d-1)E_2\left(D/2-2,1\right)\right.$$
$$\left.+\cdots+2^{d-1}E_d\left(D/2-2,1\right)\right]\Big\}^{-1}, \tag{33}$$

where $C(D)=\frac{1}{8\pi^{D/2}}\Gamma\left(\frac{D}{2}-2\right)$. It then ensues that we obtain the critical temperature as a function of L. Taking $D=3$, we have a similar situation as that of section 2. We find modified

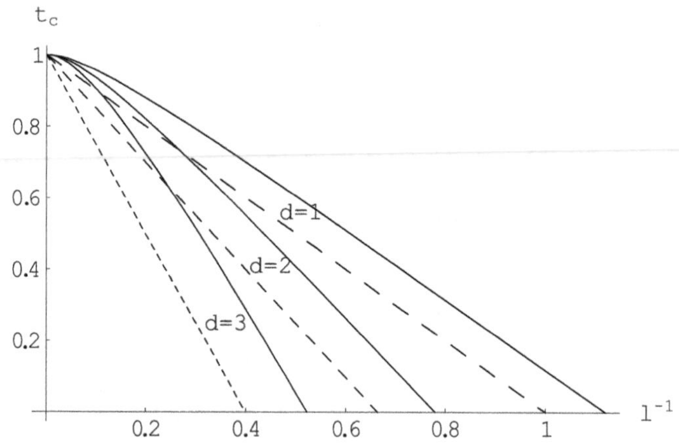

Fig. 3. Reduced transition temperature (t_c) as a function of the inverse of the reduced compactification length (l), for films ($d=1$), square wires ($d=2$) and cubic grains ($d=3$). The full and dashed lines correspond to results with and without correction of the coupling constant, respectively.

L-dependent transition temperatures, which are given by

$$T_c^{\text{film}}(L) = T_0 - \frac{48\pi C_1 \lambda}{48\pi\alpha L + \lambda\alpha L^2};$$
$$T_c^{\text{wire}}(A) = T_0 - \frac{48\pi C_2 \lambda}{48\pi\alpha\sqrt{A} + \mathcal{E}_2\lambda\alpha A}; \tag{34}$$
$$T_c^{\text{grain}}(L) = T_0 - \frac{48\pi C_1 \lambda}{48\pi\alpha V^{1/3} + \mathcal{E}_3\lambda\alpha V^{2/3}},$$

with C_1, C_2 and C_3 as before and where $\mathcal{E}_2$ and $\mathcal{E}_3$ are constants, resulting from sums involving the Bessel functions [Malbouisson et al. (2009)]. We see that the critical temperature has the same kind of dependence on the size extension L for $d = 1, 2, 3$, only constants differ in each case. The functional behavior does not depend on the number of compactified dimensions, only on the dimension of the Euclidean space, which we have computed for $D = 3$. One can also notice that the minimal size of the compact superconductor has lesser values than those computed without taking into account corrections to the coupling constant. This can be seen in Figure 3, where we have plotted the reduced transition temperature $t_c = T_c/T_0$ as a function of the inverse of the reduced compactification length $l = L/L_{\text{min}}$, where L_{min} is the corresponding minimal allowed linear extension without coupling constant boundary corrections.

5. Infrared fixed-point structure for the $\lambda\varphi^4$ model

5.1 The system in the absence of an external magnetic field

In this subsection, we study the fixed-point structure of the compactified model described by the Hamiltonian density in Eq. (29) in the large-N limit. We start from the four-point function at the critical point ($m = 0$) and for small external momenta, before compactification, which is given by

$$\Gamma_{\text{cr}}^{(4)}(p) = \frac{\lambda/N}{1 + \lambda\Pi_{\text{cr}}(p)}. \tag{35}$$

In the equation above, $\Pi_{\text{cr}}(p)$ is the one-loop four-point function at the critical point; introducing a Feynman parameter x, it is written in the form

$$\Pi_{\text{cr}}(p) = \int_0^1 dx \int \frac{d^D k}{(2\pi)^D} \frac{1}{\left[k^2 + p^2 x(1-x)\right]^2}. \tag{36}$$

Performing the Matsubara replacements (3) for d dimensions, Eq. (36) becomes ($\omega_i = 2\pi n_i/L$)

$$\Pi_{\text{cr}}(p, L) = \frac{1}{L^d} \sum_{n_1,\ldots,n_d=-\infty}^{\infty} \int_0^1 dx \int \frac{d^{D-d} q}{(2\pi)^{D-d}} \times \frac{1}{\left[q^2 + \omega_{n_1}^2 + \cdots + \omega_{n_d}^2 + p^2 x(1-x)\right]^2}, \tag{37}$$

and we define the effective L-dependent coupling constant in the large-N limit as

$$\lambda(p, L) \equiv \lim_{N\to\infty} N\Gamma_D^{(4)}(p, L) = \frac{\lambda}{1 + \lambda\Pi(p, L)}. \tag{38}$$

The sum over the n_i and the integral over q above can be treated using the formalism developed in Khanna et al. (2009) and described in section 2. We obtain

$$\Pi_{\text{cr}}(p,L) = (2\pi)^{-D/2}\int_0^1 dx\left[2^{-D/2}\left(\frac{1}{(2\pi)^2}p^2x(1-x)\right)^{D/2-2}\Gamma\left(2-\frac{D}{2}\right)\right.$$
$$+d\sum_{n=1}^{\infty}\left(\frac{\sqrt{p^2x(1-x)}}{2\pi Ln}\right)^{D/2-2}K_{D/2-2}\left(\frac{Ln}{2\pi}\sqrt{p^2x(1-x)}\right)+\cdots$$
$$+2^{d-1}\sum_{n_1,\ldots,n_d=1}^{\infty}\left(\frac{\sqrt{p^2x(1-x)}}{2\pi L\sqrt{n_1^2+\cdots+n_d^2}}\right)^{D/2-2}$$
$$\left.\times K_{D/2-2}\left(\frac{L}{2\pi}\sqrt{p^2x(1-x)}\sqrt{n_1^2+\cdots+n_d^2}\right)\right], \tag{39}$$

which, replaced in Eq. (38), gives the boundary-dependent four-point function in the large-N limit. We can write Eq. (39) in the form

$$\Pi(p,L) = A(D)|p|^{D-4} + B_d(D,L), \tag{40}$$

with the d-independent coefficient of the $|p|$-term being

$$A(D) = (2\pi)^{4-3D/2}2^{-D/2}b(D)\Gamma\left(2-\frac{D}{2}\right), \tag{41}$$

and where we have defined

$$b(D) = \int_0^1 dx\,[x(1-x)]^{D/2-2} = 2^{3-D}\sqrt{\pi}\frac{\Gamma\left(\frac{D}{2}-1\right)}{\Gamma\left(\frac{D-1}{2}\right)}, \qquad \text{for } \mathrm{Re}(D) > 2. \tag{42}$$

We remark that, for the physically interesting dimension $D = 3$, $b(3) = \pi$. This implies that $A(3) = \pi/4$.

If an infrared-stable fixed point exists for any of the models with d compactified dimensions, it is possible to determine it by a study of the infrared behavior of the Callan–Symanzik β function. Therefore, we investigate the above equations for $|p| \approx 0$. With this restriction, we may use the asymptotic formula for small values of the argument of the Bessel functions, and the expressions for B_d simplify considerably [see the reasoning leading to Eq. (13)]. The result is expressed in terms of one of the multidimensional Epstein–Hurwitz zeta functions of Eq. (14). In this limit, the p^2-dependence of the Bessel functions exactly compensates the one coming from the accompanying factors. Thus, the remaining p^2-dependence is only that of the first term of (39), which is the same for all number of compactified dimensions d. For general and detailed expressions, see Linhares et al. (2011). One can also construct analytical continuations and recurrence relations for the multidimensional Epstein functions, which permit to write them in terms of modified Bessel and Riemann zeta functions [Khanna et al. (2009); Kirsten (1994)]. We thus are able to derive expressions for each particular value of

d, from 1 to D, in the $|p| \approx 0$ limit, but let us restrict ourselves to the most expressive values, corresponding to materials in the form of a film, a wire, or a grain.

Therefore, for a film, we obtain

$$B_{d=1}(D,L) \sim (2\pi)^{-D/2} 2^{D/2-3} L^{4-D} \Gamma\left(\frac{D}{2}-2\right) \zeta(D-4). \tag{43}$$

The above expression is valid for all *odd* dimensions $D > 5$, due to the poles of the Γ and ζ functions. We can obtain an expression for smaller values of D by performing an analytic continuation of the Riemann zeta function $\zeta(D-4)$ by means of its reflection property,

$$\zeta(z) = \frac{1}{\Gamma(z/2)} \Gamma\left(\frac{1-z}{2}\right) \pi^{z-1/2} \zeta(1-z). \tag{44}$$

Then Eq. (43) leads to an expression valid for $2 < D < 4$ given by

$$B_{d=1}(D,L) = 2^{-3} \pi^{(D-9)/2} L^{4-D} \Gamma\left(\frac{5-D}{2}\right) \zeta(5-D). \tag{45}$$

For $D = 3$, we have $B_{d=1}(3,L) = L/48\pi$. For $d = 2$ and $d = 3$, similar expressions are obtained. An analysis of the singularity structure of the quantities B_d shows that their domain of existence can be extended to $2 < D < 4$ [Linhares et al. (2011)].

To discuss infrared properties of these compactified models, we insert Eq. (40) in Eq. (38) and we get the (p,L)-dependent coupling constant

$$\lambda(|p| \approx 0, D, L) \approx \frac{\lambda}{1 + \lambda\left[A(D)|p|^{D-4} + B_d(D,L)\right]}. \tag{46}$$

Let us take $|p|$ as a running scale, and define the dimensionless coupling constant

$$g = \lambda(p, D, L) |p|^{D-4}. \tag{47}$$

We recall that in these expressions p is a D-dimensional vector. The Callan-Symanzik β function controls the rate of the renormalization-group flow of the running coupling constant and a (nontrivial) fixed point of this flow is given by a (nontrivial) zero of the β function. For $|p| \approx 0$, it is obtained straightforwardly from Eq. (47),

$$\beta(g) = |p| \frac{\partial g}{\partial |p|} \approx (D-4)\left[g - A(D) g^2\right], \tag{48}$$

from which we get the infrared-stable fixed point

$$g_*(D) = \frac{1}{A(D)}. \tag{49}$$

We see that the L-dependent B_d-part of the subdiagram Π_{cr} does not play any role in this expression and, as remarked before, $A(D)$ is the same for all number of compactified dimensions, so is g_* only dependent on the space dimension.

5.2 The system with an external magnetic field

We now take the N-component Ginzburg–Landau model of the previous subsection to describe the behavior of d-confined systems, now in the presence of an external magnetic field, at leading order in $1/N$. The Hamiltonian density (29) is then modified to

$$\mathcal{H} = \left[\left(\partial_\mu - ieA_\mu^{\text{ext}}\right)\varphi_a\right]\left[\left(\partial^\mu - ieA^{\text{ext},\mu}\right)\varphi_a\right] + m^2\varphi_a\varphi_a + \frac{\lambda}{N}\left(\varphi_a\varphi_a\right)^2, \tag{50}$$

where $m^2 = \alpha(T - T_c)$, with $\alpha > 0$. For $D = 3$, from a physical point of view, such Hamiltonian is supposed to describe type-II superconductors. In this case, we assume that the external magnetic field H is parallel to the z-axis and we choose the gauge $A^{\text{ext}} = (0, xH, 0)$. In the present D-dimensional case, we assume analogously a gauge $A^{\text{ext}} = (0, x_1H, 0, 0, \ldots, 0)$, with $\{x_i\} = x_1, x_2, \ldots, x_D$, meaning that the applied external magnetic field lies on a fixed direction along one of the coordinate axis; for simplicity, in the calculations that follow, we have adopted the notation $x_1 \equiv x$, $x_2 \equiv y$. If we consider the system in unlimited space, the field φ should be written in terms of the well-known Landau-level basis,

$$\varphi(r) = \sum_{\ell=0}^{\infty}\int\frac{dp_y}{2\pi}\int\frac{d^{D-2}p}{(2\pi)^{D-2}}\tilde{\varphi}_{\ell,p_y,p}\chi_{\ell,p_y,p}(r), \tag{51}$$

where $\chi_{\ell,p_y,p}(r)$ are the Landau-level eigenfunctions given in terms of Hermite polynomials H_ℓ by

$$\chi_{\ell,p_y,p}(r) = \frac{1}{\sqrt{2^\ell \ell!}}\left(\frac{\omega}{\pi}\right)^{1/4} e^{i(p\cdot r + p_y y)} e^{-\omega(x - p_y/\omega)^2/2} H_\ell\left(\sqrt{\omega}x - \frac{p_y}{\sqrt{\omega}}\right), \tag{52}$$

with energy eigenvalues $E_\ell(|p|) = |p|^2 + (2\ell + 1)\omega + m^2$ and $\omega = eH$ is the so-called cyclotron frequency. In the above equation, p and r are $(D-2)$-dimensional vectors.

In the following, we consider only the lowest Landau level $\ell = 0$. For $D = 3$, this assumption usually corresponds to the description of superconductors in the extreme type-II limit. Under this assumption, we obtain that the effective $|\varphi|^4$ interaction in momentum space and at the critical point ($m = 0$) is written as

$$\lambda(p, L; \omega) = \frac{\lambda}{1 + \lambda\omega e^{-(1/2\omega)(p_1^2 + p_2^2)}\Pi(p, L; \omega)}, \tag{53}$$

where the single 1-loop four-point function, $\Pi(p, L; \omega)$, is given by

$$\Pi(p, L; \omega) = \frac{1}{L^d}\sum_{i=1}^{d}\sum_{n_i=-\infty}^{\infty}\int_0^1 dx \int\frac{d^{D-d-2}q}{(2\pi)^{D-d-2}} \times\left[q^2 + \omega_{n_1}^2 + \cdots + \omega_{n_d}^2 + p^2x(1-x)\right]^{-2}. \tag{54}$$

This is the same kind of expression that is encountered in the previous subsection, Eq. (37), with the only modification that $D \to D - 2$. The analysis is then performed along the same lines and we obtain, analogously,

$$\Pi(p,L;\omega) = (2\pi)^{1-D/2}\left[2^{1-D/2}\frac{1}{(2\pi)^2}c(D)\Gamma\left(3-\frac{D}{2}\right)\left(p^2\right)^{D/2-3}\right.$$
$$+\int_0^1 dx\, d\sum_{n=1}^{\infty}\left(\frac{\sqrt{p^2x(1-x)}}{2\pi Ln}\right)^{D/2-3} K_{D/2-3}\left(\frac{Ln}{2\pi}\sqrt{p^2x(1-x)}\right)$$
$$+\cdots+2^{d-1}\int_0^1 dx \sum_{n_1,\dots,n_d=1}^{\infty}\left(\frac{\sqrt{p^2x(1-x)}}{2\pi L\sqrt{n_1^2+\cdots+n_d^2}}\right)^{D/2-3}$$
$$\left.\times K_{D/2-3}\left(\frac{1}{2\pi}\sqrt{p^2x(1-x)}\sqrt{n_1^2+\cdots+n_d^2}\right)\right], \tag{55}$$

where

$$c(D) = \int_0^1 dx\,(x(1-x))^{D/2-3} = 2^{5-D}\sqrt{\pi}\frac{\Gamma\left(\frac{D}{2}-2\right)}{\Gamma\left(\frac{D-3}{2}\right)}, \qquad \text{for } \mathrm{Re}(D) > 4. \tag{56}$$

As for the infrared behavior of the β function, it suffices to study it in the neighborhood of $|p| = 0$, so that we can again use the asymptotic formula for Bessel functions for small values of the argument, as before. It turns out that in the $|p| \approx 0$ limit, the bubble Π_{cr} is written in the form

$$\Pi_{\mathrm{cr}}(|p| \approx 0, L;\omega) = A_1(D)\,|p|^{D-6} + C_d(D,L), \tag{57}$$

with

$$A_1(D) = (2\pi)^{-D/2-1}2^{1-D/2}c(D)\Gamma\left(3-\frac{D}{2}\right), \tag{58}$$

and where the quantity $C_d(D,L)$ is obtained by simply making the change $D \to D-2$ in the formula for $B_d(D,L)$ in the preceding subsection.

Let us remind Eq. (53) and define the dimensionless coupling constant

$$g^{(1)} = \omega\lambda(p_1 = p_2 = 0, D, L)|p|^{D-6}, \tag{59}$$

where we remember that in this context p is a $(D-2)$-dimensional vector. As before, we take as a running scale $|p|$ and after performing manipulations entirely analogous to those in the previous subsection and recalling Eq. (56), we have the extended domain of validity $4 < D < 6$ for the quantities $C_{d=1}(D;L)$, for all $d = 1,2,3$. We then get the β function for $|p| \approx 0$,

$$\beta(g) = |p|\frac{\partial g^{(1)}}{\partial |p|} \approx (D-6)\left[g^{(1)} - A_1(D)\left(g^{(1)}\right)^2\right], \tag{60}$$

from which the infrared-stable fixed point is obtained:

$$g_*^{(1)}(D) = \frac{1}{A_1(D)}. \tag{61}$$

6. Concluding remarks

Investigations on the dependence of the critical temperature for films with its thickness have been done in other contexts and approaches, different from the one we adopt. For instance, in Zinn-Justin (2002), an analysis of the renormalization group in finite-size geometries can be found and scaling laws have been studied. Also, such a dependence has been investigated in Asamitsu et al. (1994); Minhaj et al. (1994); Quateman (1986); Raffy et al. (1983) from both experimental and theoretical points of view, explaining this effect in terms of proximity, localization and Coulomb interaction. In particular, Quateman (1986) predicts, as our model also does, a suppression of the superconducting transition for thicknesses below a minimal value. More recently, in Shanenko et al. (2006) the thickness dependence of the critical temperature is explained in terms of a shape-dependent superconducting resonance, but no suppression of the transition is predicted or exhibited.

In this chapter, we have adopted a phenomenological approach, discussing the $(\lambda|\varphi|^4)_D$ and $(-\lambda|\varphi|^4+\eta|\varphi|^6)_D$ theories compactified in $d \leq D$ Euclidean dimensions. We have presented a general formalism which, in the framework of the Ginzburg–Landau model, is able to describe phase transitions for systems defined in spaces of arbitrary dimensions, some of them being compactified. We have focused in particular on the situations with $D = 3$ and $d = 1, 2, 3$, corresponding (in the context of condensed-matter systems) to films, wires and grains, respectively, undergoing phase transitions which may be described by Ginzburg–Landau models. This generalizes previous works dealing with first- and second-order transitions in low-dimensional systems [Abreu et al. (2005); Linhares et al. (2006); Malbouisson et al. (2002)].

We have observed the contrasting behavior of the critical temperature on the size of the system, whether the transition is first- or second-order. This may indicate that from this shape dependence one can infer the order of the transition the system undergoes.

In what a renormalization group approach is concerned, we have discussed the infrared behavior and the fixed-point structure of the compactified O(N) $\lambda\varphi^4$ in the large-N limit. We have shown that, whether in the absence or presence of an external magnetic field, the existence of an infrared-stable fixed point depends only on the space dimension D, not on the number of compactified dimensions.

7. References

Abreu, L.M.; Malbouisson, A.P.C.; Malbouisson, J.M.C. & Santana, A.E. (2003). Large-N transition temperature for superconducting films in a magnetic field, Phys. Rev. B, Vol. 67, No. 21, June 2003, 212502-1–212502-4, ISSN 0163-1829

Abreu, L.M.; Malbouisson, A.P.C. & Roditi, I. (2004). Gauge fluctuations in superconducting films, Eur. Phys. J. B, Vol. 37, No. 4, April 2004, 515–522, ISSN 1434-6028

Abreu, L.M.; de Calan, C.; Malbouisson, A.P.C.; Malbouisson,J.M.C. & Santana, A.E. (2005). Critical behavior of the compactified $\lambda\varphi^4$, J. Math. Phys., Vol. 46, No. 1, January 2005, 012304-1–012304-12, ISSN 0022-2488

Abreu, L.M.; Malbouisson, A.P.C.; Malbouisson, J.M.C. & Santana, A.E. (2009). Finite-size effects on the chiral phase diagram of four-fermion models in four dimensions, Nucl. Phys. B, Vol. 819, Nos. 1-2, September 2009, 127–138, ISSN 0550-3213

Asamitsu, A.; Iguchi, M.; Ichikawa, A. & Nishida, N. (1994). Disappearance of superconductivity in ultra thin amorphous Nb films, Physica B, Vols. 194-196, Part 2, February 1994, 1649–1650, ISSN 0163-1829

Bordag, M.; Mohideed U. & Mostepanenko, V.M. (2001). New developments in the Casimir effect, Phys. Rep., Vol. 353, No. 1-3, October 2001, 1–206, ISSN 0370-1573

Boyer, T.H. (2003). Casimir forces and boundary conditions in one dimension: Attraction, repulsion, Planck spectrum, and entropy, Am. J. Phys., Vol. 71, No. 10, October 2003, 990–998, ISSN 0002-9505

Ccapa Tira, C.; Fosco, C.D.; Malbouisson, A.P.C. & Roditi, I. (2010). Vacuum polarization for compactified QED_{4+1} in a magnetic flux background, Phys. Rev. A, Vol. 81, No. 3, March 2010, 032116-1–032116-7, ISSN 1050-2947

Coleman, S. & Weinberg, E. (1973). Radiative corrections as the origin of spontaneous symmetry breaking, Phys. Rev. D, Vol. 7, No. 6, March 1973, 1888–1910, ISSN 0556-2821

Elizalde, A. (1995). *Ten Physical Applications of Spectral Zeta-Functions*, Springer, ISBN 3-540-60230-5, Berlin

Halperin, L.; Lubensky, T.C. & Ma, S.-K. (1974). First-order phase transitions in superconductors and smectic-A liquid crystals, Phys. Rev. Lett., Vol. 32, No. 6, February 1974, 292–295, ISSN 0031-9007

Khanna, F.C.; Malbouisson, A.P.C.; Malbouisson, J.M.C. & Santana, A.E. (2009). *Thermal Quantum Field Theory: Algebraic Aspects and Applications*, World Scientific, ISBN-13 978-981-281-887-4, Singapore

Khanna, F.C.; Malbouisson, A.P.C.; Malbouisson, J.M.C. & Santana, A.E. (2010). Phase transition in the 3D massive Gross-Neveu model, EPL, Vol. 92, No. 1, October 2010, 11001-p1–11001-p6, ISSN 0295-5075

Kirsten, K. (1994). Generalized multidimensional Epstein zeta functions, J. Math. Phys., Vol. 35, No. 1, January 1994, 459–470, ISSN 0022-2488

Kleinert, H. (1989). *Gauge Fields in Condensed Matter, Vol. 1: Superflow and Vortex Lines*, World Scientific, ISBN-10 9971502100, Singapore

Kodama, J.; Itoh, M. & Hirai, H. (1983). Superconducting transition temperature versus thickness of Nb film on various substrates, J. Appl. Phys., Vol. 54, No. 7, July 1983, 4050–4053, ISSN 0021-8979

Levin, F.S. & Micha, D.A. (Eds.) (1993). *Long Range Casimir Forces*, Plenum, ISBN-10 0306443856, New York

Linhares, C.A.; Malbouisson, A.P.C.; Milla, Y.W. & Roditi, I. (2006). First-order phase transitions in superconducting films: A Euclidean model, Phys. Rev. B, Vol. 73, No. 21, June 2006, 214525-1–214525-7, ISSN 1098-0121

Linhares, C.A.; Malbouisson, A.P.C.; Milla, Y.W. & Roditi, I. (2007). Critical temperature for first-order phase transitions in critical systems, Eur. Phys. J. B, Vol. 60, No. 3, December 2007, 353–362, ISSN 1434-6028

Linhares, C.A.; Malbouisson, A.P.C. & Souza, M.L. (2011). A note on the infrared behavior of the compactified Ginzburg–Landau model in a magnetic field, EPL, Vol. 96, No. 3, November 2011, 31002-p1–31002-p6, ISSN 0295-5075

Malbouisson, A.P.C. (2002). Large-N β function for superconducting films in a magnetic field, Phys. Rev. B, Vol. 66, No. 9, September 2002, 092502-1–092502-3, ISSN 0163-1829

Malbouisson, A.P.C.; Malbouisson, J.M.C. & Santana, A.E. (2002). Spontaneous symmetry breaking in compactified $\lambda\varphi^4$ theory, Nucl. Phys. B, Vol. 631, Nos. 1-2, June 2002, 83–94, ISSN 0550-3213

Malbouisson, A.P.C.; Malbouisson, J.M.C. & Pereira, R.C. (2009). Boundary effects on the mass and coupling constant in the compactified Ginzburg–Landau model: The boundary dependent critical temperature, J. Math. Phys., Vol. 50, No. 8, August 2009, 083304-1–083304-16, ISSN 0022-2488

Milonni, P.W. (1993). *The Quantum Vacuum*, Academic, ISBN-10 0124980805, Boston

Minhaj, M.S.M.; Meepagala, S.; Chen, J.T. & Wenger, L.E. (1994). Thickness dependence on the superconducting properties of thin Nb films, Phys. Rev. B, Vol. 49, No. 21, June 1994, 15235–15240, ISSN 0163-1829

Mostepanenko, V.M. & Trunov, N.N. *The Casimir Effect and Its Applications*, Clarendon, ISBN-10 0124980805, Oxford

Pogrebnyakov, A.V.; Redwing, J.M.; Jones, J.E.; Xi, X.X.; Xu, S.Y.; Qi Li; Vaithyanathan, V. & Schlom, D.G. (2003). Thickness dependence of the properties of epitaxial MgB_2 thin films grown by hybrid physical chemical vapor deposition, Appl. Phys. Lett., Vol. 82, No. 24, June 2003, 4319–4321, ISSN 0003-6951

Quateman, J. (1986). T_c suppression and critical fields in thin superconducting Nb films, Phys. Rev. B, Vol. 34, No. 3, August 1986, 1948–1951, ISSN 0163-1829

Raffy, H.; Laibowitz, R.B.; Chaudhari, P. & Maekawa, S. (1983). Localization and interaction effects in two-dimensional W-Re films, Phys. Rev. B, Vol. 28, No. 11, December 1983, 6607–6609, ISSN 0163-1829

Seife, C. (1997). The subtle pull of emptiness, Science, Vol. 275, No. 5297, October 1997, 158, ISSN 0036-8075

Shanenko, A.A.; Croitoru, M.D.; Zgirski, M.; Peeters, J.M. & Arutyunov, K. (2006). Size-dependent enhancement of superconductivity in Al and Sn nanowires: Shape-resonance effect, Phys. Rev. B, Vol. 74, No. 5, August 2006, 052502-1–052502-4, ISSN 1098-0121

Strongin, M.; Thompson, R.; Kammerer, O.F. & Crow, J.E. (1970). Destruction of superconductivity in disordered near-monolayer films, Phys. Rev. B, Vol. 1, No. 3, February 1970, 1078–1091, ISSN 0556-2805

Zinn-Justin, J. (2002). *Quantum Field Theory and Critical Phenomena*, Clarendon, ISBN-10 0198509235, Oxford

Zgirski, M.; Riikonen, K.-P.; Touboltsev, V.; Arutyunov, K. (2005). Size Dependent Breakdown of Superconductivity in Ultranarrow Nanowires, Nano Lett., Vol. 5, No. 6, May 2005, 1029–1033, ISSN 1530-6984

Part 3

Applications in Mathematics

7

Solution of a Linearized Model of Heisenberg's Fundamental Equation

E. Brüning[1] and S. Nagamachi[2]
[1]*University of KwaZulu-Natal*
[2]*The University of Tokushima*
[1]*South Africa*
[2]*Japan*

1. Introduction

The theory of quantized fields has its roots in the quantum theory of M. Planck [Planck (1900)]. For the solution of the problem of black body radiation, he introduced the universal quantum of action to the theory of electromagnetic fields. Nonrelativistic quantum mechanics is established by W. Heisenberg [Heisenberg (1925)] and E. Schrödinger [Schrödinger (1926)]. M. Born, W. Heisenberg and P. Jordan [Born et al. (1926)] realized the quantization of electromagnetic fields, and P.A.M. Dirac [Dirac (1927)] quantized electromagnetic fields in interaction with a material system. But P. Ehrenfest noticed soon that the theory had to lead to infinities. The existence of the positron is suggested by the theory of P.A.M. Dirac [Dirac (1928; 1931)]. The discovery of the positron by C.D. Andersen (1932) established the theory of quantum electrodynamics which treats the behavior of electron, positron and electromagnetic fields. But this theory still has to lead to infinities, and these difficulties are (partially) removed by S. Tomonaga [Tomonaga (1946)], H.A. Bethe [Bethe (1947)] and J. Schwinger [Schwinger (1948)] using the subtraction formalism in perturbation theory. R.P. Feynman [Feynman (1948)] developed the method of path integral which simplifies the calculation and F.J. Dyson [Dyson (1949)] derives Feynman's prescription from Tomonaga-Schwinger theory. Thus the prescription of the subtraction formalism in electrodynamics was completely worked out. Those theories in which the infinity of each term in the perturbation series can be subtracted consistently are called renormalizable theories. But for these renormalizable theories, there are still doubts about summability of the series in perturbation theory. In electrodynamics, the expansion parameter (the coupling constant) is small and the sum of the first few terms gives an amazing agreement with experiments, but there are no proofs about the convergence of the series. In some theory of strong interaction the parameter is greater than 1, and this subtraction formalism does not work. The main question is: What is hidden behind these formal infinite series? To this question, one answer is given by the formalism of A.S. Wightman and L. Gårding [Wightman & Gårding (1964)], which is a mathematically rigorous study of quantum fields. Since the structure of the theory is axiomatic, the theory is called axiomatic quantum field theory. The axioms formulate the basic physical postulates (e.g., relativistic invariance of the state space, spectral property, unique existence of a vacuum state, Poincaré-covariance of the fields, locality or micro causality, etc., and some technical postulate: temperedness of the fields, etc.) in a mathematical language. Nowadays the strong interaction is described by the quantum chromodynamics. Its renormalizability was proven in ['t Hooft (1971)], and its asymptotic freedom discovered by [Gross & Wilczek (1973); Politzer (1973)] gave it excellent predictability as well as quantum electrodynamics. But we know nothing about its summability of perturbation series and the perturbation theory does not work in the low-energy region where the coupling constant becomes large. The situation is not so different from 1940's. The axioms show the way to reconcile the principles of quantum mechanics and those of special relativity. Quantum field theory is also important as effective theory in low-energy approximation to a deeper theory like a string theory where it is said that there is a length $\ell > 0$ (fundamental length) such that one cannot distinguish events which occur in a smaller distance than ℓ (see [Polchinski (1998)]). From this perspective, quantum field theory with a fundamental length becomes interesting.

In 1958, Heisenberg and Pauli introduced the equation

$$\gamma_{\mu}\frac{\partial}{\partial x_{\mu}}\psi(x) \pm l^{2}\gamma_{\mu}\gamma_{5} : \psi(x)\bar{\psi}(x)\gamma^{\mu}\gamma_{5}\psi(x) : = 0 \tag{1.1}$$

which was later called Heisenberg's fundamental field equation or the equation of universe and studied in [Dürr et al. (1959); Heisenberg (1966)]. The equation contains a parameter l of the dimension of length and accordingly one might speculate that this parameter can play the rôle of the fundamental length of a quantum field theory with a fundamental length. In order to verify this speculation one must solve two eminent problems:

(A) Formulate a relativistic quantum field theory with a fundament length in an axiomatic way and establish its main properties;

(B) Solve this equation with a field theory according to (A).

As a mathematical theory a solution to problem (A) has been suggested in [Brüning & Nagamachi (2004)]. According to this suggestion a *relativistic quantum field theory with a fundamental length* is a relativistic quantum field theory similar to the Gårding - Wightman theory (see [Wightman & Gårding (1964)]) where the fields are operator valued tempered ultra-hyperfunctions instead of operator valued tempered (Schwartz) distributions. The physically important aspect of this theory is that for the first time in a mathematical rigorous way, the fundamental length is realized on the level of the fields, not on the level of the geometry of the space-time on which these fields are defined. This allows to rely on the established concepts and theories of Physics based on a standard space-time. The fundamental length is introduced through the use of a class of generalized functions (tempered ultrahyperfunctions) which distinguish events in space-time only when they are separated by more than a certain length, the fundamental length ℓ. Certainly, up to now there is now no? physical evidence that this is the appropriate way of defining the fundamental length.

Concerning problem (B) we recall that nobody knows to solve Heisenberg's fundamental field equation. However there is a simplification of this equation which is solvable in the sense of classical field theory, namely the system of equations (: · : indicates the Wick product)

$$\begin{cases} (\Box + m^{2})\phi(x) = 0 \\ \left(i\gamma^{\mu}\dfrac{\partial}{\partial x^{\mu}} - M\right)\psi(x) = -2l^{2}\gamma^{\mu} : \dfrac{\partial\phi(x)}{\partial x^{\mu}}\phi(x)\psi(x) : \end{cases} \tag{1.2}$$

for a Klein-Gordon field ϕ and a spinor field ψ. It is this system of coupled equations which we discuss in the framework of [Brüning & Nagamachi (2004)], i.e., for a given Klein-Gordon field ϕ we define an operator valued tempered ultrahyperfunction ψ such that the pair (ϕ, ψ) satisfies equation (1.2). This system has been studied first by [Okubo (1961)] as a quantum field theory although this interaction is un-renormalizable in the usual sense of perturbation theory, and Green's functions are calculated. There are some interactions which look un-renormalizable but actually are renormalizable if the use of perturbations is avoided (see [Okubo (1954)]).

2. Overview

The basic idea to solve the system (1.2) is quite natural:

Take a Klein-Gordon field of mass m and suppose that we can show the following statements:

A) the Wick power series

$$\rho(x) =: e^{il^2\phi(x)^2} := \sum_{n=0}^{\infty} i^n l^{2n} : \phi(x)^{2n} : /n! \tag{2.1}$$

and

$$\rho^*(x) =: e^{-il^2\phi(x)^2} := \sum_{n=0}^{\infty} (-i)^n l^{2n} : \phi(x)^{2n} : /n!$$

are well-defined as operator-valued ultra-hyperfunctions.

B) $\rho(x)$ satisfies

$$\frac{\partial}{\partial x^\mu}\rho(x) = 2il^2 : \frac{\partial\phi(x)}{\partial x^\mu}\phi(x)e^{il^2\phi(x)^2} := 2il^2 : \frac{\partial\phi(x)}{\partial x^\mu}\phi(x)\rho(x) : . \tag{2.2}$$

C) the free Dirac field $\psi_0(x)$ is a multiplier for the field ρ and thus we can define the field

$$\psi(x) = \rho(x)\psi_0(x). \tag{2.3}$$

D) Show that the field ψ defined in C) is indeed al relativistic quantum field with a fundamental length.

E) Calculate

$$\begin{aligned}\left(i\gamma^\mu \frac{\partial}{\partial x^\mu} - M\right)\psi(x) &= \rho(x)\left[\left(i\gamma^\mu \frac{\partial}{\partial x^\mu} - M\right)\psi_0(x)\right] + \gamma^\mu \frac{\partial\rho(x)}{\partial x^\mu}\psi_0(x)\\ &= -2l^2\gamma^\mu : \frac{\partial\phi(x)}{\partial x^\mu}\phi(x)\rho(x)\psi_0(x) := -2l^2\gamma^\mu : \frac{\partial\phi(x)}{\partial x^\mu}\phi(x)\psi(x) : .\end{aligned}$$

Thus, if A) – E) hold, the operator-valued ultra-hyperfunctions $\phi(x), \psi(x)$ satisfy the system of equations (1.2).

Note that the above Wick power series do not converge in the sense tempered distributions. They even do not converge in the sense Fourier hyperfunctions (see [Ito (1988); Nagamachi (1981a;b)]). But as we show they converge in the sense of tempered ultra-hyperfunctions.

Remark 2.1.

The definition of a relativistic quantum field with a fundamental length has been proposed in [Brüning & Nagamachi (2004)] and there the functional characterization has been derived for the case of a scalar field. The functional characterization for a spinor field is given in [Brüning & Nagamachi (2008)].

Naturally the localization properties of a relativistic quantum field with a fundament length $\ell > 0$ are very different from those of a standard quantum field. According to their definition these fields do not distinguish events in space-time which are separated by $\ell' < \ell$ (see Remark 3.11).

According to these new localization properties the counter part of the "locality or causality condition for standard fields" looks quite different. It is called "extended causality" and the verification of this condition is the major difficulty in verifying that the fields ρ and ψ as introduced above are indeed relativistic quantum fields with a fundamental length.

Remark 2.2. The key to our approach is the use of tempered ultra-hyperfunctions (see [Morimoto (1970; 1975a;b)]) and their localization properties (see [Nagamachi & Brüning (2003)]). This has first been suggested in [Brüning & Nagamachi (2004)]. This class of generalized functions is not too well known and therefore we will briefly explain what tempered ultra-hyperfunctions are and that and how these localization properties arise.

3. Tempered ultra-hyperfunctions

For any subset A of $\mathbb{R}^n$, denote by $T(A) = \mathbb{R}^n + iA \subset \mathbb{C}^n$ the tubular set with base A. For a convex compact set K of $\mathbb{R}^n$, $\mathcal{T}_b(T(K))$ is, by definition, the space of all continuous functions f on $T(K)$ which are holomorphic in the interior of $T(K)$ and which satisfy

$$\|f\|^{T(K),j} = \sup\{|z^p f(z)|; z \in T(K), |p| \leq j\} < \infty, \; j = 0,1,\ldots \tag{3.1}$$

where $p = (p_1,\ldots,p_n)$ and $z^p = z_1^{p_1}\cdots z_n^{p_n}$. $\mathcal{T}_b(T(K))$ is a Fréchet space with the semi-norms $\|f\|^{T(K),j}$. If $K_1 \subset K_2$ are two compact convex sets, we have the canonical mapping:

$$\mathcal{T}_b(T(K_2)) \to \mathcal{T}_b(T(K_1)). \tag{3.2}$$

For a convex open set O in $\mathbb{R}^n$ we define

$$\mathcal{T}(T(O)) = \lim_{\leftarrow} \mathcal{T}_b(T(K)), \tag{3.3}$$

where K runs through the convex compact sets contained in O and the projective limit is taken following the restriction mappings (3.2).

Definition 3.1. *A* **tempered ultra-hyperfunction** *is by definition a continuous linear functional on* $\mathcal{T}(T(\mathbb{R}^n))$.

The Fourier transformation $\mathcal{F}$ is well defined on $\mathcal{T}(T(\mathbb{R}^n))$ by the standard formula (3.8). In order to determine the range of $\mathcal{F}$ on $\mathcal{T}(T(\mathbb{R}^n))$ we introduce another function space.

The gauge functional h_K of a compact convex set $K \subset \mathbb{R}^n$ is defined by

$$h_K(x) = \sup\{\langle x,\xi\rangle; \xi \in K\}. \tag{3.4}$$

For a convex compact set K of $\mathbb{R}^n$, denote by $H_b(\mathbb{R}^n;K)$ the space of all C^∞ functions f on $\mathbb{R}^n$ which satisfy, for $j = 0,1,\ldots,$

$$\|f\|_{K,j} = \sup\{\exp(h_K(x))|D^p f(x)|; x \in \mathbb{R}^n, |p| \leq j\} < \infty. \tag{3.5}$$

Equipped with the system of semi-norms $\|f\|_{K,j}$, $H_b(\mathbb{R}^n;K)$ is a Fréchet space. If $K_1 \subset K_2$ are two compact convex sets, then $h_{K_1} \leq h_{K_2}$ and thus one has the canonical mappings:

$$H_b(\mathbb{R}^n;K_2) \to H_b(\mathbb{R}^n;K_1). \tag{3.6}$$

For a convex open set $O \subset \mathbb{R}^n$ the space $H(\mathbb{R}^n;O)$ is the projective limit of the spaces $H_b(\mathbb{R}^n;K)$ along the restriction mappings (3.6), i.e.,

$$H(\mathbb{R}^n;O) = \lim_{\leftarrow} H_b(\mathbb{R}^n;K), \tag{3.7}$$

where K runs through the convex compact sets contained in O.

In order to relate the space $H(\mathbb{R}^n;\mathbb{R}^n)$ to the Schwartz space $\mathcal{S}(\mathbb{R}^n)$ we derive a more direct characterization of $H(\mathbb{R}^n;\mathbb{R}^n)$. Observe that for any convex compact set $K \subset \mathbb{R}^n$ there is a number $k > 0$ such that $K \subseteq [-k,k]^n$. For the sets $K = [-k,k]^n$ the gauge function h_K is easily determined:

$$h_K(x) = \sup\{\langle x,\xi\rangle;\ \xi \in K\} = k\sum_{i=1}^{n} |x_i|,$$

and the system of continuous norms takes the form, using the notation $|x| = \sum_{i=1}^n |x_i|$,

$$\begin{aligned}\|f\|_{K,j} &= \sup\{\exp(h_K(x))|D^p f(x)|;\ |p| \leq j,\ x \in \mathbb{R}^n\}\\ &= \sup\{e^{k|x|}|D^p f(x)|;\ |p| \leq j,\ x \in \mathbb{R}^n\}.\end{aligned}$$

Thus, the space $H(\mathbb{R}^n;\mathbb{R}^n)$ can be defined as the projective limit of the spaces $H_b(\mathbb{R}^n;K)$ along the restriction mappings (3.6), where $K = [-k,k]^n$, $0 < k < \infty$. Accordingly, the space $H(\mathbb{R}^n;\mathbb{R}^n)$ is the space of all C^∞-functions on $\mathbb{R}^n$ which, together with all derivatives, decrease faster than any (linear) exponential. An easy consequence is

Corollary 3.2. *1. The space $H(\mathbb{R}^n;\mathbb{R}^n)$ is continuously embedded into the Schwartz space $\mathcal{S}(\mathbb{R}^n)$;*

2. The elements of $\mathcal{S}(\mathbb{R}^n)$ are multipliers for the space $H(\mathbb{R}^n;\mathbb{R}^n)$, and for each $g \in \mathcal{S}(\mathbb{R}^n)$ the map $f \mapsto gf$ is a continuous linear map of $H(\mathbb{R}^n;\mathbb{R}^n)$ into itself.

Proof: See [Hasumi (1961); Morimoto (1975b)]. □

The following theorem collects the basic facts about the spaces introduced above.

Theorem 3.3. *For the spaces introduced above the following statements hold, for any convex compact set K respectively convex open set O.*

1. *The space of $\mathcal{D}(\mathbb{R}^n)$ all C^∞ functions with compact support is dense in $H(\mathbb{R}^n;O)$.*
2. *The space $H(\mathbb{R}^n;\mathbb{R}^n)$ is dense in $H(\mathbb{R}^n;O)$ and in $H(\mathbb{R}^n;K)$.*
3. *$H(\mathbb{R}^m;\mathbb{R}^m) \otimes H(\mathbb{R}^n;\mathbb{R}^n)$ is dense in $H(\mathbb{R}^{m+n};\mathbb{R}^{m+n})$.*

Proof: For the proof of the first two items we refer to [Hasumi (1961); Morimoto (1975b)]. The proof the last item can be found in [Brüning & Nagamachi (2004)]. □

Proposition 3.4. *The Fourier transformation* $f \mapsto \tilde{f} \equiv \mathcal{F}f$,

$$\tilde{f}(p) = (2\pi)^{-n/2} \int_{\mathbb{R}^n} f(z) e^{i\langle p,z\rangle} dz \tag{3.8}$$

is a topological isomorphism between the spaces $\mathcal{T}(T(O))$ *and* $H(\mathbb{R}^n;O)$, *for any open convex nonempty set* $O \subset \mathbb{R}^n$. *The inverse transformation is*

$$f(z) = \bar{\mathcal{F}}\tilde{f} = (2\pi)^{-n/2} \int_{\mathbb{R}^n} \tilde{f}(p) e^{-i\langle p,z\rangle} dp. \tag{3.9}$$

Proof: See [Hasumi (1961); Morimoto (1975b)]. □

Proposition 3.5. *Let* $O \subset \mathbb{R}^n$ *be a nonempty convex open subset. Then the spaces* $H(\mathbb{R}^n;O)$ *and* $\mathcal{T}(T(O))$ *are nuclear Fréchet spaces and thus, in particular, reflexive.*

Proof: In the case of $O = \mathbb{R}^n$ Hasumi [Hasumi (1961)] proved this result, and his proof is valid in the general case. A sketch of the proof for $H(\mathbb{R}^n;O)$ is provided in [Brüning & Nagamachi (2004)]. □

Theorem 3.6 (Corollary of Theorem 34.1 of [Treves (1967)]). *Let* E *be a Fréchet space,* E_1 *a metrizable space,* G *a locally convex space. Then a separately continuous bilinear map of* $E \times E_1$ *into* G *is continuous.*

Theorem 3.7 (Kernel theorem for ultra-hyperfunctions). *Let* M *be a separately continuous multi-linear map of* $[\mathcal{T}(T(\mathbb{R}^4))]^n$ *into a Banach space* G. *Then there is a unique continuous linear map* F *of* $\mathcal{T}(T(\mathbb{R}^{4n}))$ *into* G *such that, for all* $f_i \in \mathcal{T}(T(\mathbb{R}^4))$, $i = 1, \ldots, n$,

$$M(f_1, \ldots, f_n) = F(f_1 \otimes \cdots \otimes f_n).$$

Proof: The proof is quite involved and lengthy. Details are again given in [Brüning & Nagamachi (2004)]. □

For an open set V in $\mathbb{R}^n$ and a positive number ϵ introduce the set V^ϵ defined by

$$V^\epsilon = \{z \in \mathbb{C}^n; \exists x \in V, |\mathrm{Re}\, z - x| < \epsilon, |\mathrm{Im}\, z|_\beta < \epsilon\},$$

where $|y|_\beta$ is a norm of $\mathbb{R}^n$ satisfying $|y|_\beta \geq |y|$ for the Euclidean norm $|y|$. Let K_p be the closure of $V^{\epsilon/(1+1/p)}$ in $\mathbb{C}^n$ and $L_p = \{w \in \mathbb{C}^m; |\mathrm{Im}\, w| \leq p\}$. Denote $U = V^\epsilon \times \mathbb{C}^m$ and $M_p = K_p \times L_p$. $\mathcal{T}_b(M_p)$ is, by definition, the space of all continuous functions f on M_p which are holomorphic in the interior of M_p and satisfy, for $k = 1, 2, \ldots$,

$$\|f\|^{M_p,k} = \sup\{|z^s w^t f(z,w)|; (z,w) \in M_p, |s| + |t| \leq k\} < \infty;$$

$\mathcal{T}_b(M_p)$ is a Fréchet space with the seminorms $\|f\|^{M_p,k}$.

If $k < m$, then we have the canonical mappings:

$$\mathcal{T}_b(M_m) \to \mathcal{T}_b(M_k). \tag{3.10}$$

We define

$$\mathcal{T}(U) = \lim_{\leftarrow} \mathcal{T}_b(M_m), \tag{3.11}$$

where the projective limit is taken following the restriction mappings (3.10).

Theorem 3.8. *$\mathcal{T}(T(\mathbb{R}^{n+m}))$ is dense in $\mathcal{T}(U)$.*

Proof: The proof is similar to the proofs of Proposition 2.4 of [Nishimura & Nagamachi (1990)] and Proposition 9.1.2 of [Hörmander (1983)]. For more details we refer to the appendix of [Brüning & Nagamachi (2004)]. □

Theorem 3.9. *Let V be a closed convex cone and K a convex compact set in $\mathbb{R}^n$. Define a function $h_{K,V}(\xi)$, $\xi \in \mathbb{R}^n$, and a set V_K^0 as follows (see Equation (3.4) for the definition of h_K):*

$$h_{K,V}(\xi) = \sup_{x \in V} h_K(x) - \langle x, \xi \rangle, \quad \text{and} \quad V_K^0 = \{\xi \in \mathbb{R}^n; h_{K,V}(\xi) < \infty\}.$$

Then for every $\mu \in H(\mathbb{R}^n;O)'$ with support in the cone V there is a function

$$\hat{\mu}(\zeta) = \langle \mu, e^{i\langle \cdot, \zeta \rangle} \rangle \tag{3.12}$$

with the following properties: $\hat{\mu}$ is well defined and holomorphic in the interior of $\mathbb{R}^n \times iV_K^0$ and satisfies there the following estimate, for a suitable $K \subset O$.

$$|\hat{\mu}(\zeta)| \le C(1+|\zeta|)^j \exp(h_{K,V}(\operatorname{Im}\zeta)). \tag{3.13}$$

$\hat{\mu}$ is called the Laplace transform of the tempered ultra-hyperfunction μ.

Proof: See [Brüning & Nagamachi (2004)]. □

Remark 3.10. Let $|x|_\infty = \max\{|x^0|, |\boldsymbol{x}|\}$ be a norm in $\mathbb{R}^4$ and $\bar{V}_+$ the closed forward light-cone in $\mathbb{R}^4$. Abbreviate $V = \bar{V}_+^n$ and for $\ell_i > 0$ introduce $h_K(x) = \sum_{i=1}^n \ell_i |x_i|_\infty$. Then we estimate

$$h_{K,V}(\xi) = \sup_{x_i \in \bar{V}_+} \sum_{i=1}^n (\ell_i |x_i|_\infty - \langle x_i, \xi_i \rangle) \le \sum_{i=1}^n \sup_{x_i \in \bar{V}_+} (\ell_i |x_i|_\infty - \langle x_i, \xi_i \rangle).$$

Let V_+ be the open forward light-cone. It follows

$$\sup_{x \in \bar{V}_+} -\langle x, \eta \rangle < \infty$$

for $\eta \in V_+$. Let $\xi_i = \eta_i + (\ell_i, \mathbf{o}) \in V_+ + (\ell_i, \mathbf{o})$. Since $|x|_\infty = x^0$ in $\bar{V}_+$, we find

$$\sup_{x_i \in \bar{V}_+} (\ell_i |x_i|_\infty - \langle x_i, \xi_i \rangle) = \sup_{x_i \in \bar{V}_+} (\ell_i x_i^0 - \langle x_i, \eta_i \rangle - x_i^0 \ell_i) = \sup_{x_i \in \bar{V}_+} -\langle x_i, \eta_i \rangle < \infty.$$

Thus the set

$$V_+(\ell_1, \ldots, \ell_n) = \{(\xi_1, \ldots, \xi_n) \in \mathbb{R}^{4n}; \xi_i \in V_+ + (\ell_i, \mathbf{o})\} \tag{3.14}$$

is contained in V_K^0.

The reason why we use tempered ultra-hyperfunctions for the formulation of relativistic quantum field theory with a fundamental length is illustrated in the following remark.

Remark 3.11. If $f(z)$ is a holomorphic function in the strip $|\mathrm{Im}\, z| < \ell$ around the real axis, then, for $|a| < \ell$, we have

$$\langle \sum_{n=0}^{\infty} \frac{a^n}{n!} \delta^{(n)}(x), f(x) \rangle = \sum_{n=0}^{\infty} \frac{(-a)^n}{n!} f^{(n)}(0) = f(0-a) = \langle \delta(x+a), f(x) \rangle, \tag{3.15}$$

that is, as an equation for functionals defined on the function space $\mathcal{T}(T(-\ell,\ell))$ whose elements are holomorphic functions in $T(-\ell,\ell) = \mathbb{R} + i(-\ell,\ell) \subset \mathbb{C}$, the identity

$$\sum_{n=0}^{\infty} \frac{a^n}{n!} \delta^{(n)}(x) = \delta(x+a)$$

holds, i.e., the sequence of generalized functions $S_N = \sum_{n=0}^{N} \frac{a^n}{n!} \delta^{(n)}(x)$ with support $\{0\}$ converges (weakly, in the dual space of $\mathcal{T}(T(-\ell,\ell))$) to the generalized function $\delta(x+a)$ with support $\{-a\}$, as $N \to \infty$. However, if $|a| > \ell$, then this sequence does not converge in $\mathcal{T}(T(-\ell,\ell))'$. This phenomenon can be understood as follows. If $|a| < \ell$, then elements in $\mathcal{T}(T(-\ell,\ell))'$ do not distinguish between the points $\{0\}$ and $\{-a\}$, but if $|a| > \ell$ then elements in $\mathcal{T}(T(-\ell,\ell))'$ can distinguish between the points $\{0\}$ and $\{-a\}$. Since $|a| < \ell$ is arbitrary, one can say that elements in $\mathcal{T}(T(-\ell,\ell))'$ do not distinguish between points which are separated by less than ℓ.

Remark 3.12. Let $U = (-\ell,\ell) + i(-\ell,\ell)$. Then the functional (3.15) is considered to be a functional on $\mathcal{T}(U)$ (the space of holomorphic functions on U), i.e., it is continuously extendable to $\mathcal{T}(U)$. If a functional μ on $\mathcal{T}(T(\mathbb{R}))$ is continuously extendable to $\mathcal{T}(U)$, one says that a carrier of μ is contained in U. The notion of carrier for tempered ultra-hyperfunctions is the counterpart of the notion of support for distributions. We can recognize the similarity of these notions: If a distribution $\mu \in \mathcal{S}'(\mathbb{R})$ is continuously extendable to $\mathcal{E}(U)$ for some open set $U \subset \mathbb{R}$, then we know that the support of μ is contained in U.

4. Relativistic quantum fields with a fundamental length and their functional characterization

4.1 Wightman's Axioms for relativistic quantum fields with a fundamental length

In Wightman's scheme, the concept of a relativistic quantum field $\phi^{(\kappa)}$ of type κ plays a fundamental role. Such a field, for example a scalar, tensor or spinor field, has a finite number of Lorentz components $\phi_j^{(\kappa)}$ $(j = 1, \ldots, r_\kappa)$.

The field components $\phi_j^{(\kappa)}(x)$ are operator-valued tempered ultra-hyperfunctions, i.e., for $f \in \mathcal{T}(T(\mathbb{R}^4))$,

$$\phi_j^{(\kappa)}(f) = \int \phi_j^{(\kappa)}(x) f(x) d^4x$$

are densely defined linear operators in a complex Hilbert space $\mathcal{H}$. They are not assumed to be bounded.

Here we state Wightman's axioms for the ultra-hyperfunction quantum field theory [Brüning & Nagamachi (2008)]. For the neutral scalar fields, these axioms are the axioms listed in [Brüning & Nagamachi (2004)].

W.I. **Relativistic invariance and state space**: There is a complex Hilbert space $\mathcal{H}$ with positive metric in which a unitary representation $U(a, A)$ of the Poinaré spinor group $\mathcal{P}_0$ acts. $(a, A) \mapsto U(a, A)$ is weakly continuous.

W.II. **Spectral property**: The spectrum Σ of the energy-momentum operator P which generates the translations in this representation, i.e., $e^{iaP} = U(a, 1)$, is contained in the closed forward light cone
$\bar{V}_+ = \{p = (p^0, \dots, p^3) \in \mathbb{R}^4; p^0 \geq |\boldsymbol{p}|\}$.

W.III. **Existence and uniqueness of the vacuum**: In $\mathcal{H}$ there exists unit vector Φ_0 (called the vacuum vector) which is unique up to a phase factor and which is invariant under all space-time translations $U(a, 1)$, $a \in \mathbb{R}^4$.

W.IV. **Fields as operator-valued tempered ultra-hyperfunctions**: The components $\phi_j^{(\kappa)}$ of the quantum field $\phi^{(\kappa)}$ are operator-valued generalized functions $\phi_j^{(\kappa)}(x)$ over the space $\mathcal{T}(T(\mathbb{R}^4))$ with common dense domain $\mathcal{D}$; i.e., for all $\Psi \in \mathcal{D}$ and all $\Phi \in \mathcal{H}$,

$$\mathcal{T}(T(\mathbb{R}^4)) \ni f \to (\Phi, \phi_j^{(\kappa)}(f)\Psi) \in \mathbb{C}$$

is a tempered ultra-hyperfunction. It is supposed that the vacuum vector Φ_0 is contained in $\mathcal{D}$ and that $\mathcal{D}$ is invariant under the action of the operators $\phi_j^{(\kappa)}(f)$ and $U(a, A)$, i.e.,

$$\phi_j^{(\kappa)}(f)\mathcal{D} \subset \mathcal{D},\ U(a, A)\mathcal{D} \subset \mathcal{D}.$$

Moreover it is assumed that there exist indices $\bar{\kappa}$, $\bar{\jmath}$ such that $\phi_{\bar{\jmath}}^{(\bar{\kappa})}(\bar{f}) \subset \phi_j^{(\kappa)}(f)^*$ where $*$ indicates the Hilbert space adjoint of the operator in question.

W.V. **Poincaré-covariance of the fields**: According to the type of the field, there is a finite dimensional real or complex matrix representation $V^{(\kappa)}(A)$ of $SL(2, \mathbb{C})$ such that

$$U(a, A)\phi_j^{(\kappa)}(x)U(a, A)^{-1} = \sum_{\ell} V_{j,\ell}^{(\kappa)}(A^{-1})\phi_\ell^{(\kappa)}(\Lambda(A)x + a),$$

i.e., for any $f \in \mathcal{T}(T(\mathbb{R}^4))$ and $\Psi \in \mathcal{D}$,

$$U(a, A)\phi_j^{(\kappa)}(f)U(a, A)^{-1}\Psi = \sum_{\ell} V_{j,\ell}^{(\kappa)}(A^{-1})\phi_\ell^{(\kappa)}(f_{(a,A)})\Psi,$$

where $f_{(a,A)}(x) = f(\Lambda(A)^{-1}(x - a))$. We have $V^{(\kappa)}(-1) = \pm 1$. If $V^{(\kappa)}(-1) = 1$, then the field is called a tensor field. If $V^{(\kappa)}(-1) = -1$, then the field is called a spinor field.

W.VI. **Extended causality or extended local commutativity**: Any two field components $\phi_j^{(\kappa)}(x)$ and $\phi_l^{(\kappa')}(y)$ either commute or anti-commute if the space-like distance between x

and y is greater than ℓ: In some Lorentz frame[1] , for any $\ell' > \ell$ and arbitrary elements Φ, Ψ in $\mathcal{D}$,

a) the functionals

$$\mathcal{T}(T(\mathbb{R}^4)) \otimes \mathcal{T}(T(\mathbb{R}^4)) \ni f \otimes g \rightarrow (\Phi, \phi_j^{(\kappa)}(f)\phi_l^{(\kappa')}(g)\Psi)$$

and

$$\mathcal{T}(T(\mathbb{R}^4)) \otimes \mathcal{T}(T(\mathbb{R}^4)) \ni f \otimes g \rightarrow (\Phi, \phi_l^{(\kappa')}(g)\phi_j^{(\kappa)}(f)\Psi)$$

can be extended continuously to $\mathcal{T}(T(L^{\ell'}))$, where

$$T(L^{\ell}) = \{(z_1, z_2) \in \mathbf{C}^{4\cdot 2}; |\mathrm{Im}\, z_1 - \mathrm{Im}\, z_2|_1 < \ell\},$$

with $|y|_1 = |y^0| + \sqrt{\sum_{i=1}^{3}(y^i)^2}$, and moreover,

b) the carrier of the functional

$$f \otimes g \rightarrow (\Phi, [\phi_j^{(\kappa)}(f), \phi_l^{(\kappa')}(g)]_{\mp}\Psi)$$

on $\mathcal{T}(T(\mathbb{R}^4)) \otimes \mathcal{T}(T(\mathbb{R}^4))$ is contained in the set

$$W^{\ell'} = \{(z_1, z_2) \in \mathbf{C}^{4\cdot 2}; z_1 - z_2 \in V^{\ell'}\},$$

where

$$V^{\ell} = \{z \in \mathbf{C}^4; \exists x \in V, |\mathrm{Re}\, z - x| < \ell, |\mathrm{Im}\, z|_1 < \ell\}$$

with $|y| = \sqrt{\sum_{i=0}^{3}(y^i)^2}$ is a complex neighborhood of light cone V, i.e., this functional can be extended continuously to $\mathcal{T}(W^{\ell'})$.

W.VII. **Cyclicity of the vacuum**: The set $\mathcal{D}_0$ of finite linear combinations of vectors of the form

$$\phi_{j_1}^{(\kappa_1)}(f_1) \cdots \phi_{j_n}^{(\kappa_n)}(f_n)\Phi_0,\ f_j \in \mathcal{T}(T(\mathbb{R}^4))\ (n = 0, 1, \ldots)$$

is dense in $\mathcal{H}$.

Remark 4.1. Condition a) of axiom W.VI expresses the fact that if the distance between x and y is greater than ℓ then x and y are distinguishable (see Remark 3.11). Condition b) of axiom W.VI corresponds to the locality condition of ordinary quantum field theory (see Remark 3.12).

4.2 Main properties of the system of vacuum expectation values

A vector-valued generalized function $\Phi_{\underline{\mu}_n}^{(\underline{\kappa}_n)}(f)$ is defined as follows: First, for $g(x_1, \ldots, x_n) = f_1(x_1) \cdots f_n(x_n)$, $f_j \in \mathcal{T}(T(\mathbb{R}^4))$, define $\Phi_{\mu_1 \ldots \mu_n}^{(\kappa_1 \ldots \kappa_n)}(g)$ by:

$$\Phi_{\mu_1 \ldots \mu_n}^{(\kappa_1 \ldots \kappa_n)}(g) = \phi_{\mu_1}^{(\kappa_1)}(f_1) \cdots \phi_{\mu_j}^{(\kappa_j)}(f_j) \cdots \phi_{\mu_n}^{(\kappa_n)}(f_n)\Phi_0.$$

[1] In [Nagamachi & Brüning (2010)] it is shown that the fundamental length does actually not dependent on the Lorentz frame.

This mapping is naturally extended to $\mathcal{T}(T(\mathbb{R}^4))^{\otimes n}$ by linearity. Then, by the same argument as Proposition 4.1 of [Brüning & Nagamachi (2004)]and using Theorem 3.7, $\Phi^{(\kappa_1\ldots\kappa_n)}_{\mu_1\ldots\mu_n}(g)$ is extended to a continuous mapping

$$\mathcal{T}(T(\mathbb{R}^{4n})) \ni f \to \Phi^{(\kappa_1\ldots\kappa_n)}_{\mu_1\ldots\mu_n}(f) \in \mathcal{H}.$$

The Wightman (generalized) function $\mathcal{W}^{(\kappa_1\ldots\kappa_n)}_{\mu_1\ldots\mu_n}(f)$ is defined by

$$\mathcal{T}(T(\mathbb{R}^{4n})) \ni f \to \mathcal{W}^{(\kappa_1\ldots\kappa_n)}_{\mu_1\ldots\mu_n}(f) = (\Phi_0, \Phi^{(\kappa_1\ldots\kappa_n)}_{\mu_1\ldots\mu_n}(f)) \in \mathbb{C}.$$

With the definition of the Fourier transform $\tilde{\Phi}^{(\underline{\kappa}_n)}_{\underline{\mu}_n}$ of $\Phi^{(\underline{\kappa}_n)}_{\underline{\mu}_n}$ by

$$\Phi^{(\underline{\kappa}_n)}_{\underline{\mu}_n}(f) = \tilde{\Phi}^{(\underline{\kappa}_n)}_{(\underline{\mu}_n)}(\tilde{f}).$$

we find

$$U(a,1)\tilde{\Phi}^{(\underline{\kappa}_n)}_{\underline{\mu}_n}(\tilde{f}) = U(a,1)\Phi^{(\underline{\kappa}_n)}_{\underline{\mu}_n}(f) = \Phi^{(\underline{\kappa}_n)}_{\underline{\mu}_n}(f_{(a,1)}) = \tilde{\Phi}^{(\underline{\kappa}_n)}_{\underline{\mu}_n}\left(\tilde{f}e^{[i(\sum_{k=1}^{n} p_k a)]}\right).$$

According to the standard strategy we use this identity to determine support properties of the Fourier transforms of the field operators. For $h \in \mathcal{T}(T(\mathbb{R}^4))$ calculate

$$(2\pi)^2\tilde{h}(P)\tilde{\Phi}^{(\underline{\kappa}_n)}_{\underline{\mu}_n}(\tilde{f}) = \int_{\mathbb{R}^4} h(a)U(a,1)da\tilde{\Phi}^{(\underline{\kappa}_n)}_{\underline{\mu}_n}(\tilde{f})$$

$$= (2\pi)^2\langle\tilde{\Phi}^{(\underline{\kappa}_n)}_{\underline{\mu}_n}(p_1,\ldots,p_n), \tilde{h}(p_1+\cdots+p_n)\cdot\tilde{f}(p_1,\ldots,p_n)\rangle.$$

Let χ_n be the linear mapping defined by

$$(p_1,\ldots,p_n) = \chi_n(q_0,\ldots,q_{n-1}),\ p_k = q_{k-1} - q_k (k = 1,\ldots,n-1),\ p_n = q_{n-1}.$$

The inverse mapping χ_n^{-1} is:

$$q_k = \sum_{j=k+1}^{n} p_j \quad (k = 0,\ldots,n-1).$$

Define $\tilde{Z}^{(\underline{\kappa}_n)}_{\underline{\mu}_n}$ by

$$\tilde{Z}^{(\underline{\kappa}_n)}_{\underline{\mu}_n}(\tilde{f}\circ\chi_n) = \tilde{\Phi}^{(\underline{\kappa}_n)}_{\underline{\mu}_n}(\tilde{f}).$$

Then

$$\tilde{Z}^{(\underline{\kappa}_n)}_{\underline{\mu}_n}(\tilde{g}) = \tilde{\Phi}^{(\underline{\kappa}_n)}_{\underline{\mu}_n}(\tilde{g}\circ\chi_n^{-1}).$$

In particular, for $\tilde{g}_2 \in H(\mathbb{R}^{4(n-1)};\mathbb{R}^{4(n-1)})$ and $\tilde{g}_1 \in H(\mathbb{R}^4;\mathbb{R}^4)$ we find

$$\tilde{h}(P)\tilde{Z}^{(\underline{\kappa}_n)}_{\underline{\mu}_n}(\tilde{g}_1\otimes\tilde{g}_2)) = \tilde{Z}^{(\underline{\kappa}_n)}_{\underline{\mu}_n}(\tilde{h}\cdot\tilde{g}_1\otimes\tilde{g}_2)) = \tilde{g}_1(P)\tilde{Z}^{(\underline{\kappa}_n)}_{\underline{\mu}_n}(\tilde{h}\otimes\tilde{g}_2)).$$

These identities show that the vector-valued generalized function

$$H(\mathbb{R}^4;\mathbb{R}^4) \ni \tilde{g}_1 \to \tilde{Z}^{(\underline{\kappa}_n)}_{\underline{\mu}_n}(\tilde{g}_1 \otimes \tilde{g}_2)) \in \mathcal{H}$$

has its support contained in the spectrum Σ of energy-momentum operator P (see Proposition 4.5 of [Brüning & Nagamachi (2004)]), moreover we can define a functional $\tilde{W}^{(\underline{\kappa}_n)}_{\underline{\mu}_n}$ by

$$(2\pi)^2\tilde{g}_1(0)\tilde{W}^{(\underline{\kappa}_n)}_{\underline{\mu}_n}(\tilde{g}_2) = (\Phi_0, \tilde{Z}^{(\underline{\kappa}_n)}_{\underline{\mu}_n}(\tilde{g}_1 \otimes \tilde{g}_2)),$$

and we have

$$(\tilde{Z}^{(\underline{\kappa}_m)}_{\underline{\mu}_m}(\tilde{g}_1), \tilde{Z}^{(\underline{\kappa}_n)}_{\underline{\mu}_n}(\tilde{g}_2)) = (2\pi)^2\langle \tilde{W}^{(\underline{\kappa}_{m+n})}_{\underline{\mu}_{m+n}}(q_1,\ldots,q_{m+n-1}), \bar{\tilde{g}}_1(q_m,\ldots,q_1)\tilde{g}_2(q_m,\ldots,q_{m+n-1})\rangle.$$

This identity implies that the support of $\tilde{W}^{(\underline{\kappa}_n)}_{\underline{\mu}_n}(q_1,\ldots,q_{n-1})$ is contained in Σ^{n-1} (see Proposition 4.6 of [Brüning & Nagamachi (2004)]). Moreover, the equality

$$(\tilde{Z}^{(\underline{\kappa}_n)}_{\underline{\mu}_n}(\tilde{g}), \tilde{Z}^{(\underline{\kappa}_n)}_{\underline{\mu}_n}(\tilde{g})) = (2\pi)^2\langle \tilde{W}^{(\underline{\kappa}_{2n})}_{\underline{\mu}_{2n}}(q_1,\ldots,q_{2n-1}), \bar{\tilde{g}}(q_n,\ldots,q_1)\tilde{g}(q_n,\ldots,q_{2n-1})\rangle$$

shows that the support of $\tilde{Z}^{(\underline{\kappa}_n)}_{\underline{\mu}_n}(q_0,\ldots,q_{n-1})$ is contained in Σ^n.

From this support property it follows that $\tilde{Z}^{(\underline{\kappa}_n)}_{\underline{\mu}_n}(\tilde{g})$ exists for a much wider class of test functions $\tilde{g}$ than was originally considered. For example, the function

$$\tilde{g}_\zeta(q) = (2\pi)^{-2n}e^{i[\sum_{j=0}^{n-1} q_j\zeta_j]}, \quad \operatorname{Im}\zeta_j \in V_+ + \ell_j(1,\mathbf{o})$$

belongs to the class of test functions for sufficiently large ℓ_j. We investigate the region of holomorphy of the following function

$$\begin{aligned}
&\langle \tilde{W}^{(\underline{\kappa}_{2n})}_{\underline{\mu}_{2n}}(q_1,\ldots,q_{2n-1}), \tilde{g}^*_{\zeta'}(q_1,\ldots,q_n)\tilde{g}_\zeta(q_n,\ldots,q_{2n-1})\rangle \\
&= \frac{1}{(2\pi)^{4n}}\langle \tilde{W}^{(\underline{\kappa}_{2n})}_{\underline{\mu}_{2n}}(q_1,\ldots,q_{2n-1}), e^{-i[\sum_{j=1}^{n} q_{n+1-j}\bar{\zeta}'_{j-1}]}e^{i[\sum_{k=1}^{n} q_{n+k-1}\zeta_{k-1}]}\rangle \\
&= W^{(\underline{\kappa}_{2n})}_{\underline{\mu}_{2n}}(-\bar{\zeta}'_{n-1},\ldots,-\bar{\zeta}'_0+\zeta_0,\ldots,\zeta_{n-1}).
\end{aligned}$$

Now recall the following proposition.

Proposition 4.2 (Proposition 4.7 of [Brüning & Nagamachi (2004)]). *There exist decreasing functions $R_{ij}(r)$ defined for $\ell < r$ such that $W^{(\underline{\kappa}_{2n})}_{\underline{\mu}_{2n}}(\zeta_1,\ldots,\zeta_{2n-1})$ is holomorphic in*

$$\bigcup_{i=1}^{2n-1}\{\zeta \in \mathbb{C}^{4(2n-1)}; \operatorname{Im}\zeta_i \in V_+ + (\ell',\mathbf{o}), \operatorname{Im}\zeta_j \in V_+ + (R_{ij}(\ell'),\mathbf{o}), \ell < \ell', j \neq i\}.$$

This proposition shows that $Z^{(\underline{\kappa}_n)}_{\underline{\mu}_n}(\zeta_0,\ldots,\zeta_{n-1})$ is holomorphic in the domain $\operatorname{Im}\zeta_0 \in V_+ + (\ell,\mathbf{o})/2$ and $\operatorname{Im}\zeta_k \in V_+ + (\ell_k,\mathbf{o})$ for sufficiently large ℓ_k for $k = 1,\ldots,n$.

Note that

$$(\tilde{g}_\zeta \circ \chi_n^{-1})(p_1,\dots,p_n) = (2\pi)^{-2n} e^{i\langle \zeta, \chi_n^{-1} p\rangle} = (2\pi)^{-2n} e^{i\langle \chi_n^{-1T}\zeta, p\rangle} = (2\pi)^{-2n} e^{i\langle z,p\rangle},$$

where $z = \chi_n^{-1T}\zeta$ and $\zeta = \chi_n^T z$, that is,

$$\zeta_0 = z_1,\ \zeta_j = z_{j+1} - z_j\ (j = 1,\dots,n-1),\ z_1 = \zeta_0,\ z_j = \sum_{k=0}^{j-1} \zeta_k\ (j = 2,\dots,n).$$

Therefore we get

$$Z^{(\underline{\kappa}_n)}_{\underline{\mu}_n}(\zeta_0,\dots,\zeta_{n-1}) = \tilde{Z}^{(\underline{\kappa}_n)}_{\underline{\mu}_n}(\tilde{g}_\zeta) = \tilde{\Phi}^{(\underline{\kappa}_n)}_{\underline{\mu}_n}(\tilde{g}_\zeta \circ \chi_n^{-1}) = \Phi^{(\underline{\kappa}_n)}_{\underline{\mu}_n}(z_1,\dots,z_n),$$

and

$$\Phi^{(\underline{\kappa}_n)}_{\underline{\mu}_n}(f) = \int \Phi^{(\underline{\kappa}_n)}_{\underline{\mu}_n}(x_1 + i\ell_0,\dots,x_n + i\sum_{k=1}^{n}\ell_{k-1}) f(x_1 + i\ell_0,\dots,x_n + i\sum_{k=1}^{n}\ell_{k-1}) dx_1\cdots dx_n,$$

where $\ell_0 = \ell/2 + \epsilon$ for any $\epsilon > 0$.

Note that the Poincaré group acts on $\tilde{g}_\zeta(q)$ as

$$(a, A) : \tilde{g}_\zeta(q) \to \tilde{g}_\zeta(\Lambda(A)^{-1}q)e^{iaq_0} = (2\pi)^{-2n} e^{i[\sum_{j=0}^{n-1} \Lambda(A)^{-1}q_j\zeta_j]} e^{iaq_0}$$

$$= (2\pi)^{-2n} \exp i[\sum_{j=0}^{n-1} q_j \Lambda(A)\zeta_j] e^{iaq_0} = \tilde{g}_{\Lambda(A)\zeta}(q) \exp iaq_0.$$

Then the formula of covariance

$$U(a, A)\Phi^{(\underline{\kappa}_n)}_{\underline{\mu}_n}(f) = \sum_{\nu_1,\dots,\nu_n} \prod_{j=1}^{n} V^{(\kappa_j)}_{\mu_j,\nu_j}(A^{-1}) \Phi^{(\kappa_1\dots\kappa_n)}_{\nu_1\dots\nu_n}(f_{(a,A)})$$

implies the following simple formula of covariance in the domain of holomorphy of $\Phi^{(\underline{\kappa}_n)}_{\underline{\mu}_n}(z_1,\dots,z_n)$ in complex space:

$$U(a, A)\Phi^{(\underline{\kappa}_n)}_{\underline{\mu}_n}(z_1,\dots,z_n)$$

$$= \sum_{\nu_1,\dots,\nu_n} \prod_{j=1}^{n} V^{(\kappa_j)}_{\mu_j,\nu_j}(A^{-1}) \Phi^{(\kappa_1\dots\kappa_n)}_{\nu_1\dots\nu_n}(\Lambda(A)z_1 + a,\dots,\Lambda(A)z_n + a). \qquad (4.1)$$

4.3 Functional characterization of fundamental length quantum fields

The analysis of the previous sections has shown that the sequence of vacuum expectation values of an ultra-hyperfunction quantum field theory has a number of specific properties. In analogy to standard quantum field theory we single out a set of properties of these vacuum expectation values which actually characterizes an ultra-hyperfunction quantum field theory up to isomorphisms. For the use in Section 5 (and because of space restrictions), we state them only in case of a scalar field.

Properties of UHQFT functionals:

(R1) $\mathcal{W}_0 = 1$, $\mathcal{W}_n \in \mathcal{T}(T(\mathbb{R}^{4n}))'$ for $n \geq 1$, and $\mathcal{W}_n(f^*) = \overline{\mathcal{W}_n(f)}$, for all $f \in \mathcal{T}(T(\mathbb{R}^{4n})) \equiv E(n)$, where $f^*(z_1, \ldots, z_n) = \overline{f(\bar{z}_n, \ldots, \bar{z}_1)}$.

(R2) For the Fourier transform $\tilde{\mathcal{W}}_n \in H(\mathbb{R}^{4n}; \mathbb{R}^{4n})'$ of $\mathcal{W}_n$, there exists $\tilde{W}_{n-1} \in H(\mathbb{R}^{4(n-1)}; \mathbb{R}^{4(n-1)})'$ such that

$$\tilde{\mathcal{W}}_n \circ \chi_n(q_0, \ldots, q_{n-1}) = (2\pi)^2 \delta(q_0) \tilde{W}_{n-1}(q_1, \ldots, q_{n-1})$$

and $\operatorname{supp} \tilde{W}_{n-1} \subset \Sigma^{n-1}$.

(R3) For a space-like vector $\boldsymbol{a} \in \mathbb{R}^4$ and $g_n \in E(n)$ introduce, for all $\lambda > 0$,

$$g_{n,\lambda}(x_1, \ldots, x_n) = g_n(x_1 - \lambda \boldsymbol{a}, \ldots, x_n - \lambda \boldsymbol{a}).$$

Then, for every $f_m \in E(m)$ and $g_n \in E(n)$ as $\lambda \to \infty$,

$$\mathcal{W}_{m+n}(f_m \otimes g_{n,\lambda}) \to \mathcal{W}_m(f_m)\mathcal{W}_n(g_n).$$

(R4) For any finite set $f_0, f_1, \ldots, f_N$ of test functions such that $f_0 \in \mathbb{C}$, $f_n \in \mathcal{T}(T(\mathbb{R}^{4n}))$ for $1 \leq n \leq N$, one has

$$\sum_{m,n=0}^{N} \mathcal{W}_{m+n}(f_m^* \otimes f_n) \geq 0.$$

(R5) $\mathcal{W}_n(f) = \mathcal{W}_n(f_{(a,\Lambda)})$ for all $(a, \Lambda) \in \mathcal{P}_+^\uparrow$, all $f \in \mathcal{T}(T(\mathbb{R}^{4n}))$, and all $n = 1, 2, \ldots$.

(R6) For all $n = 2, 3, \ldots$ and all $i = 1, \ldots, n-1$ denote

$$L_i^\ell = \{x = (x_1, \ldots, x_n) \in \mathbb{R}^{4n}; |x_i - x_{i+1}|_1 < \ell\},$$
$$W_i^\ell = \{(z_1, \ldots, z_n) \in \mathbb{C}^{4n}; z_i - z_{i+1} \in V^\ell\}.$$

Then, for any $\ell' > \ell$,

(i) $\mathcal{W}_n \in \mathcal{T}(T(\mathbb{R}^{4n}))'$ belongs to $\mathcal{T}(T(L_i^{\ell'}))'$ and

(ii) $\mathcal{W}_n \circ c_i^n$ belongs to $\mathcal{T}(W_i^{\ell'})'$,

where

$$(\mathcal{W}_n \circ c_i^n)(f) = \mathcal{W}_n(c_i^n(f)),$$
$$c_i^n(f)(x_1, \ldots, x_n) = f(x_1, \ldots, x_i, x_{i+1}, \ldots, x_n) - f(x_1, \ldots, x_{i+1}, x_i, \ldots, x_n).$$

The derivation of the properties (R1) - (R6) is found in [Brüning & Nagamachi (2004)].

Theorem 4.3 (reconstruction theorem). *To a given sequence $(\mathcal{W}_n)_{n \in \mathbb{N}}$ of tempered ultra-hyperfunctions satisfying the conditions (R1) - (R6), there corresponds a neutral scalar field $A(f)$ which obeys all the axioms W.I - W.VII and has the given tempered ultra-hyperfunctions as vacuum expectation values. The field A is unique up to isomorphisms.*

Proof: The proof of the theorem is found in [Brüning & Nagamachi (2004)].
□

5. $: \exp(\mathrm{i}l^2\phi(x)^2):$ as a fundamental length quantum field

We are going to construct models of relativistic quantum fields with a fundamental length by constructing a sequence of n-point functionals which satisfies conditions (R1) – (R6) and then applying the reconstruction theorem (Theorem 4.3). Our starting point are the well-known results of Jaffe [Jaffe (1965)] on formal Wick power series of free fields. If we consider the power series of a free field ϕ

$$\rho^{(i)}(x) = \sum_{n=0}^{\infty} a_n^{(i)} \frac{:\phi(x)^n:}{n!}, \tag{5.1}$$

then we have the following theorem.

Theorem 5.1 (Theorem A.1 of [Jaffe (1965)]). *As a formal power series*

$$(\Phi_0, \rho^{(1)}(x_1)\cdots\rho^{(n)}(x_n)\Phi_0) = \sum_{r_{ij}=0;\,1\le i<j\le n}^{\infty} \frac{A(R)T^R}{R!} \tag{5.2}$$

$$r_{ij} = r_{ji},\ r_{ii} = 0,\ R_i = \sum_{j=1}^{n} r_{ij},\ A(R) = \prod_{j=1}^{n} a_{R_j}^{(j)}$$

$$R! = \prod_{1\le i<j\le n} (r_{ij})!, \quad T^R = \prod_{1\le i<j\le n} (t_{ij})^{r_{ij}} \tag{5.3}$$

$$t_{ij} = (\Phi_0, \phi(x_i)\phi(x_j)\Phi_0) = D_m^{(-)}(x_i - x_j).$$

Therefore

$$(\Phi_0, \rho^{(i)}(x)\rho^{(i)}(y)\Phi_0) = \sum_{n=0}^{\infty} \frac{a_n^{(i)2}}{n!} D_m^{(-)}(x-y)^n,$$

$$D_m^{(-)}(x) = (2\pi)^{-3} \int_{\boldsymbol{R}^3} [2\omega(\boldsymbol{k})]^{-1} e^{-i\omega(\boldsymbol{k})x^0} e^{i\boldsymbol{k}\cdot\boldsymbol{x}} d\boldsymbol{k}$$

$$(k\cdot x = k^0x^0 - \boldsymbol{k}\cdot\boldsymbol{x},\ \omega(\boldsymbol{k}) = \sqrt{\boldsymbol{k}^2 + m^2}).$$

If the coefficients $\{a_n^{(i)}\}$ satisfy $\lim_{n\to\infty}[|a_n^{(i)}|^2/n!]^{1/n} = 0$ then the series (5.1) defines a hyperfunction quantum field (see [Nagamachi & Mugibayashi (1986)]).

Now we assume that for some $\sigma > 0$

$$\limsup_{n\to\infty} [|a_n^{(i)}|^2/n!]^{1/n} = \sigma. \tag{5.4}$$

For example, consider

$$\rho(x) =: e^{g\phi(x)^2} := \sum_{n=0}^{\infty} g^n \frac{:\phi(x)^{2n}:}{n!} = \sum_{n=0}^{\infty} g^n \frac{(2n)!}{n!} \frac{:\phi(x)^{2n}:}{(2n)!}. \tag{5.5}$$

Then

$$\sigma = \lim_{n\to\infty} \left[|g^n|^2 \frac{(2n)!}{(n!)^2} \right]^{1/2n} = 2|g| \tag{5.6}$$

and

$$(\Phi_0, \rho(x)\rho(y)\Phi_0) = \sum_{n=0}^{\infty} \left(g^n \frac{(2n)!}{n!}\right)^2 \frac{1}{(2n)!} D_m^{(-)}(x-y)^{2n}.$$

Since

$$(1-x)^{-\alpha} = 1 + \alpha x + \frac{\alpha(\alpha+1)}{2!}x^2 + \ldots + \frac{\alpha(\alpha+1)\cdots(\alpha+n-1)}{n!}x^n + \ldots,$$

and for $\alpha = 1/2$

$$\frac{\alpha(\alpha+1)\cdots(\alpha+n-1)}{n!} = \frac{(2n)!}{4^n n!}\frac{1}{n!},$$

we get, in the sense of formal power series,

$$(\Phi_0, \rho(x)\rho(y)\Phi_0) = [1 - 4g^2 D_m^{(-)}(x-y)^2]^{-1/2}. \tag{5.7}$$

Now we investigate the convergence of this power series, in the sense of tempered ultra-hyperfunctions. To this end consider the power series

$$\sum_{r_{ij}=0; 1\le i<j\le n}^{\infty} \frac{A(R)Z^R}{R!} \tag{5.8}$$

in the variables z_{ij} $(1 \le i < j \le n)$, where $Z^R = \prod_{1\le i<j\le n}(z_{ij})^{r_{ij}}$. Let

$$\|R\| = \sum_{1\le i<j\le n} r_{ij} \tag{5.9}$$

and $t_{ij}(1 \le i < j \le n)$ be positive constants. Suppose

$$\limsup_{\|R\|\to\infty} \left[\frac{|A(R)|T^R}{R!}\right]^{1/\|R\|} \le 1.$$

Then the fact that the series (5.8) converges if $|z_{ij}| < t_{ij}$ $(1 \le i < j \le n)$ follows from the following theorem of Lemire.

Theorem 5.2. *The associate convergence radii* $(r_1, \ldots, r_n)$ *of a series* $\sum a_{\nu_1,\ldots,\nu_n} z_1^{\nu_1} \cdots z_n^{\nu_n}$ *satisfy*

$$\limsup_{\nu_1+\cdots+\nu_n\to\infty} [|a_{\nu_1,\ldots,\nu_n}| r_1^{\nu_1} \cdots r_n^{\nu_n}]^{1/(\nu_1+\cdots+\nu_n)} = 1.$$

The multinomial theorem implies

$$R_i! \prod_{j=1}^{n} \frac{t_{ij}^{r_{ij}}}{(r_{ij})!} \le \left(\sum_{j=1}^{n} t_{ij}\right)^{R_i}.$$

and according to equations (5.3) and (5.9) we know

$$\sum_{i=1}^{n} R_i = 2\|R\|, \quad \prod_{i=1}^{n}\prod_{j=1}^{n}(r_{ij})! = (R!)^2, \quad \prod_{i=1}^{n}\prod_{j=1}^{n} t_{ij}^{r_{ij}} = (T^R)^2;$$

hence

$$\left[\frac{|A(R)|T^R}{R!}\right]^2 = \frac{\prod_{i=1}^n |a_{R_i}^{(i)}|^2 (T^R)^2}{(R!)^2} = \prod_{i=1}^n \left(|a_{R_i}^{(i)}|^2 \prod_{j=1}^n \frac{t_{ij}^{r_{ij}}}{(r_{ij})!} \right)$$

and

$$\left[\frac{|A(R)|T^R}{R!}\right]^{1/\|R\|} = \prod_{i=1}^n \left[|a_{R_i}^{(i)}|^2 \prod_{j=1}^n \frac{t_{ij}^{r_{ij}}}{(r_{ij})!} \right]^{1/2\|R\|} = \prod_{i=1}^n \left[\frac{|a_{R_i}^{(i)}|^2}{R_i!} R_i! \prod_{j=1}^n \frac{t_{ij}^{r_{ij}}}{(r_{ij})!} \right]^{1/2\|R\|}$$

$$\leq \prod_{i=1}^n \left[\frac{|a_{R_i}^{(i)}|^2}{R_i!} \left(\sum_{j=1}^n t_{ij} \right)^{R_i} \right]^{1/2\|R\|} = \prod_{i=1}^n \left[\left(\frac{|a_{R_i}^{(i)}|^2}{R_i!} \right)^{1/R_i} \left(\sum_{j=1}^n t_{ij} \right) \right]^{R_i/2\|R\|}.$$

Suppose that $t_{kk+1} < 1/\sigma$, and the other t_{ij}'s are so small that

$$\sum_{1 \leq i < j \leq n} t_{ij} < \frac{1}{\sigma}.$$

This then implies

$$\limsup_{\|R\| \to \infty} \prod_{i=1}^n \left[\left(\frac{|a_{R_i}^{(i)}|^2}{R_i!} \right)^{1/R_i} \left(\sum_{j=1}^n t_{ij} \right) \right]^{R_i/2\|R\|} \leq 1$$

and the power series (5.8) is convergent for $|z_{ij}| < t_{ij} (1 \leq i < j \leq n)$. This shows the convergence of the vacuum expectation value

$$(\Phi_0, \rho^{(1)}(x_1) \cdots \rho^{(n)}(x_n) \Phi_0)$$

in the sense of tempered ultra-hyperfunctions.

Now we consider the case of $m = 0$ for simplicity. In this case the growth of the two-point function of the free field is easier to estimate. In the case of $m > 0$, see Proposition 8.3. Recall

$$D_0^{(-)}(x) = \lim_{\epsilon \to +0} (2\pi)^{-2} [(x^0 - i\epsilon)^2 - \boldsymbol{x}^2]^{-1},$$

$$|(x^0 - i\epsilon)^2 - \boldsymbol{x}^2| = |x^2 - \epsilon^2 - 2i\epsilon x^0|.$$

We claim that we can find $\epsilon \geq 0$ such that

$$|(2\pi)^{-2} [(x^0 - i\epsilon)^2 - \boldsymbol{x}^2]^{-1}| < 1/\sigma = 1/(2|g|),$$

where we used the relation (5.6). For $x^2 \leq 0$, $|x^2 - \epsilon^2 - 2i\epsilon x^0| \geq |x^2 - \epsilon^2| \geq |x^2| + \epsilon^2$, and for $x^2 \geq 0$, $|x^2 - \epsilon^2 - 2ix^0| \geq |x^2 - \epsilon^2 - 2i\epsilon\sqrt{x^2}| = x^2 + \epsilon^2$, and $(|x^2| + \epsilon^2)^{-1} < (2\pi)^2/\sigma$ is equivalent to $\epsilon^2 > \sigma/(2\pi)^2 - |x^2|$. Choose a number $r' > \sqrt{\sigma}/(2\pi)$ and define

$$\epsilon(x) = \sqrt{\max\{r'^2 - |x^2|, 0\}}. \tag{5.10}$$

For such a choice one has

$$|(2\pi)^{-2}[(x^0 - i\epsilon(x))^2 - \boldsymbol{x}^2]^{-1}| < 1/\sigma.$$

Finally we fix the fundamental length for these models:

$$\ell = \sqrt{\sigma}/(2\pi) = \sqrt{2|g|}/(2\pi) = l/(\sqrt{2}\pi), \tag{5.11}$$

where we used the relation $g = il^2$ for $l > 0$. It is easily seen that for any $\ell' > \ell$ there exist $\epsilon(x)$ such that

$$\{(x^0 + i\epsilon(x), x^1, x^2, x^3); x \in \mathbb{R}^4\} \subset V^{\ell'}.$$

Therefore, for any $\ell' > \ell$ there exists $R > 0$ such that (in formal but suggestive notation)

$$W_{n-1}(\zeta) = \mathcal{W}_n(z) = (\Phi_0, \rho(z_1) \cdots \rho(z_k)\rho(z_{k+1}) \cdots \rho(z_n)\Phi_0)$$

is a well-defined holomorphic function for

$$\mathrm{Im}\,\zeta_k = \mathrm{Im}\,(z_{k+1} - z_j) \in V_+ + (\ell', 0, 0, 0)$$

and

$$\mathrm{Im}\,\zeta_j = \mathrm{Im}\,(z_{j+1} - z_j) \in V_+ + (R, 0, 0, 0), \ (j \neq k). \tag{5.12}$$

This implies that $\mathcal{W}_n$ satisfies the condition (i) of the axiom (R6). That is, the mapping

$$\mathcal{T}(T(\mathbb{R}^{4n})) \ni f \to \mathcal{W}(f) = \int_{\prod_{i=0}^{n-1}\Gamma_i} W_{n-1}(\zeta) g(\zeta) d\zeta_0 \cdots d\zeta_{n-1}$$

is continuous and can be extended continuously to

$$\mathcal{T}(T(L_k^{\ell'})) \ni f \to \mathcal{W}(f) = \int_{\prod_{i=0}^{n-1}\Gamma_i} W_{n-1}(\zeta) g(\zeta) d\zeta_0 \cdots d\zeta_{n-1},$$

where, with $\epsilon(x)$ according to (5.10) and R sufficiently large,

$$\Gamma_k = \{(x^0 + i\epsilon(x), x^1, x^2, x^3); x \in \mathbb{R}^4\}, \ \Gamma_j = \{(x^0 + iR, x^1, x^2, x^3); x \in \mathbb{R}^4\}$$

and $g(\zeta) = f(\zeta_0, \zeta_0 + \zeta_1, \ldots, \zeta_0 + \cdots + \zeta_{n-1})$. Now consider the formula

$$\begin{aligned}\prod_{1 \le i < j \le n} (t_{i,j})^{r_{i,j}} = (t_{k,k+1})^{r_{k,k+1}} & \prod_{1 \le i < j \le n, i \neq k, j \neq k+1} (t_{i,j})^{r_{i,j}} \\ & \times \prod_{1 \le i < k} (t_{i,k})^{r_{i,k}} \prod_{k+1 < j \le n} (t_{k,j})^{r_{k,j}} \\ & \times \prod_{1 \le i < k} (t_{i,k+1})^{r_{i,k+1}} \prod_{k+1 < j \le n} (t_{k+1,j})^{r_{k+1,j}}.\end{aligned}$$

The transposition of x_k and x_{k+1} causes the transposition of $(t_{k,k+1})^{r_{k,k+1}}$ and $(t_{k+1,k})^{r_{k+1,k}}$ in the first line, and the transposition of the second line and the third line. If x_k and x_{k+1} are space-like separated, then $t_{k,k+1} = t_{k+1,k}$. The function

$$W_{n-1}^k(\zeta) = (\Phi_0, \rho(z_1) \cdots \rho(z_{k+1})\rho(z_k) \cdots \rho(z_n)\Phi_0)$$

is also holomorphic in a domain defined by (5.12) and

$$-\mathrm{Im}\,\zeta_k \in V_+ + (\ell', 0, 0, 0).$$

Moreover, if ζ_k lies in $\mathbb{R}^4 \backslash V^{\ell'}$, the functions $W_{n-1}(\zeta)$ and $W^k_{n-1}(\zeta)$ are well-defined and coincide. Thus we have

$$(\mathcal{W}_n \circ c^n_k)(f) = \int_{\prod_{i=0}^{n-1}\Gamma_i} W_{n-1}(\zeta)g(\zeta)d\zeta_0 \cdots d\zeta_{n-1} - \int_{-\Gamma_k \prod_{i\neq k}\Gamma_i} W^k_{n-1}(\zeta)g(\zeta)d\zeta_0 \cdots d\zeta_{n-1}$$

$$= \int_{\Gamma^{\ell'}_k \prod_{i\neq k}\Gamma_i} W_{n-1}(\zeta)g(\zeta)d\zeta_0 \cdots d\zeta_{n-1} - \int_{-\Gamma^{\ell'}_k \prod_{i\neq k}\Gamma_i} W^k_{n-1}(\zeta)g(\zeta)d\zeta_0 \cdots d\zeta_{n-1},$$

where

$$\Gamma^{\ell}_k = \{(x^0 + i\epsilon(x), x^1, x^2, x^3); x \in \mathbb{R}^4 \cap V^{\ell}\}$$

and we used the fact that $W_{n-1}(\zeta)$ and $W^k_{n-1}(\zeta)$ coincides for $\zeta_k \in \mathbb{R}^4 \backslash V^{\ell'}$. The above formula shows that the functional $\mathcal{W}_n \circ c^n_k$ belongs to $\mathcal{T}(T(W^{\ell'}_k))'$ for any $\ell' > \ell$ which shows condition (ii) of axiom (R6). We can show that $\mathcal{W}_n$'s satisfy the axioms (R1), (R3), (R4) and (R5) in a similar way as [Brüning & Nagamachi (2001)] where it is shown that if the coefficients $\{a^{(i)}_n\}$ satisfy $\lim_{n\to\infty}[|a^{(i)}_n|^2/n!]^{1/n} = 0$ then the series (5.1) define hyperfunction quantum fields. There, a Wick polynomial $\rho_N(x)$ is introduced as a truncation of $\rho(x)$,

$$\rho_N(x) = \sum_{n=0}^{N} g^n \frac{:\phi(x)^{2n}:}{n!}.$$

Then the Wightman functions $\mathcal{W}^N_n(x) = (\Phi_0, \rho_N(x_1)\cdots\rho_N(x_n)\Phi_0)$ for $\rho_N(x)$ satisfy all the standard Wightman axioms, and they converge weakly to $\mathcal{W}_n(x) = (\Phi_0, \rho(x_1)\cdots\rho(x_n)\Phi_0)$ as $N \to \infty$ in the sense of tempered ultra-hyperfunctions. Thus they satisfy the above axioms. The proof of the spectral condition (R2) is easier than in the case of hyperfunction quantum field theory because $\tilde{W}^N_{n-1}(q)$ and $\tilde{W}_{n-1}(q)$ are distributions, and $\tilde{W}^N_{n-1}(q)$ converge weakly to $\tilde{W}_{n-1}(q)$ as $N \to \infty$ in the sense of distributions. Since the limit in the sense of distributions preserves the support, (R2) is valid for $\rho(x)$. Accordingly we formulate the main result of the section.

Theorem 5.3 (Existence of fields with fundamental length). *For a free field ϕ of mass $m \geq 0$ the Wick power series (5.1) (or more specifically (5.5)) define ultra-hyperfunction quantum fields with a fundamental length ℓ given by equations (5.4) and (5.11).*

Remark 5.4. We explained this only in the case of $m = 0$, but the above theorem is valid for $m \geq 0$. For details we have to refer to [Brüning & Nagamachi (2008)].

For the explicit form of $(\Phi_0, \rho^{(1)}(x_1)\cdots\rho^{(n)}(x_n)\Phi_0)$, we have the following proposition.

Proposition 5.5. *Abbreviate*

$$\rho^{(j)}(x_j) =: e^{-r_j i l^2 \phi(x_j)^2}:$$

with $r_j = \pm 1$. Then the vacuum expectation values of these fields are given by

$$(\Phi_0, \rho^{(1)}(x_1)\cdots\rho^{(n)}(x_n)\Phi_0) = (\det A)^{-1/2}, \tag{5.13}$$

where A is the $n \times n$ symmetric matrix whose entries $a_{j,k}$ are given by

$$a_{j,k} = a_{k,j} = 2h_{r_j}h_{r_k}l^2 D_m^{(-)}(x_j - x_k)$$

for $h_{\pm 1} = e^{\pm i\pi/4}$, $j < k$ *and* $a_{j,j} = 1$.

Proof. The proof is given in [Brüning & Nagamachi (2008)].

□

6. Proof of equation 2.2

In order to prove statement B) we need some further properties of Wick products. Thus we begin by recalling some basic facts about Wick products of free fields which are then used to derive this statement.

Let $\mathcal{H}$ be the Hilbert space defined by

$$\mathcal{H} = \oplus_{n=0}^{\infty} \mathcal{H}_n.$$

Here, $\mathcal{H}_n$ is the set of symmetric square-integrable functions on the direct product of the momentum space hyperboloids

$$\xi_k^2 = m^2,\ \xi_k^0 > 0,\ k = 1, \ldots, n \tag{6.1}$$

with respect to the Lorentz invariant measure $\prod_{k=1}^{n} d\Omega_m(\xi_k)$, given by

$$d\Omega_m(\xi) = \frac{d\xi^1 d\xi^2 d\xi^3}{\sqrt{\sum_{k=1}^{3}(\xi^k)^2 + m^2}}.$$

In the fundamental paper [Wightman & Gårding (1964)], we find the following quite general formula (3.44) for the definition of Wick products of a free field ϕ of mass m as operators in $\mathcal{H}$: For $f \in \mathcal{S}(\mathbb{R}^4)$ and $\Phi \in \mathcal{H}$ one has:

$$(: D^{\alpha^{(1)}}\phi D^{\alpha^{(2)}}\phi \cdots D^{\alpha^{(l)}}\phi : (f)\Phi)^{(n)}(\xi_1, \ldots, \xi_n) \tag{3.44}$$

$$= \frac{\pi^{l/2}}{(2\pi)^{2(l-1)}} \sum_{j=0}^{l} \left[\frac{(n-l+2j)!}{n!}\right]^{1/2} \int \cdots \int \left(\prod_{k=1}^{j} d\Omega_m(\eta_k)\right) \times$$

$$\sum_{1 \le k_1 < k_2 < \ldots < k_{l-j} \le n} (j!)^{-1} \sum_{P} P\left((-i\eta_1)^{\alpha^{(1)}} \cdots (-i\eta_j)^{\alpha^{(j)}} (i\xi_{k_1})^{\alpha^{(j+1)}} \cdots\right.$$

$$\left.\cdots (i\xi_{k_{l-j}})^{\alpha^{(l)}}\ \tilde{f}\left(\sum_{r=1}^{j}\eta_r - \sum_{r=1}^{l-j}\xi_{k_r}\right)\right) \Phi^{(n-l+2j)}(\eta_1, \ldots, \eta_j, \xi_1, \ldots, \hat{\xi}_{k_1}, \ldots, \hat{\xi}_{k_{l-j}}, \ldots, \xi_n),$$

where in the summation $\sum_{j=0}^{l}$, only those terms are to be retained for which $n - l + 2j \geq 0$, and the sum $\sum_P$ is over all permutation P of the variables $\eta_1, \ldots, \eta_j, (-\xi_{k_1}), \ldots, (-\xi_{k_{l-j}})$. We reconsider this formula in the sense of operator-valued ultra-hyperfunctions. Let $|\beta| = 1$ and $|\alpha^{(1)}| = |\alpha^{(2)}| = \ldots = |\alpha^{(l)}| = 0$. Then we have from (3.44)

$$
\begin{aligned}
(: \phi^l : (-D^\beta f)\Phi)^{(n)}(\xi_1, \ldots, \xi_n) &= \frac{\pi^{l/2}}{(2\pi)^{2(l-1)}} \sum_{j=0}^{l} \left[\frac{(n-l+2j)!}{n!}\right]^{1/2} \int \cdots \int \left(\prod_{k=1}^{j} d\Omega_m(\eta_k)\right) \\
&\times \sum_{1 \leq k_1 < k_2 < \ldots < k_{l-j} \leq n} (j!)^{-1} \sum_P P\left(i\left(\sum_{r=1}^{j} \eta_r - \sum_{r=1}^{l-j} \xi_{k_r}\right)^\beta \tilde{f}\left(\sum_{r=1}^{j} \eta_r - \sum_{r=1}^{l-j} \xi_{k_r}\right)\right) \times \\
&\times \Phi^{(n-l+2j)}(\eta_1, \ldots, \eta_j, \xi_1, \ldots, \hat{\xi}_{k_1}, \ldots, \hat{\xi}_{k_{l-j}}, \ldots, \xi_n). \\
&= \frac{\pi^{l/2}}{(2\pi)^{2(l-1)}} \sum_{j=0}^{l} \left[\frac{(n-l+2j)!}{n!}\right]^{1/2} \int \cdots \int \left(\prod_{k=1}^{j} d\Omega_m(\eta_k)\right) \\
&\times \sum_{1 \leq k_1 < k_2 < \cdots < k_{l-j} \leq n} (j!)^{-1} \sum_P P\left(l(i\eta_1)^\beta \tilde{f}\left(\sum_{r=1}^{j} \eta_r - \sum_{r=1}^{l-j} \xi_{k_r}\right)\right) \\
&\times \Phi^{(n-l+2j)}(\eta_1, \ldots, \eta_j, \xi_1, \ldots, \hat{\xi}_{k_1}, \ldots, \hat{\xi}_{k_{l-j}}, \ldots, \xi_n).
\end{aligned}
$$

Observe that

$$\sum_P P(\eta_i) = \sum_P P(-\xi_{k_r})$$

for any i and r. This implies for $|\beta| = 1$,

$$\sum_P P\left((\eta_i)^\beta\right) = \sum_P P\left((-\xi_{k_r})^\beta\right)$$

and therefore

$$\sum_P P\left(i\left(\sum_{r=1}^{j} \eta_r - \sum_{r=1}^{l-j} \xi_{k_r}\right)^\beta \tilde{f}\left(\sum_{r=1}^{j} \eta_r - \sum_{r=1}^{l-j} \xi_{k_r}\right)\right) = \sum_P P\left(l(i\eta_1)^\beta \tilde{f}\left(\sum_{r=1}^{j} \eta_r - \sum_{r=1}^{l-j} \xi_{k_r}\right)\right).$$

On the other hand, we also have from (3.44), for $|\alpha^{(1)}| = 1$, and $|\alpha^{(2)}| = \ldots = |\alpha^{(l)}| = 0$

$$
\begin{aligned}
&(: (D^{\alpha^{(1)}}\phi)\phi^{l-1} : (f)\Phi)^{(n)}(\xi_1, \ldots, \xi_n) \\
&= \frac{\pi^{l/2}}{(2\pi)^{2(l-1)}} \sum_{j=0}^{l} \left[\frac{(n-l+2j)!}{n!}\right]^{1/2} \int \cdots \int \left(\prod_{k=1}^{j} d\Omega_m(\eta_k)\right) \\
&\times \sum_{1 \leq k_1 < k_2 < \ldots < k_{l-j} \leq n} (j!)^{-1} \sum_P P\left((i\eta_1)^{\alpha^{(1)}} \tilde{f}\left(\sum_{r=1}^{j} \eta_r - \sum_{r=1}^{l-j} \xi_{k_r}\right)\right)
\end{aligned}
$$

$$\times \Phi^{(n-l+2j)}(\eta_1,\dots,\eta_j,\xi_1,\dots,\hat{\xi}_{k_1},\dots,\hat{\xi}_{k_{l-j}},\dots,\xi_n).$$

This shows that

$$(:\phi^l:(-D^{\alpha^{(1)}}f)\Phi)^{(n)} = l(:(D^{\alpha^{(1)}}\phi)\phi^{l-1}:(f)\Phi)^{(n)}, \tag{6.2}$$

that is,

$$D^{\alpha^{(1)}}:\phi(x)^l := l:(D^{\alpha^{(1)}}\phi(x))\phi^{l-1}(x):. \tag{6.3}$$

Let $\mathcal{D}_0$ be the set generated by the vectors of the form

$$\rho^{(1)}(f_1)\cdots\rho^{(n)}(f_n)\Phi_0, f_k \in \mathcal{T}(T(\mathbb{R}^4)),$$

where $\rho^{(k)}(x)$ is one of $\phi(x)$, $\rho(x)$ and $\rho^*(x)$, and $\Phi \in \mathcal{D}_0$. Then it follows from the weak conveargence of

$$\rho(-D^{\alpha^{(1)}}f)\Phi =: e^{ig\phi^2}:(-D^{\alpha^{(1)}}f)\Phi = \sum_{l=0}^{\infty}\frac{(ig)^l}{l!}:\phi^{2l}:(-D^{\alpha^{(1)}}f)\Phi$$

which we have seen in the previous section that the above series is also strongly convergent, and by (6.2)

$$:\phi^{2l}:(-D^{\alpha^{(1)}}f)\Phi = l:(D^{\alpha^{(1)}}\phi)\phi^{l-1}:(f)\Phi.$$

This shows that

$$\sum_{l=0}^{\infty}\frac{(ig)^l}{l!}:\phi^{2l}:(-D^{\alpha^{(1)}}f)\Phi = \sum_{l=1}^{\infty}\frac{(ig)^l}{(l-1)!}2:(D^{\alpha^{(1)}}\phi)\phi\phi^{2(l-1)}:(f)\Phi$$

$$= \sum_{l=0}^{\infty}2(ig)\frac{(ig)^l}{l!}:(D^{\alpha^{(1)}}\phi)\phi\phi^{2l}:(f)\Phi.$$

We write the last expression as

$$= 2(ig):(D^{\alpha^{(1)}}\phi)\phi\sum_{l=0}^{\infty}\frac{(ig)^l}{l!}\phi^{2l}:(f)\Phi = 2ig:(D^{\alpha^{(1)}}\phi)\phi\rho:(f)\Phi.$$

That is, the formal expression (which is difficult to give a direct meaning)

$$2ig:(D^{\alpha^{(1)}}\phi(x))\phi(x)(:e^{ig\phi(x)^2}:):\Phi = 2ig:(D^{\alpha^{(1)}}\phi(x))\phi(x)\sum_{l=0}^{\infty}\frac{(ig)^l}{l!}:\phi^{2l}(x)::\Phi$$

should be understood to be

$$\sum_{l=0}^{\infty}2ig:(D^{\alpha^{(1)}}\phi(x))\phi(x)\frac{(ig)^l}{l!}\phi^{2l}(x):\Phi = \sum_{l=1}^{\infty}2:(D^{\alpha^{(1)}}\phi(x))\frac{(ig)^l}{(l-1)!}\phi^{2l-1}(x):\Phi.$$

Then by (6.3), the above expression equals

$$\sum_{l=1}^{\infty}\frac{(ig)^l}{l!}D^{\alpha^{(1)}}:\phi^{2l}(x):\Phi,$$

and this is equal to

$$D^{\alpha^{(1)}} \sum_{l=1}^{\infty} \frac{(ig)^l}{l!} : \phi^{2l}(x) : \Phi = D^{\alpha^{(1)}} \rho(x)\Phi$$

in the sense of generalized functions. In the above understanding, we have

$$D^{\alpha^{(1)}} \rho(x)\Phi = 2ig : (D^{\alpha^{(1)}} \phi(x))\phi(x)\rho(x) : \Phi, \tag{6.4}$$

that is, if the Wick product

$$: (D^{\alpha^{(1)}} \phi(x))\phi(x)\rho(x) :$$

is defined by the Wick power series

$$\sum_{l=0}^{\infty} 2ig : (D^{\alpha^{(1)}} \phi(x))\phi(x) \frac{(ig)^l}{l!} \phi^{2l}(x) :,$$

then (6.4) holds and (2.2) follows.

7. Multiplier

As stated at the end of Section 5, $\{\mathcal{H}, \Phi_0, U(a,\Lambda), \phi(x), \rho(x), \rho^*(x)\}$ satisfies the axioms of UHFQFT (= ultrahyperfunction quantum field theory). Let $\rho^{(\kappa)}(x) = \rho(x)$ and $\rho^{(\bar{\kappa})}(x) = \rho^*(x)$. Then, as we learned in Section 4.2, the vector-valued function $\rho^{(\lambda_1)}(z_1) \cdots \rho^{(\lambda_n)}(z_n)\Phi_0$ is holomorphic in

$$\{(z_1, \ldots, z_n) \in \mathbb{C}^{4n}; \mathrm{Im}\, z_1 \in V_+ + (\ell/2, \mathbf{o}), \mathrm{Im}\,(z_{j+1} - z_j) \in V_+ + (\ell_j, \mathbf{o})\}$$

for some $\ell_j > \ell > 0$ $(j = 1, \ldots, n-1)$, where $\rho^{(\lambda)}(x)$ is one of $\rho^{(\kappa)}(x)$, $\rho^{(\bar{\kappa})}(x)$ and $\phi(x)$. Let $\psi_{0,\alpha}^{(\kappa)}(x) = \psi_{0,\alpha}(x)$ and $\psi_{0,\bar{\alpha}}^{(\bar{\kappa})}(x) = \bar{\psi}_{0,\bar{\alpha}}(x)$ be free Dirac fields of mass M. Then the system

$$\{\mathcal{K}, \Psi_0, V(a,\Lambda), \psi_{0,\alpha}^{(\kappa)}(x), \psi_{0,\bar{\alpha}}^{(\bar{\kappa})}(x)\}$$

satisfies the axioms of standard quantum field theory in terms of tempered distributions (and consequently, that of UHFQFT), and therefore $\psi_{0,\beta_1}^{(\lambda_1)}(z_1) \cdots \psi_{0,\beta_n}^{(\lambda_n)}(z_n)\Psi_0$ is holomorphic in

$$\{(z_1, \ldots, z_n) \in \mathbb{C}^{4n}; \mathrm{Im}\, z_1 \in V_+, \mathrm{Im}\,(z_j - z_{j-1}) \in V_+\},$$

where $\lambda = \kappa$, $\beta = \alpha$ or $\lambda = \bar{\kappa}$, $\beta = \bar{\alpha}$. Therefore, $\rho(z)\Phi$ for $\Phi = \rho^{(\lambda_2)}(f_2) \cdots \rho^{(\lambda_n)}(f_n)\Phi_0$, $f_j \in \mathcal{T}(T(\mathbb{R}^4))$ is holomorphic in

$$\{z \in \mathbb{C}^4; \mathrm{Im}\, z \in V_+ + (\ell/2, \mathbf{o})\}$$

and $\psi_{0,\alpha_1}(z)\Psi$ for $\Psi = \psi_{0,\beta_2}^{(\lambda_2)}(g_2) \cdots \psi_{0,\beta_n}^{(\lambda_n)}(g_n)\Psi_0$, $g_j \in \mathcal{S}(\mathbb{R}^4)$ is holomorphic there too. The composite system

$$\{\mathcal{H} \otimes \mathcal{K}, \Phi_0 \otimes \Psi_0, U(a,\Lambda) \otimes V(a,\Lambda), \phi(x) \otimes I_{\mathcal{K}}, \rho(x) \otimes I_{\mathcal{K}}, \\ \rho^*(x) \otimes I_{\mathcal{K}}, I_{\mathcal{H}} \otimes \psi_{0,\alpha}(y), I_{\mathcal{H}} \otimes \bar{\psi}_{0,\bar{\alpha}}(y)\}$$

is the tensor product of two systems and thus satisfies all the axioms of UHFQFT. Although the tensor product is well-defined, the pointwise product is not necessarily well-defined for generalized (vector-valued) functions. In the category of distributions, the following theorem is well-known:

Theorem 7.1 (Theorem 8.2.10 of [Hörmander (1983)]). *If $u, v \in \mathcal{D}'(X)$ then the product uv can be defined as the pullback of the tensor product $u \otimes v$ by the diagonal map $\delta : X \to X \times X$ unless $(x, \xi) \in WF(u)$ and $(x, -\xi) \in WF(v)$.*

In our case, the condition that $\rho(z)\Phi$ and $\psi_{0,\alpha_1}(z)\Psi$ have the common domain of holomorphy,

$$\{z \in \mathbb{C}^4; \operatorname{Im} z \in V_+ + (\ell/2, \mathbf{o})\},$$

which corresponds to the condition of the wave front sets $WF(u)$ and $WF(v)$ of distributions, implies that the product $(\rho\psi_{0,\alpha})(f)$ is well-defined by the formula

$$(\rho\psi_{0,\alpha})(f)(\Phi \otimes \Psi) = \int_{\Gamma_N} f(z)\rho(z)\Phi \otimes \psi_{0,\alpha}(z)\Psi dz, \quad \Gamma_N = \{z \in \mathbb{C}^4; z = x + i(N, \mathbf{o})\}$$

for suitable $N > 0$. Thus the field $\psi_0(x)$ is a multiplier of the field $\rho(x)$. Similarly one can show that $\frac{\partial}{\partial x^\mu}\psi_{0,\alpha}$ is a multiplier for $\rho(x)$ and then we calculate

$$\begin{aligned}(\frac{\partial}{\partial x^\mu}(\rho\psi_{0,\alpha}))(f)\Phi \otimes \Psi &= (\rho\psi_{0,\alpha})(-\frac{\partial}{\partial x^\mu} f)\Phi \otimes \Psi = \int_{\Gamma_N} (-\frac{\partial}{\partial x^\mu} f(z))\rho(z)\Phi \otimes \psi_{0,\alpha}(z)\Psi dz \\ &= \int_{\Gamma_N} f(z)\{\rho(z)\Phi \otimes \frac{\partial}{\partial x^\mu}\psi_{0,\alpha}(z)\Psi + \frac{\partial}{\partial x^\mu}\rho(z)\Phi \otimes \psi_{0,\alpha}(z)\Psi\} dz \\ &= (\rho\frac{\partial}{\partial x^\mu}\psi_{0,\alpha})(f)\Phi \otimes \Psi + ((\frac{\partial}{\partial x^\mu}\rho)\psi_{0,\alpha})(f)\Phi \otimes \Psi.\end{aligned}$$

This gives

$$\frac{\partial}{\partial x^\mu}(\rho(x)\psi_{0,\alpha}(x))(\Phi \otimes \Psi) = \rho(x)\frac{\partial}{\partial x^\mu}\psi_{0,\alpha}(x)\Phi \otimes \Psi + (\frac{\partial}{\partial x^\mu}\rho(x))\psi_{0,\alpha}(x)\Phi \otimes \Psi.$$

Let $\psi(x) = \rho(x)\psi_0(x)$ and $\bar{\psi}(x) = \rho^*(x)\bar{\psi}_0(x)$. We can easily see that the fields $\psi(x), \bar{\psi}(x), \phi(x)$ satisfy the axioms of UHFQFT except for the condition of extended causality, which is proven in the next section. In fact, W.I - W.V follow from those of the systems $\{\mathcal{H}, \Phi_0, U(a, \Lambda), \phi(x), \rho(x), \rho^*(x)\}$ and $\{\mathcal{K}, \Psi_0, V(a, \Lambda), \psi_{0,\alpha}^{(\kappa)}(x), \psi_{0,\bar{\alpha}}^{(\bar{\kappa})}(x)\}$ (for W.V the relation (4.1) is used). For W.VI, we have only to restrict the Hilbert space $\mathcal{H} \otimes \mathcal{K}$ to the subspace generated by

$$\phi_{j_1}^{(\kappa_1)}(f_1) \cdots \phi_{j_n}^{(\kappa_n)}(f_n)\Phi_0 \otimes \Psi_0, \ f_j \in \mathcal{T}(T(\mathbb{R}^4)) \ (n = 0, 1, \ldots),$$

where $\phi_j^{(\kappa)}(x)$ is $\psi_\alpha(x) = (\rho(x) \otimes I_{\mathcal{K}}) \cdot (I_{\mathcal{H}} \otimes \psi_{0,\alpha}(x)) = \rho(x) \otimes \psi_{0,\alpha}(x)$ or $\bar{\psi}_{\bar{\alpha}}(x) = (\rho^*(x) \otimes I_{\mathcal{K}}) \cdot (I_{\mathcal{H}} \otimes \bar{\psi}_{0,\bar{\alpha}}(x)) = \rho^*(x) \otimes \bar{\psi}_{0,\bar{\alpha}}(x)$ or $\phi(x) \otimes I_{\mathcal{K}}$.

8. Fundamental length quantum fields

In this section we are going to prove the condition of extended causality (the axiom W.VI). In a first step we prove that Axiom W.VI is equivalent to a condition **R6** for the Wightman functionals. Then we proceed to verify condition **R6**.

Proposition 8.1. *Assuming the validity of the other axioms, the axiom of extended causality WVI is equivalent to the following condition*

R6 *For all $n = 2, 3, \ldots$ and all $i = 1, \ldots, n-1$ denote*

$$
\begin{aligned}
L_i^\ell &= \{x = (x_1, \ldots, x_n) \in \mathbb{R}^{4n}; |x_i - x_{i+1}|_1 < \ell\}, \\
W_i^\ell &= \{x = (z_1, \ldots, z_n) \in \mathbb{C}^{4n}; z_i - z_{i+1} \in V^\ell\}, \\
V^\ell &= \{z \in \mathbb{C}^4; \exists x \in V, |\mathrm{Re}\, z - x| < \ell, |\mathrm{Im}\, z|_1 < \ell\}.
\end{aligned} \tag{8.1}
$$

Then, for any $\ell' > \ell$,

a) the functional

$$\mathcal{T}(T(\mathbb{R}^{4n})) \ni f \to \mathcal{W}_{\mu_1 \ldots \mu_n}^{(\kappa_1 \ldots \kappa_n)}(f) \in \mathbb{C}$$

is extended continuously to $\mathcal{T}(T(L_i^{\ell'}))$, and

b) the functional on $\mathcal{T}(T(\mathbb{R}^{4n}))$

$$f \to \mathcal{W}_{\mu_1 \ldots \mu_j \mu_{j+1} \ldots \mu_n}^{(\kappa_1 \ldots \kappa_j \kappa_{j+1} \ldots \kappa_n)}(f) + \mathcal{W}_{\mu_1 \ldots \mu_{j+1} \mu_j \ldots \mu_n}^{(\kappa_1 \ldots \kappa_{j+1} \kappa_{j+} \ldots \kappa_n)}(f) \in \mathbb{C}$$

is extended continuously to $\mathcal{T}(W_i^{\ell'})$.

Proof. Since the spinor/tensor indices do not play a role in this statement the proof given in [Brüning & Nagamachi (2004)] for the scalar case applies (see Propositions 4.3, 4.4 and Theorem 5.1 of [Brüning & Nagamachi (2004)]). □

Proposition 8.2. *The Wightman functions $\mathcal{W}_\alpha^r$ as given in [Nagamachi & Brüning (2008)] or in formula (8.6) below satisfy condition* **R6**.

Proof. The determinant $\det A$ of (5.13) can be expressed as

$$\det A = 1 + P_n(a_{j,k}) \tag{8.2}$$

where $P_n(a_{j,k})$ is the sum of homogeneous polynomials of degrees $m = 2, \cdots, n$ in the entries $a_{j,k}$, $1 \le j < k \le n$ with integer coefficients.

Introduce

$$Q_{n,j}(a_{i,k}) = \sum_{\substack{(i,k,\ldots,l) \neq (1,2,\ldots,n) \\ (i,k,\ldots,l) \neq (1,2,\ldots,j+1,j,\ldots,n)}} \mathrm{sgn}\,(i,k,\ldots,l) a_{1,j} a_{2,k} \cdots a_{n,l} \tag{8.3}$$

and denote by $\sigma(j+1,j)$ the permutation $(1,\ldots,j-1,j,j+1,\ldots,n) \longrightarrow (1,\ldots,j-1,j+1,j,\ldots,n)$. Then we have

$$P_n(a_{i,k}) = \text{sgn}\,(\sigma(j+1,j))a_{1,1}a_{2,2}\cdots a_{j-1,j-1}a_{j,j+1}a_{j+1,j}a_{j+1,j+1}\cdots a_{n,n}$$
$$+ Q_{n,j}(a_{i,k}) = -a_{j,j+1}^2 + Q_{n,j}(a_{i,k}) = \pm 4l^2 D_m^{(-)}(z_j - z_{j+1})^2 + Q_{n,j}(a_{i,k}).$$

Hence we can rewrite (8.2) as

$$\det A = 1 + P_n(a_{i,k}) = 1 \pm 4l^4 D_m^{(-)}(z_j - z_{j+1})^2 + Q_{n,j}(a_{i,k}).$$

It is clear from (8.2), (8.3) and the details provided about the polynomial P_n that each term of $Q_{n,j}(a_{i,k})$ contains products of 2-points functions $D_m^{(-)}$ at arguments different from $z_j - z_{j+1}$.

Assume

$$y_{j+1}^0 - y_j^0 > \ell = l/(\sqrt{2}\pi). \tag{8.4}$$

Then we have $|4l^4 D_m^{(-)}(z_j - z_{j+1})^2| < 1$ by the estimate

$$|D_m^{(-)}(x^0 - i\epsilon, \boldsymbol{x})| \le (2\pi\epsilon)^{-2} \quad \text{for all } x \in \mathbb{R}^4. \tag{8.5}$$

If we choose the arguments $y_k^0 - y_i^0$ $(i < k)$ in these 2-points functions sufficiently large, $Q_{n,j}(a_{i,k})$ becomes very small; and for these points z_j the determinant $(\det A(z))^{-1/2}$ is holomorphic and the function $(\det A(z))^{-1/2}\mathcal{W}_{0,\alpha}^r(z_1,\ldots,z_n)$ defines a functional in $\mathcal{T}(T(L_j^{\ell'}))'$ for any $\ell' > \ell$ by the formula

$$\mathcal{W}_\alpha^r(f) = \int_{\prod_{j=1}^n \Gamma_j} (\det A(z))^{-1/2}\mathcal{W}_{0,\alpha}^r(z_1,\ldots,z_n)f(z)dz \tag{8.6}$$

for all $f \in \mathcal{T}(T(L_j^{\ell'}))$, where $\Gamma_j = \mathbb{R}^4 + i(y_j^0,0,0,0)$ and

$$\mathcal{W}_{0,\alpha}^r(z_1,\ldots,z_n) = (\Psi_0, \psi_{0,\alpha_1}^{(r_1)}(z_1)\cdots\psi_{0,\alpha_n}^{(r_n)}(z_n)\Psi_0)$$

is the Wightman function of free Dirac field. In fact, for $\ell' > \ell$, we choose $\ell' > y_{j+1}^0 - y_j^0 > \ell$ and the other $y_k^0 - y_i^0$ sufficiently large so that $(\det A(z))^{-1/2}$ is a bounded function of x. Then the corresponding integration path $\prod_{j=1}^n \Gamma_j$ of (8.6) is contained in

$$T(L_j^{\ell'}) = \{z = x + iy \in \mathbb{C}^{4n}; |y_j - y_{j+1}|_1 < \ell'\},$$

where $|y|_1 = |y^0| + |\boldsymbol{y}|$. We conclude that the functional defined by

$$(\det A(z))^{-1/2}\mathcal{W}_{0,\alpha}^r(z_1,\ldots,z_n)$$

satisfies condition a) of **R6**.

The transposition of z_j and z_{j+1} causes the change of $a_{j,j+1} = a_{j+1,j}$:

$$D_m^{(-)}(z_j - z_{j+1}) \to D_m^{(-)}(z_{j+1} - z_j)$$

and for an index k with $j < k \neq j+1$ the change

$$a_{j,k} = a_{k,j} = D_m^{(-)}(z_j - z_k) \to D_m^{(-)}(z_{j+1} - z_k) = a_{j+1,k} = a_{k,j+1},$$

$$a_{j+1,k} = a_{k,j+1} = D_m^{(-)}(z_{j+1} - z_k) \to D_m^{(-)}(z_j - z_k) = a_{j,k} = a_{k,j},$$

results while for an index k with $j > k \neq j+1$ the change is

$$a_{j,k} = a_{k,j} = D_m^{(-)}(z_k - z_j) \to D_m^{(-)}(z_k - z_{j+1}) = a_{j+1,k} = a_{k,j+1},$$

$$a_{j+1,k} = a_{k,j+1} = D_m^{(-)}(z_k - z_{j+1}) \to D_m^{(-)}(z_k - z_j) = a_{j,k} = a_{k,j}.$$

We consider the matrix $B = (b_{i,j})$ obtained from A by the change of j-th and $(j+1)$-th rows and j-th and $(j+1)$-th columns. Then we have $\det A = \det B$. Next we consider the matrix $C = (c_{j,k})$ obtained from B by changing only $b_{j,j+1} = b_{j+1,j} = a_{j,j+1} = a_{j+1,j}$, i.e., $c_{j,j+1} = c_{j+1,j} = D_m^{(-)}(z_{j+1} - z_j)$. If x_j and x_{j+1} are space-like separated, then $D_m^{(-)}(x_j - x_{j+1})$ is analytic (space-like points x are Jost points of $D_m^{(-)}(x)$) and $D_m^{(-)}(x_j - x_{j+1}) = D_m^{(-)}(x_{j+1} - x_j)$. Therefore for space-like separated x_j, x_{j+1} ($y_j^0 - y_{j+1}^0 = 0$) and the other $y_k^0 - y_i^0$ sufficiently large, we have $\det A = \det C$. Note that $\mathcal{W}_{0,\alpha}^r(z_1, \ldots, z_n)$ is also expressed by the sum of the products of the two-point functions of the free Dirac field as in the scalar case, and for space-like separated x_j, x_{j+1} ($y_j^0 - y_{j+1}^0 = 0$) and the other $y_k^0 - y_i^0$ positive, one has

$$\mathcal{W}_{0,\alpha}^r(z_1, \ldots, x_j, x_{j+1}, \ldots, z_n) = -\mathcal{W}_{0,\alpha}^r(z_1, \ldots, x_{j+1}, x_j, \ldots, z_n).$$

In order to proceed, we need some estimate for $D_m^{(-)}(x_j - x_{j+1})$.

Proposition 8.3 (Corollary 2.4 of [Brüning & Nagamachi (2008)]). *Denote by* $\mathrm{dist}\,(x, \bar{V})$ *the distance between* x *and the closed light cone* $\bar{V}$*, and for* $\ell > 0$*,*

$$V_\ell = \{x \in \mathbb{R}^4; \mathrm{dist}\,(x, V) < \ell\}.$$

Define $\epsilon_\ell(x)$ *by* $\epsilon_\ell(x) = \ell$ *if* $\mathrm{dist}\,(x, \bar{V}) < \ell/\sqrt{2}$, $\epsilon_\ell(x) = \sqrt{2\ell^2 - 2\mathrm{dist}\,(x, \bar{V})^2}$ *if* $\ell/\sqrt{2} \leq \mathrm{dist}\,(x, \bar{V}) < \ell$ *and* $\epsilon_\ell(x) = 0$ *if* $\mathrm{dist}\,(x, \bar{V}) \geq \ell$. *Then* $0 \leq \epsilon_\ell(x) \leq \ell$ *and* $\mathrm{supp}\,\epsilon_\ell(x) \subset \bar{V}_\ell$. *Let* $\ell = l/(\sqrt{2}\pi)$ *and assume* $ml < 2$. *Then, if* $\ell'' > \ell$*, the estimate*

$$2l^2|D_m^{(-)}(x^0 - i\epsilon_{\ell''}(x), \boldsymbol{x})| < 1$$

holds.

For any $\ell' > \ell$, we choose $\ell < \ell'' < \ell'$. Let $\epsilon(x) = \epsilon_{\ell''}(x)$ and $a_{j,j+1} = D_m^{(-)}(x_j - x_{j+1} + i\epsilon(x_j - x_{j+1}))$ and for other $a_{i,k}$, $y_k^0 - y_i^0$ sufficiently large. Then $(\det A(x))^{-1/2}$ and $(\det C(x))^{-1/2}$ are well-defined continuous functions of x and $(\det A(x))^{-1/2} = (\det C(x))^{-1/2}$ if $x_j - x_{j+1} \in \mathbb{R}^4 \backslash V^{\ell'}$. Denote

$$\mathcal{W}_\alpha^r(z_1, \ldots, z_n) = (\det A(z))^{-1/2} \mathcal{W}_{0,\alpha}^r(z_1, \ldots, z_n)$$

and

$$\mathcal{W}_\alpha^{r,j}(z) = \mathcal{W}_{\alpha'}^{r'}(z'), z' = (z_1, \ldots, z_{j+1}, z_j, \ldots, z_n),$$

$$r' = (r_1, \ldots, r_{j+1}, r_j, \ldots, _n), \alpha' = (\alpha_1, \ldots, \alpha_{j+1}, \alpha_j, \ldots, \alpha_n).$$

Then, by deforming the path $\Gamma_j \times \Gamma_{j+1}$ in Eq. (8.6) into $G_{j,j+1}$, we can write

$$\mathcal{W}^r_\alpha(f) + \mathcal{W}^{r,j}_\alpha(f) = \int_{G_{j,j+1}\prod_{i\neq j,j+1}\Gamma_i} \mathcal{W}^r_\alpha(z) f(z) dz + \int_{G_{j+1,j}\prod_{i\neq j,j+1}\Gamma_i} \mathcal{W}^{r,j}_\alpha(z) f(z) dz,$$

where $y^0_j = y^0_{j+1}$ and

$$G_{j,j+1} = \{(x^0_j + iy^0_j - i\epsilon(x_j - x_{j+1}), \boldsymbol{x}_j, x^0_{j+1} + iy^0_{j+1}, \boldsymbol{x}_{j+1}); (x_j, x_{j+1}) \in \mathbb{R}^{2\cdot 4}\},$$

$$G_{j+1,j} = \{(x^0_j + iy^0_j, \boldsymbol{x}_j, x^0_{j+1} + iy^0_{j+1} - i\epsilon(x_{j+1} - x_j), \boldsymbol{x}_{j+1}); (x_j, x_{j+1}) \in \mathbb{R}^{2\cdot 4}\}.$$

Since $\mathcal{W}^r_\alpha(z) + \mathcal{W}^{r,j}_\alpha(z) = 0$ for $x_j - x_{j+1} \in \mathbb{R}^4 \backslash V^{\ell'}$,

$$\mathcal{W}^r_\alpha(f) + \mathcal{W}^{r,j}_\alpha(f) = \int_{G^{\ell'}_{j,j+1}\prod_{i\neq j,j+1}\Gamma_i} \mathcal{W}^r_\alpha(z) f(z) dz + \int_{G^{\ell'}_{j+1,j}\prod_{i\neq j,j+1}\Gamma_i} \mathcal{W}^{r,j}_\alpha(z) f(z) dz,$$

where

$$G^{\ell'}_{j,j+1} = \{(x^0_j + iy^0_j - i\epsilon(x_j - x_{j+1}), \boldsymbol{x}_j, x^0_{j+1} + iy^0_{j+1}, \boldsymbol{x}_{j+1}); (x_j, x_{j+1}) \in \mathbb{R}^{2\cdot 4} \cap V^{\ell'}\},$$

$$G^{\ell'}_{j+1,j} = \{(x^0_j + iy^0_j, \boldsymbol{x}_j, x^0_{j+1} + iy^0_{j+1} - i\epsilon(x_{j+1} - x_j), \boldsymbol{x}_{j+1}); (x_j, x_{j+1}) \in \mathbb{R}^{2\cdot 4} \cap V^{\ell'}\}.$$

Since $G^{\ell'}_{j,j+1}\prod_{i\neq j,j+1}\Gamma_i, G^{\ell'}_{j+1,j}\prod_{i\neq j,j+1}\Gamma_i \subset W^{\ell'}_j$, this shows that

$$\mathcal{T}(W^{\ell'}_j) \ni f \to \mathcal{W}^r_\alpha(f) + \mathcal{W}^{r,j}_\alpha(f) \in \mathbb{C}$$

is continuous and satisfies the condition b) of **R6**. □

Remark 8.4. In our previous paper [Brüning & Nagamachi (2004)], we defined a complex neighbourhood $V^{\ell'}$ by

$$V^{\ell'} = \{z \in \mathbb{C}^4; \exists x \in V; |\mathrm{Re}\, z - x| + |\mathrm{Im}\, z|_1 < \ell'\}. \tag{8.7}$$

But we found that to treat the present model, the neighbourhood (8.1) is convenient, and by this change of the ℓ'-neighbourhood of V, our theory [Brüning & Nagamachi (2004)] is not affected.

9. Conclusion

In the space of operator-valued tempered ultrahyperfunctions we have solved Heisenberg's linearized equation (1.2). This equation contains a parameter l which has the dimension of a length and it has been found that this parameter l is proportional to the fundamental length ℓ of our recently developed relativistic quantum field theory with a fundamental length. We found (see (5.11), (8.4) and Proposition 8.3)

$$\ell = l/(\sqrt{2}\pi).$$

The use of tempered ultrahyperfunctions was unavoidable, at least in our solution strategy. In this sense we conclude that Heisenberg's linearized equation only has a solution in the framework of relativistic quantum field theory with a fundamental length, not in the framework of ordinary relativistic quantum field theory.

10. References

Bethe, H. (1947). The electromagnetic shift of energy levels, *Phys. Rev.* 72: 339.

Born, M., Heisenberg, W. & Jordan, P. (1926). Zur Quantenmechnik, II, *Z. Phys.* 35: 557.

Brüning, E. & Nagamachi, S. (2001). Kernel theorem for Fourier hyperfunction, *J. Math. Univ. Tokushima* .

Brüning, E. & Nagamachi, S. (2004). Relativistic quantum field theory with a fundamental length, *J. Math. Phys.* 45: 2199–2231.

Brüning, E. & Nagamachi, S. (2008). Solution of a linearized model of Heisenber's fundamental equation II, *J. Math. Phys.* 49: 052304–1 – 052304–22.

Dirac, P. (1927). The quantum theory of emission and absorption of radiation, *Proc. Roy. Soc.* 114: 243.

Dirac, P. (1928). The quantum theory of the electron, *Proc. Roy. Soc.* 117: 610.

Dirac, P. (1931). Quantized singularities in the electromagnetic field, *Proc. Roy. Soc.* 133: 60.

Dürr, H.-P., Heisenberg, W., Mitter, H., Schlieder, S. & Yamazaki, K. (1959). Zur Theorie der Elementarteilchen, *Z. Naturf.* 14a: 441–485.

Dyson, F. (1949). The radiation theories of Tomonaga, Schwinger and Feynman, *Phys. Rev.* 75: 486, 1736.

Feynman, R. (1948). A relativistic cut-off for classical electrodynamics, *Phys. Rev.* 74: 939, 1430.

Gross, D. & Wilczek, F. (1973). Ultraviolet Behavior of Non-Abelian Gauge Theories, *Phys. Rev. Lett.* 30: 1343–1346.

Hasumi, M. (1961). Note on the n-dimensional tempered ultra-distributions, *Tohoku Math. J.* 13: 94–104.

Heisenberg, W. (1925). Über die quantentheoretische Umdeutung kinematischer und mechnischer Beziehungen, *Zeitsch. f. Phys.* 33: 879.

Heisenberg, W. (1966). *Introduction to the Unified Field Theory of Elementary Particles*, John Wiley & Sons.

Hörmander, L. (1983). *The analysis of linear partial differential operators I*, Vol. 256 of *Grundlehren der mathematischen Wissenschaften*, Springer-Verlag, Berlin Heidelberg New York Tokyo.

Ito, Y. (1988). Fourier hyperfunctions of general type, *Journal of Mathematics of Kyoto University* 28(2): 213–265.

Jaffe, A. (1965). Entire functions of the free field, *Ann. Phys.* 32: 127–156.

Morimoto, M. (1970). Sur les ultradistributions cohomologiques, *Ann. Inst. Fourier* 19: 129–153.

Morimoto, M. (1975a). Convolutors for ultrahyperfunctions, *International Symposium on Mathematical Problems in Theoretical Physics*, Vol. 39 of *Lecture Notes in Phys.*, Springer, Berlin, pp. 49–54.

Morimoto, M. (1975b). Theory of Tempered Ultrahyyperfunctions I, *Proc. Japan Acad.* 51: 87–81.

Nagamachi, S. (1981a). The theory of vector valued Fourier hyperfunctions of mixed type. I, *Publ. RIMS, Kyoto Univ.* 17: 25–63.

Nagamachi, S. (1981b). The theory of vector valued Fourier hyperfunctions of mixed type. II, *Publ. RIMS, Kyoto Univ.* 17: 65–93.

Nagamachi, S. & Brüning, E. (2003). Localization in Quantum Field Theory, *in* J. Arafune (ed.), *A Garden of Quanta*, World Scientific, Singapore.

Nagamachi, S. & Brüning, E. (2008). Solutions of a linearized model of Heisenberg's fundamental equation I, arXiv:0804.1663 [math-ph].

Nagamachi, S. & Brüning, E. (2010). Frame independence of the fundamental length in relativistic quantum field theory, *J. Math. Phys.* 51: 022305–1 – 022305–18.

Nagamachi, S. & Mugibayashi, N. (1986). Hyperfunctions and renormalization, *J. Math. Phys.* 27: 832–839.

Nishimura, T. & Nagamachi, S. (1990). On supports of Fourier hyperfunctions, *Math. Japonica* 35: 293–313.

Okubo, S. (1954). Note on the Second Kind Interaction, *Progr. Theor. Phys.* 11: 80–94.

Okubo, S. (1961). Green's Function in Some Covariant Soluble Problems, *Nuovo Cimeno* 19: 574–585.

Planck, M. (1900). Zur Theorie des Gesetzes der Energieverteilung in Normalspektrum, *Verh. Deutsch. Phys. Ges.* 2: 237.

Polchinski, J. (1998). *String theory I*, Cambridge University Press, Cambridge.

Politzer, H. (1973). Reliable Perturbative Results for Strong Interactions?, *Phys. Rev. Lett.* 30: 1346–1349.

Schrödinger, E. (1926). Quantisierung als Eigenwertproblem (Dritte Mitteilung), *Ann. d. Phys.* 80: 437.

Schwinger, J. (1948). Quantum electrodynamics, I, *Phys. Rev.* 74: 1439.

't Hooft, G. (1971). Renormalization of massless Yang-Mills fields, *Nucl. Phys.* B33: 173–199.

Tomonaga, S. (1946). On a relativistically invariant formulation of the quantum theory of wave fields I, *Prog. Theoret. Phys.* 1: 27.

Treves, F. (1967). *Topological Vector Spaces, Distributions and Kernels*, Pure and Applied Mathematics, Academic Press, New York London.

Wightman, A. S. & Gårding, L. (1964). Fields as operator-valued distributions in relativistic quantum theory., *Arkiv för Fysik* 28: 129–184.

8

Topological Singularity of Fermion Current in Abelian Gauge Theory

Ai-Dong Bao and Shi-Shu Wu

Department of Physics Northeast Normal University Chang Chun

Center for Theoretical Physics Department of Physics Jilin University Chang Chun

People's Republic of China

1. Introduction

Since calculations in a quantized theory are often plagued by divergencies , we have to impose a regularization scheme in order to eliminate the singularity from ill-field functions such as the infrared-divergence problem in quantum electrodynamics[1,2]. Among these divergence phenomena, a type of divergence appears in product of quantum operators taken at the same time. The product of the local operators is often singular, which can destroy symmetries of conservation currents and classical equations of motions. For instance, in derivation of the conservation equation for the axial vector current, the differential equation of motion for fermion fields can not be used to reduce fermion-photon vertex in quantum electrodynamics in the most straightforward way[3]. Through standard perturbative treatments of singularity in the operator product of fermion current, a closer look at the manipulations reveals some subtleties. These unavoidable divergences reflect the feature that the symmetry of the classical theory may be ruined in quantum theory by quantum anomaly. Usually, we deal with the divergence of the anomaly by examining the "triangle graph", which consist of an internal fermion loop connected to two vector fields and to one axial vector field. These triangle graphs give rise to anomaly in orders of perturbation theory, which arise from momentum-space integrals. In consequence, the anomaly is affected by interactions of gauge bosons attached to the fermion loop[4,5].

From the path-integral viewpoint, the anomaly term associated with the axial-vector current can be understood in the path-integral formulation of quantum gauge theory as a consequence of the fact that the functional measure is not invariant with respect to the relevant local group transformations on the fields[6]. The chiral anomaly responds to local group transformations but vanishes for a class of space-independent transformations. One discovers that the anomaly has a topological character given by the index of a Dirac operator in a gauge background[1]. The mathematical explanation of such a anomaly is directly related to Atiyah-Singer index theorem. The theorem relates the number of chirality zero modes of Dirac operator to the topological charge of the gauge field. Furthermore, a number of papers have addressed the relationships among chiral anomalies,the geometry of the space of vector potentials and families of Dirac operators[7,8,9]. The authors showed that the first characteristic class of the index bundle for the Dirac operator is related to anomalies. In particular, in the gravitation case,the

Abelian and non-Abelian chiral anomalies in $2n$ dimensional spacetime is determined by a variation of gauged effective action in differential geometric form which satisfies the Wess-Zumino consistency condition. Obviously, the use of the family's index theorem in the study of gravitational anomalies has the advantage over using Feynman diagram methods, which enables us to obtain the anomalies without having to calculate Feynman diagrams[10]. This implies that singularity in fermion currents connects to the non-perturbative effects of anomalies in the presences of topologically non-trivial field configurations. This sort of problem also implies the index of Dirac operator relates to quantization of the space integral of the anomaly ,and the supertrace of the kernel function of the square of a Dirac operator is the quantization of the topological character of the corresponding connection on a manifold[11]. From this point of view, it is natural how to find a means to distinguish the divergence of various fermion currents coupling with gauge field becomes extremely important. In this paper we present an exposition that the topological singularity of operator product of various fermion current in gauge theory can be described in terms of Atiyah-Segal-Singer index theorem for Dirac operator.

2. Transformation property of fermion current

Now we are in position to specify a transformation of operator product of fermion current in Abelian quantum field to determine the property of the current. According to the form of fermion current coupling with gauge field $B_\mu(x)$ in Dirac equation, the fermion current coupling with gauge field is defined as

$$I^{[\mu\nu]} = \bar{\psi}(x)\gamma^\mu \Gamma^{[\nu]}\psi(x) \tag{1}$$

where $\Gamma^{[\nu]}$ indicates the corresponding Dirac matrix, which is also an element of Dirac algebra, $\psi(x)$ and $\bar{\psi}(x)$ is Dirac spinors.

One easily verifies that these fermion currents possess bilinear covariant property under Lorentz transformation. Some of the "products of currents" itself have quantum field theoretical meaning, which is subject to infinitesimal change of varibles along the symmetry direction. To do this, in the broad sense, a set of the changing variables of matter field in quantum theory can be transformed as [3]

$$\begin{aligned} \psi'_\alpha(x) &= \psi_\alpha(x) + \delta\psi_\alpha(x) \\ \delta\psi_\alpha(x) &= \theta_{[\nu]}(x) H_\alpha^{[\nu]}(\psi_\beta, \psi_{\beta\lambda}) + \partial_\mu \theta_{[\nu]}(x) h_\alpha^{\mu[\nu]}(\psi_\beta, \psi_{\beta\lambda}) \end{aligned} \tag{2}$$

where the matrices $H_\alpha^{[\nu]}, h_\alpha^{\mu[\nu]}$ are the functions composed of both Dirac matrix and field variables, the group parameter $\theta_{[\nu]}(x)$ is a real function.

In the light of Eq,(2), the infinitesimal local transformation of fermion fields in Abelian gauge theory is taken tentatively as [12]

$$\begin{aligned} \psi'(x) &= e^{-i\theta_{[\nu]}(x)\Gamma^{[\nu]}}\psi(x) \\ \bar{\psi}'(x) &= \bar{\psi}(x)\gamma^0 e^{i\theta_{[\nu]}(x)\Gamma^{[\nu]+}}\gamma^0 \end{aligned} \tag{3}$$

In fact, there is no contradiction between Eq.(2) and Eq.(3). In the case of QED, the origin of the Wald-Takahashi identities relative to femion currents lies in the gauge invariance of the generating functional of QED. We know that the generating functional $Z\left(B_{\mu}(x),\bar{\eta},\eta\right)$ remains the same. This is because changing variables in an integral never affects its value. In this sense, we can consider a gauge transformation to be more general type of field redefinition.This is the transfornation Eq.(2) as a guiding rule.

Thus, to discuss the property of the fermion current, we need to choose a comparatively simple transformation on the fields for fermion currents. This is transformation Eq.(3), in which the term coupling with the spacetime derivative of $\theta_{[v]}(x)$ is not included.

To expand the fermion current carefully, let us introduce a complete set of eigenstates ϕ_n of the Dirac operator $iD = i\gamma^{\mu}\left(\partial_{\mu} - igB_{\mu}(x)\right)$ as follows

$$iD\phi_n(x) = \lambda_n\phi_n(x) \tag{4}$$

here real λ_n is the discrete eigenvalues. This means the fermion fields can be decomposed in this set of eigenstates

$$\begin{aligned} \psi'(x) &= \sum_n c_n'\phi_n(x) \\ \bar{\psi}'(x) &= \sum_n \bar{c}_n'^{+}\phi_n(x) \\ \delta_{nm} &= \int d^4x\phi_n^{+}\phi_m \qquad (1 \le n,m \le d) \end{aligned} \tag{5}$$

where d stands for the eigenstate space dimension. It is true that the anticommuting coefficients $c_n', \bar{c}_n'$ of Dirac spins $\psi'(x)$ and $\bar{\psi}'(x)$ span a Grassmann algebra.

Thus under the transformation Eq.(3), the fermion current Eq.(1) is changed into

$$\begin{aligned} I^{[\mu v]'} &= \bar{\psi}'(x)\gamma^{\mu}\Gamma^{[v]}\psi'(x) \\ &= \sum_{nm} \bar{c}_n' c_m' \phi_n^{+}\gamma^{\mu}\Gamma^{[v]}\phi_m \end{aligned} \tag{6}$$

To find the relation between the fermion current and the corresponding integral measure, let's construct the power function $I^{[\mu v]d}$ of the fermion current over $\psi(x)$ and $\bar{\psi}(x)$:

$$\begin{aligned} I^{[\mu v]d} &= \left[\bar{\psi}(x)\gamma^{\mu}\Gamma^{[v]}\psi(x)\right]^{d} \\ &= \sum_{n_1m_1} \bar{c}_{n_1}c_{m_1}\phi_{n_1}^{+}\gamma^{\mu}\Gamma^{[v]}\phi_{m_1} \sum_{n_2m_2} \bar{c}_{n_2}c_{m_2}\phi_{n_2}^{+}\gamma^{\mu}\Gamma^{[v]}\phi_{m_2} \cdots \sum_{n_dm_d} \bar{c}_{n_d}c_{m_d}\phi_{n_d}^{+}\gamma^{\mu}\Gamma^{[v]}\phi_{m_d} \end{aligned} \tag{7}$$

where d denotes power of the operator function $I^{[\mu v]}(x)$, its value is equal to the dimension of the eigenspace of the regulation operator(see Eq.(4)). In like manner, due to the transformation Eq.(3), the power function $I^{[\mu v]'}(x)$ over $\psi'(x)$ and $\bar{\psi}'(x)$ can be expressed as

$$I^{[\mu\nu]'d}(x)=\left[\bar{\psi}'(x)\gamma^{\mu}\Gamma^{[\nu]}\psi'(x)\right]^{d}$$
$$=\sum_{n_1m_1}\bar{c}'_{n_1}c'_{m_1}\phi^{+}_{n_1}\gamma^{\mu}\Gamma^{[\nu]}\phi_{m_1}\sum_{n_2m_2}\bar{c}'_{n_2}c'_{n_2}\phi^{+}_{n_2}\gamma^{\mu}\Gamma^{[\nu]}\phi_{m_2}\cdots\sum_{n_dm_d}\bar{c}'_{n_d}c'_{m_d}\phi^{+}_{n_d}\gamma^{\mu}\Gamma^{[\nu]}\phi_{m_d} \tag{8}$$

Obviously in the light of the property of the Grassmann algebra, the expression (8) implies that $I^{[\mu\nu]'d}(x)$ links with the operator product of the expansion coefficients $c'_n,\bar{c}'_n$.

In order to understand the meaning of the operator product composed of c'_n and $\bar{c}'_n$ in the expression $I^{[\mu\nu]'d}(x)$, we now recall the algebraic property of functional measure in the path integral formulation. Under the transformation Eq.(3), the corresponding functional measure $d\mu'_{[\nu]}$ over $\psi'(x)$ and $\bar{\psi}'(x)$ has the transformation property

$$\begin{aligned} d\mu'_{[\nu]} &= \prod_{n} d\bar{c}'_n \prod_{m} dc'_m \\ &= J_{[\nu]}d\mu \end{aligned} \tag{9}$$

where $d\mu'$ is the functional measure over $\psi'(x)$ and $\bar{\psi}'(x)$, $J_{[\nu]}$ is the Jacobian determinant of the corresponding transformation of the fermion variables, which is a key quantity for anomaly current. Then we see that the Jacobian of differential operator $d\mu$ connects with the transformation of the operator function μ, which is defined as the operator product of the anticommuting coefficient $c_n,\bar{c}_m$ of $\psi(x)$ and $\bar{\psi}(x)$

$$\mu=\prod_{n}\bar{c}_n\prod_{m}c_m \tag{10}$$

After the transformation Eq.(3), the expression of the Grassmann operator function μ is changed explicitly into

$$\begin{aligned} \mu'_{[\nu]} &= \prod_{n}\bar{c}'_n\prod_{m}c'_m \\ &= \prod_{n}\sum_{m}\int d^4x\phi^{+}_n(x)e^{-i\theta_{[\nu]}(x)\Gamma^{[\nu]}}\phi_m\bar{c}_m\prod_{l}\sum_{s}\int d^4x\phi^{+}_l e^{i\theta_{[\nu]}(x)\Gamma^{[\nu]}}\phi_s c_s \\ &= \tilde{J}_{[\nu]}\mu \qquad (1\le n,m,l,s\le d) \end{aligned} \tag{11}$$

where $\tilde{J}_{[\nu]}$ is the Jacobian determinant of the transformations Eq.(3).

From the properties of the Grassmann algebra we find the identity

$$\tilde{J}_{[\nu]}=J^{-1}_{[\nu]} \tag{12}$$

It shows that the Jacobian $\tilde{J}_{[\nu]}$ corresponding to the operator product μ connects with the inverse of the Jacobian of the corresponding transformation of the integral measure.

Having clarified the relation between the operator product of fermion current and the integral measure, we therefore see that the power function of fermion current Eq.(8) can be rewritten as

$$\begin{aligned}
I^{[\mu\nu]'d}(x) &= \left[\bar{\psi}'(x)\gamma^{\mu}\Gamma^{[\nu]}\psi'(x)\right]^{d} \\
&= \sum_{n_1 m_1}\bar{c}'_{n_1}c'_{m_1}\phi^{+}_{n_1}\gamma^{\mu}\Gamma^{[\nu]}\phi_{m_1}\sum_{n_2 m_2}\bar{c}'_{n_2}c'_{n_2}\phi^{+}_{n_2}\gamma^{\mu}\Gamma^{[\nu]}\phi_{m_2}\cdots\sum_{n_d m_d}\bar{c}'_{n_d}c'_{m_d}\phi^{+}_{n_d}\gamma^{\mu}\Gamma^{[\nu]}\phi_{m_d} \\
&= \sum_{\substack{n_1\cdots n_d \\ m_1\cdots m_d}}\bar{c}'_{n_1}c'_{m_1}\bar{c}'_{n_2}c'_{n_2}\bar{c}'_{n_d}c'_{m_d}\phi^{+}_{n_1}\gamma^{\mu}\Gamma^{[\nu]}\phi_{m_1}\phi^{+}_{n_2}\gamma^{\mu}\Gamma^{[\nu]}\phi_{m_2}\cdots\phi^{+}_{n_d}\gamma^{\mu}\Gamma^{[\nu]}\phi_{m_d} \\
&= \mu'_{[\nu]}\sum_{\substack{n_1\cdots n_{dD} \\ m_1\cdots m_d}}\phi^{+}_{n_1}\gamma^{\mu}\Gamma^{[\nu]}\phi_{m_1}\phi^{+}_{n_2}\gamma^{\mu}\Gamma^{[\nu]}\phi_{m_2}\cdots\phi^{+}_{n_{dd}}\gamma^{\mu}\Gamma^{[\nu]}\phi_{m_d} \\
&= J^{-1}_{[\nu]}\mu\sum_{\substack{n_1\cdots n_d \\ m_1\cdots m_d}}\phi^{+}_{n_1}\gamma^{\mu}\Gamma^{[\nu]}\phi_{m_1}\phi^{+}_{n_2}\gamma^{\mu}\Gamma^{[\nu]}\phi_{m_2}\cdots\phi^{+}_{n_d}\gamma^{\mu}\Gamma^{[\nu]}\phi_{m_d} \\
&= J^{-1}_{[\nu]}I^{[\mu\nu]d}(x)
\end{aligned} \tag{13}$$

This is the desired result, which is just the transformation identity of fermion current. The expression illustrates that the property of fermion current is presented in its functional Jacobian through the operator product function μ. In other words, the mathematical property of fermion current can be characterized by the nature of Jacobian of functional integral measure due to transformation of fermion field. For instance, the process of regularization of Jacobian in the functional measure gives rise to the anomaly for the axial vector current. Its topological property of the anomaly term is described by Atiyah-Singer index theorem for Dirac operator.

3. Topological property of fermion current

From the topological viewpoint, the non-perturbative effect of the Abelian anomaly associated with Ward-Takahashi type's identity for axial vector current is related to the topological character in the presence of topologically non-trivial field configuration [14,15]. The topological exposition of the quantum anomaly for the current is addressed by Atiyah-Singer index theorem in a gauge background .The Abelian anomaly term can be derived by using path integral formulation in Euclidean spacetime E^4. Of particular interest is applications of the index theorem for Dirac operator, because it can be used to discuss the topological property of fermion current.

First restricting ourselves to the axial-vector current case, we calculate the analytical index and topological index for Dirac operator. For the following discussion, think of Euclidean spacetime E^4 as a four-dimensional space-time. The spacetime manifold E^4 is compacted into 4-dimensional sphere manifold S^4.

If E is a vector bundle on S^4, let $\Gamma\left(S^4,E\right)$ be the space of smooth section of E. Thus In terms of γ^5, the Dirac spin space ϕ is decomposed into two subspace $\phi_+\left(E^+\right)$ and $\phi_-\left(E^-\right)$, in which the eigenvectors of Dirac operator can be chosen eigenvectors of definite chirality with eigenvalue ± 1. That is

$$iD\phi_n = 0\,,\quad \gamma^5\phi_n = \pm\phi_n \tag{14}$$

Naturally the analytical index definition $Ind_a D$ of D to be integer is defined as [11,16,17]

$$\begin{aligned} Ind_a D &= \dim\left(Ker\left(D^+\right)\right) - \dim\left(Ker\left(D^-\right)\right) \\ &= n_+ - n_- \end{aligned} \tag{15}$$

where $D^\pm$ is the restriction of D to $\Gamma\left(S^4, E^\pm\right)$, $\dim\left(Ker\left(D^+\right)\right)$ is the dimension of the kernel $Ker\left(D^+\right)$, here $n_\pm$ are the numbers of zero modes of the Dirac operator.

In the mean time, in explicitly performing the regulariaztion for anomaly function, we find a local expression giving the analytical index of the Dirac operator

$$\begin{aligned} Ind_a D &= \int d^4x \sum_n \phi_n^+(x) \Gamma^{[5]} \phi_n(x) \\ &= \int d^4x A^{[5]}(x) \\ &= -\frac{1}{32\pi^2} \int d^4x \varepsilon^{\mu\nu\rho\sigma} Trace\left(F_{\mu\nu} F_{\rho\sigma}\right) \\ &= -\frac{4}{32\pi^2} \int d^4x Trace \partial_\mu \left(\varepsilon^{\mu\nu\rho\sigma} B_\nu \partial_\rho B_\sigma\right) \end{aligned} \tag{16}$$

with "trace" here denoting a trace only over indices labeling the various fermion species, which could consider as a regularization of the trace in function space. $F_{\mu\nu}$ is the field strength, and $A^{[5]}(x)$ is the anomaly function.

This formula shows that the index of D is given in terms of the restriction to the diagonal of the kernel of D^2 at arbitrarily large masses, which we know by the asymptotic expansion of the kernel to be given by local formula in the curvature of the connection. In the above equation, the following identity is used

$$\varepsilon^{\mu\nu\rho\sigma}\left(F_{\mu\nu} F_{\rho\sigma}\right) = 4\partial_\mu\left(\varepsilon^{\mu\nu\rho\sigma} B_\nu \partial_\rho B_\sigma\right) \tag{17}$$

The fact reveals the important information that the integral of the local anomaly function $A^{[5]}(x)$ can not change smoothly under variations in the gauge field. Since the anomaly function in Eq.(16) can be written as a total derivative, the space integral of the anomaly function depends only the behavior of the gauge field at the boundaries. That is, there exists a number of singularities of the E vector bundle on the manifold.

Since D is an elliptic operator, the Atiyah-Singer local index theorem gives a formula for the topological index $Ind_t D$ of D in terms of the Chern character of ChE and the $\hat{A}$ -genus of E^4 [18,19]

$$Ind_a D = Ind_t D \tag{18}$$

So that [11]

$$Ind_t D = \int_M \hat{A}(M) Ch(E) \tag{19}$$

Here the more natural definition of $\hat{A}$ -genus form with respect the Riemannian curvature R is

$$\hat{A}(M)=\det{}^{1/2}\left(\frac{R/2}{\sinh\left(R/2\right)}\right) \tag{20}$$

and the Chern character form is

$$Ch(E)=Str\left(e^{-\nabla^2}\right) \tag{21}$$

Substituting this and Eq.(20) into Eq.(19) yields

$$Ind_t D=-\frac{4}{32\pi^2}\int d^4x Trace\partial_\mu\left(\varepsilon^{\mu\nu\rho\sigma}B_\nu\partial_\rho B_\sigma\right) \tag{22}$$

Obviously, the quantity on the right of the Eq.(22) is kown as Chern-Pontrjagin term.

Since the anomaly function $A^{[5]}(x)$ does not vanish in the Jacobian determinant J, we have the explicit identity

$$Ind_a D=\int d^4x\frac{1}{2i}\frac{\delta J}{\delta\theta}\bigg|_{\theta=0} \tag{23}$$

The gauge group parameter $\theta_{[\nu]}(x)$ is independent of the topological index.

The Abelian anomaly arisen from axial vector current is a local quantity, because it is a consequence of short distance singularities, which does violate the chiral symmetry. That is, the topological singularity of operator product of fermion current Eq.(13) is presented by Jacobian of the measure due to the transformation of fermion fields.

Now we turn to calculations of anomaly function in the transformation of measure for other fermion currents. Due to the Eq.(3),the anomaly function $A^{[\nu]}(x)$ corresponding to the defined fermion currents Eq.(1) is given by the formula(see appendix)

$$\begin{aligned}A^{[\nu]}(x)&=\sum_n\phi_n^{\ +}(x)\Gamma^{[\nu]}\phi_n(x)\\&=\langle x|\Gamma^{[\nu]}|x\rangle\end{aligned} \tag{24}$$

In virtue of path -integral technique, the anomaly emerging from the functional measure for the fermions arises from regularization of the covariant derivative acting on the fermions. Thus the anomaly function becomes

$$A^{[\nu]}(x)=\lim_{M\to\infty}\langle x|\Gamma^{[\nu]}f\left(\frac{-D^2}{M^2}\right)|x\rangle \tag{25}$$

with regulator $f\left(\frac{-D^2}{M^2}\right)=e^{-\frac{D^2}{M^2}}$.

We noticed that the quantity $\langle x|\Gamma^{[\nu]}f\left(\frac{-D^2}{M^2}\right)|x\rangle$ is a kernel function of Dirac operator, which is also a section of the endorphisms $End(E)$ on Euclidean manifold E^4.

In analogy to the axial-vector current case, we have a generalization of the local index theorem for D which enable us to calculate the $\langle x|\Gamma^{[\nu]}f\left(\frac{-D^2}{M^2}\right)|x\rangle$ for arbitrary $\Gamma^{[\nu]}$ .Here the element $\Gamma^{[\nu]}$ (Dirac matrix) lies in a compact group,which is a representation of Clifford algebra.

According to the local Atiyah-Segal-Singer index theorem, the existence of an asymptotic expansion for $\langle x|\Gamma^{[\nu]}f\left(\frac{-D^2}{M^2}\right)|x\rangle$ at $M\to\infty$ is presented by the index $Ind\left(\Gamma^{[\nu]},D\right)$ of Dirac operator D on the manifold as follows

$$\begin{aligned} Ind\left(\Gamma^{[\nu]},D\right) &= \lim_{M\to\infty}\int d^4x\langle x|\Gamma^{[\nu]}f\left(\frac{-D^2}{M^2}\right)|x\rangle \\ &= \int d^4xA^{[\nu]}(x) \end{aligned} \tag{26}$$

Here the kernel function in Eq.(26) is no other than the anomaly function for Jacobian of functional measure.

This is what we do. It is shown that the index theorem for the operator $\Gamma^{[\nu]}e^{-D^2/M^2}$ associated to the square of Dirac operator D is a statement about the relationship between the kernel and the topological character of the associated connection in gauge theory.

Fortunately, the asymptotic expansion of the kernel of the operator $\Gamma^{[\nu]}e^{-D^2/M^2}$ on the right of the Eq.(26) can evaluated by Fujikawa's method(see appendix) for these fermion currents.

The computation shows that anomaly functions for many fermion currents vanish. In other words, there is no topological singularity for these currents.

4. An Example of Abelian gauge theory

As one see from the following quantum electrodynamics case, the above argument provides a approach to examine the topological singularity in the operator products of the fermion current. Let us consider an Abelian gauge field to show our argument. The Lagrangian density based on the minimal coupling ansatz for quantum electrodynamics is the following [4]

$$L_{eff} = \bar{\psi}(x)\gamma^\mu\left(\partial_\mu - igB_\mu(x)\right)\psi(x) - \bar{\psi}(x)m\psi(x) - \frac{1}{4}F^{\mu\nu}F_{\mu\nu} - \frac{1}{2\xi}\left(\partial^\mu B_\mu\right)^2 \tag{27}$$

where g and m denote, respectively, the charge and mass of the electron. In this case, the gauge field is just the photon field $B_\mu(x)$, ξ is the gauge constant.

Then the generating functional of QED is given by

$$Z\left(J_{\mu},\overline{\eta},\eta\right)=\int D\left(B_{\mu},\overline{\psi},\psi\right)e^{i\int d^4x\left[L+B_{\mu}J^{\mu}+\overline{\eta}\psi+\overline{\psi}\eta\right]} \tag{28}$$

Noted that the path integral method allows us to generalize WT identies relative to fermion currents, because the functional $Z\left(B_{\mu},\overline{\eta},\eta\right)$ is gauge invariant under gauge tranformations. This means that the fact holds irrespecttive whether these re-naming field variables take the form of symmetry of action, or not. Therefore, this implicitly expect that a anomaly might take place at the quantum measure, which is the failure of a class symmetry to survive the process of quantization and regularization. Normally, under gauge transformation, the determinant in the measure transformation is descarded because it appears to be a constant. However, closer analysis of this term shows that it is actually divergent and here requires regularization.

Following Fujikawa's prescription[6,20], the regularization procedure for the variation of the integral measure can provide access to a wider class of such anomaly objects. In order to analyze the topology property of various fermion current, the anomaly functions $A^{[\nu]}(x)$ corresponding to Dirac matrix $\Gamma^{[\nu]}$ in the transformation Eq.(3) can be written as the limit of a manifestly convergent integral[14]

$$A^{[\nu]}(x)=\lim_{M\to\infty}\int\frac{d^4k}{(2\pi)^4}e^{-ikx}\Gamma^{[\nu]}f\left(\frac{(iD)^2}{M^2}\right)e^{ikx} \tag{29}$$

$$\overline{A}^{[\nu]}(x)=\lim_{M\to\infty}\int\frac{d^4k}{(2\pi)^4}e^{-ikx}\gamma^0\Gamma^{[\nu]+}\gamma^0 f\left(\frac{(iD)^2}{M^2}\right)e^{ikx} \tag{30}$$

Thus the Jacobian $J_{[\nu]}$ of measure can be put in the form

$$J_{[\nu]}=e^{i\int d^4xA^{[\nu]}(x)\theta_{[\nu]}(x)}e^{-i\int d^4x\overline{A}^{[\nu]}(x)\theta_{[\nu]}(x)} \tag{31}$$

The corresponding Jacobian $J_{[\nu]}$ is evaluated below for various fermion currents

i. $\Gamma^{[1]}=I \qquad (unit-matrix);$

$$A^{[1]}(x)=\frac{-i}{2^9\pi^2}tr\left(F_{\mu\nu}F_{\lambda\rho}\right)\left(g_{\mu\rho}g_{\nu\lambda}-g_{\mu\lambda}g_{\nu\rho}\right)\ \text{and}\ J_{[1]}=1. \tag{32}$$

ii. $\Gamma^{[5]}=\gamma^{[5]};\ A^{[5]}(x)=\frac{i}{16\pi^2}tr\left(F_{\mu\nu}\tilde{F}_{\rho\sigma}\right)$ with $\tilde{F}_{\rho\sigma}(x)=-\frac{i}{2}\varepsilon_{\mu\nu\rho\sigma}F_{\mu\nu}$;

$$J_{[5]}=e^{-\int d^4x\theta_5(x)\left(\frac{i}{16\pi^2}tr\left(F_{\mu\nu}\tilde{F}_{\rho\sigma}\right)\right)} \tag{33}$$

iii. $\Gamma^{[\nu]} = \gamma^{\mu}$;

$$A^{[\mu]}(x) = 0 \text{ and } J_{[\mu]} = 1. \tag{34}$$

iv. $\Gamma^{[\mu 5]} = \gamma^{\mu}\gamma_5$

$$A^{[\mu 5]}(x) = 0 \text{ and } J_{[\mu 5]} = 1. \tag{35}$$

v. $\Gamma^{[\lambda\mu\nu 5]} = \varepsilon^{\lambda\mu\nu\rho}\gamma_{\rho}\gamma_5$;

$$A^{[\lambda\mu\nu 5]}(x) = 0 \text{ and } J_{[\lambda\mu\nu]} = 1. \tag{36}$$

vi. $\Gamma^{[\mu\nu]} = \sigma^{\mu\nu}$; $A^{[\mu\nu]}(x) = 0$

$$J_{[\mu\nu]} = 1 \tag{37}$$

vii. $\Gamma^{[\mu\nu 5]} = \sigma^{\mu\nu}\gamma_5 \quad A^{[\mu\nu 5]}(x) = -\overline{A}^{[\mu\nu 5]}(x)$

$$J_{[\mu\nu 5]} = 1 \tag{38}$$

The above computation shows that the anomaly function in Jacobian vanishes for many fermion currents,(in other words, they cancel each other in regularization).

For simplicity, we take a two-dimensional eigenspace $(d = 2)$ of the regulation operator $e^{-\frac{D^2}{M^2}}$ as an following example to illustrate in detail the link between the Jacobian of the integral measure and the topological property of the fermion current. In the path-integral formulation of gauge theory, the functional measure over $\psi'(x)$ and $\overline{\psi}'(x)$. can be read off

$$d\mu'_{[\nu]} = d\overline{c}'_1 d\overline{c}'_2 dc'_1 dc'_2 \tag{39}$$

and operator product μ' in Eq.(9) is given by

$$\begin{aligned} \mu'_{[\nu]} &= \overline{c}'_1\overline{c}'_2 c'_1 c'_2 \\ &= \tilde{J}_{[\nu]}\mu \end{aligned} \tag{40}$$

with the Grassmann expansion coefficients of $\psi'(x)$ and $\overline{\psi}'(x)$

$$\begin{aligned}
c_1' &= c_1 \int d^4x \phi_1^+ e^{-i\theta_{[\nu]}(x)\Gamma^{[\nu]}} \phi_1 + c_2 \int d^4x \phi_1^+ e^{-i\theta_{[\nu]}(x)\Gamma^{[\nu]}} \phi_2 \\
c_2' &= c_1 \int d^4x \phi_2^+ e^{-i\theta_{[\nu]}(x)\Gamma^{[\nu]}} \phi_1 + c_2 \int d^4x \phi_2^+ e^{-i\theta_{[\nu]}(x)\Gamma^{[\nu]}} \phi_2 \\
\overline{c}_1' &= \overline{c}_1 \int d^4x \phi_1^+ \gamma^0 e^{i\theta_{[\nu]}(x)\Gamma^{[\nu]+}} \gamma^0 \phi_1 + \overline{c}_2 \int d^4x \phi_1^+ \gamma^0 e^{-i\theta_{[\nu]}(x)\Gamma^{[\nu]+}} \gamma^0 \phi_2 \\
\overline{c}_2' &= \overline{c}_1 \int d^4x \phi_1^+ \gamma^0 e^{i\theta_{[\nu]}(x)\Gamma^{[\nu]+}} \gamma^0 \phi_2 + \overline{c}_2 \int d^4x \phi_2^+ \gamma^0 e^{-i\theta_{[\nu]}(x)\Gamma^{[\nu]+}} \gamma^0 \phi_2
\end{aligned} \tag{41}$$

The corresponding expansion of the fermion current can be performed in the eigenspace

$$\begin{aligned}
I^{[\mu\nu]'}(x) &= \left(\overline{c}_1' \phi_1^+ + \overline{c}_2' \phi_2^+\right) \gamma^\mu \Gamma^{[\nu]} \left(c_1' \phi_1 + c_2' \phi_2\right) \\
&= \overline{c}_1' c_1' \phi_1^+ \gamma^\mu \Gamma^{[\nu]} \phi_1 + \overline{c}_1' c_2' \phi_1^+ \gamma^\mu \Gamma^{[\nu]} \phi_2 + \overline{c}_2' c_1' \phi_2^+ \gamma^\mu \Gamma^{[\nu]} \phi_1 + \overline{c}_2' c_2' \phi_2^+ \gamma^\mu \Gamma^{[\nu]} \phi_2
\end{aligned} \tag{42}$$

In accordance with Eq.(11), the square of the fermion current $I^{[\mu\nu]'}(x)$ is written out

$$\begin{aligned}
I^{[\mu\nu]'2}(x) &= \left(\overline{c}_1' \phi_1^+ + \overline{c}_2' \phi_2^+\right) \gamma^\mu \Gamma^{[\nu]} \left(c_1' \phi_1 + c_2' \phi_2\right) \left(\overline{c}_1' \phi_1^+ + \overline{c}_2' \phi_2^+\right) \gamma^\mu \Gamma^{[\nu]} \left(c_1' \phi_1 + c_2' \phi_2\right) \\
&= \overline{c}_1' \overline{c}_1' c_1' c_1' \phi_1^+ \gamma^\mu \Gamma^{[\nu]} \phi_1 \phi_1^+ \gamma^\mu \Gamma^{[\nu]} \phi_1 + \overline{c}_2' \overline{c}_1' c_1' c_1' \phi_2^+ \gamma^\mu \Gamma^{[\nu]} \phi_1 \phi_1^+ \gamma^\mu \Gamma^{[\nu]} \phi_1 + \cdots \\
&+ \overline{c}_2' \overline{c}_2' c_2' c_2' \phi_2^+ \gamma^\mu \Gamma^{[\nu]} \phi_2 \phi_2^+ \gamma^\mu \Gamma^{[\nu]} \phi_2
\end{aligned} \tag{43}$$

Further by taking the property of Grassmann algebra into account, the calculus leads straightforwardly to the following result

$$\begin{aligned}
I^{[\mu\nu]'2}(x) &= 2\overline{c}_1' \overline{c}_2' c_1' c_1' \phi_1^+ \gamma^\mu \Gamma^{[\nu]} \phi_1 \phi_2^+ \gamma^\mu \Gamma^{[\nu]} \phi_2 + 2\overline{c}_1' \overline{c}_2' c_1' c_2' \phi_1^+ \gamma^\mu \Gamma^{[\nu]} \phi_2 \phi_2^+ \gamma^\mu \Gamma^{[\nu]} \phi_1 \\
&= -2\mu_\nu' \phi_1^+ \gamma^\mu \Gamma^{[\nu]} \phi_1 \phi_2^+ \gamma^\mu \Gamma^{[\nu]} \phi_2 + 2\mu_\nu' \phi_1^+ \gamma^\mu \Gamma^{[\nu]} \phi_2 \phi_2^+ \gamma^\mu \Gamma^{[\nu]} \phi_1 \\
&= -2\tilde{J}_{[\nu]} \mu \phi_1^+ \gamma^\mu \Gamma^{[\nu]} \phi_1 \phi_2^+ \gamma^\mu \Gamma^{[\nu]} \phi_2 + 2\tilde{J}_{[\nu]} \mu \phi_1^+ \gamma^\mu \Gamma^{[\nu]} \phi_2 \phi_2^+ \gamma^\mu \Gamma^{[\nu]} \phi_1 \\
&= -2J_{[\nu]}^{-1} \mu \phi_1^+ \gamma^\mu \Gamma^{[\nu]} \phi_1 \phi_2^+ \gamma^\mu \Gamma^{[\nu]} \phi_2 + 2J_{[\nu]}^{-1} \mu \phi_1^+ \gamma^\mu \Gamma^{[\nu]} \phi_2 \phi_2^+ \gamma^\mu \Gamma^{[\nu]} \phi_1 \\
&= J_{[\nu]}^{-1} \left(-2\mu \phi_1^+ \gamma^\mu \Gamma^{[\nu]} \phi_1 \phi_2^+ \gamma^\mu \Gamma^{[\nu]} \phi_2 + 2\mu \phi_1^+ \gamma^\mu \Gamma^{[\nu]} \phi_2 \phi_2^+ \gamma^\mu \Gamma^{[\nu]} \phi_1\right) \\
&= J_{[\nu]}^{-1} I^{[\mu\nu]2}(x)
\end{aligned} \tag{44}$$

From this example, we see clearly that the topological property of products of the fermion currents $I^{[\mu\nu]2}(x)$ ($i.e. I^{[\mu\nu]}(x)$) relates with the corresponding Jacobian factor. Combining the evaluated Jacobians of various fermion currents Eq. (32~38) and the Atiyah-Singer local index theorem Eq.(26), we find that only axial-vector current $I^{[\mu 5]}(x) = \overline{\psi}(x) i\gamma^\mu \gamma^5 \psi(x)$ has topological singularity, which is coincident with previous analysis made by perturbational methods.

5. Conclusion

We have presented that the topological singularity in operator product of various fermion currents coupling to a gauge field is characterized by the topological properties of anomaly function in a quantum gauge background in terms of Atiyah-Singer index theorem for Dirac operator. The anomaly functions corresponding to various fermion currents have been evaluated through the calculus of the kernel of Dirac operator.

As the above illustration, the topological singularity of various fermion currents coupling gauge field is indeed understand on Atiyah-Singer index theorem in quantum field theory as a consequence of the fact that the Jacobian of integration measure possesses anomaly terms. That is, the kernel of the Dirac operator may have short distance singularities.

No doubt, the singularity of the fermion current has to be considered when dealing with reduction of the interaction vertex by using the Dirac differential equation of motion in the Dyson-schwinger equation. Also in the non-abelian gauge case, the relation of anommalies in conservation o general axial currents to the indes of the Dirac operator in a gauge background need to discussed further. In addition, the property of quantum anomaly associated with Ward-Takahashi relation plays an important role in the nonperturbative study of gauge theories, such as the dynamical chiral symmetry breaking.

6. Appendix: Regularization of measure for anomaly

a. regularization prescription for low-rank tensor current ($\bar{\psi}(x)\gamma^{\mu}\Gamma^{[\nu]}\psi(x)$)

Following Fujikawa's method, we work out a calculation of Jacobian of anomaly function in the transformation of measure due to the Eq.(3). The change in functional measure is in the form

$$d\mu \to d\mu' = \prod_n d\bar{c}_n' \prod_n dc_n' = \left(\det f_{nm}\right)^{-1}\left(\det f'_{nm}\right)^{-1} \prod_m d\bar{c}_m \prod_m dc_m \qquad \text{(A1)}$$

with

$$f_{nm} = \int d^4x \phi_n^{+}(x) e^{-i\theta_{[\nu]}(x)\Gamma^{\{\nu\}}} \phi_m(x). \qquad \text{(A2)}$$

The corresponding Jacobian factor is given explicitly by

$$\left(\det f'_{nm}\right)^{-1} = e^{-i\int d^4x \gamma^0 A^{[\nu]+}(x)\gamma^0 \theta_{[\nu]}(x)}, \qquad \text{(A3)}$$

where anomaly function $A^{[\nu]}(x)$ denotes the trace of Dirac matrix $\Gamma^{[\nu]}$ in the function space above

$$A^{[\nu]}(x) = \sum_n \varphi_n^{+}(x)\Gamma^{[\nu]}\varphi_n(x). \qquad \text{(A4)}$$

Based on the use of path integrals in Euclidean space, regularization of the anomaly function is achieved by inserting the convergent factor $f\left(-\frac{D^2}{M^2}\right) = e^{-D^2/M^2}$ and taking the limit as $M \to \infty$. To do this the above anomaly function can be written as the limit of a manifestly convergent integral

$$A^{[\nu]}(x) = \lim_{M\to\infty}\int\frac{d^4k}{(2\pi)^4}e^{-ikx}\Gamma^{[\nu]}f\left(\frac{-D^2}{M^2}\right)e^{ikx}$$
$$= \lim_{M\to\infty}\frac{1}{32M^2}Tr\left[-\Gamma^{[\nu]}\left(\left[\gamma^{\mu},\gamma^{\nu}\right]F_{\mu\nu}\right)^2\right]\int\frac{d^4k}{(2\pi)^4}f''\left(\frac{k^2}{M^2}\right) \quad (A5)$$

In terms of the trace over Dirac indices, when we expand the regularization operator $f\left((-D^2/M^2\right)$ in D, the terms with less than four gamma matrices evaluate to zero. In the light of the limit over M (mass), terms with more than four D s will also drop out.

In the last step, we have used the operator identities

$$\left(\gamma^{\mu}D_{\mu}\right)^2 = D_{\mu}D_{\mu} - i\frac{1}{4}\left[\gamma^{\mu},\gamma^{\nu}\right]F_{\mu\nu} \quad (A6)$$

$$f\left(\frac{\left(i\gamma^{\mu}D_{\mu}\right)^2}{M^2}\right) = \sum_n\frac{1}{n!}f^{(n)}\left(\frac{D_{\mu}^2}{M^2}\right)\left(-i\frac{1}{4M^2}\left[\gamma^{\mu},\gamma^{\nu}\right]F_{\mu\nu}\right)^n \quad (A7)$$

In the case of $\Gamma^{[\nu]} = \gamma^5$ for the chiral anomaly, the $A^{[5]}(x)$ becomes into

$$A^{[5]}(x) = \lim_{M\to\infty}\frac{1}{32M^2}Tr\left[-\gamma^5\left(\left[\gamma^{\mu},\gamma^{\nu}\right]F_{\mu\nu}\right)^2\right]\int\frac{d^4k}{(2\pi)^4}f''\left(\frac{k^2}{M^2}\right)$$
$$= -\frac{1}{32M^2}\varepsilon^{\mu\nu\rho\sigma}Tr\left[F_{\mu\nu}F_{\rho\sigma}\right] \quad (A8)$$

It is just the well-known result.

b. regularization description for high rank tensor current

In a similar way, we discuss the regularization of the anomaly faction $A^{[\nu]}(x)$ for high rank Dirac matrix transformation.The expression of the anomaly function $A^{[\alpha\beta]}(x)$ for the case of Eq.(37) can be put in the regulating form

$$A^{[\alpha\beta]}(x) = \lim_{M\to\infty}\int\frac{d^4k}{(2\pi)^4}e^{-ikx}\sigma^{\alpha\beta}f\left(\frac{-D^2}{M^2}\right)e^{ikx}$$
$$= \lim_{M\to\infty}\int\frac{d^4k}{(2\pi)^4}e^{-ikx}\sigma^{\alpha\beta}\sum_n\frac{1}{n!}f^{(n)}\left(\frac{D_{\mu}{}^2}{M^2}\right)\left(-i\frac{1}{4M^2}\left[\gamma^{\mu},\gamma^{\nu}\right]F_{\mu\nu}\right)^n e^{ikx}$$
$$= \lim_{M\to\infty}\int\frac{d^4k}{(2\pi)^4}e^{-ikx}Tr\left[\sigma^{\alpha\beta}f^{(2)}\left(\frac{-D^2}{M^2}\right)\left(\frac{i}{4M^2}\left[\gamma^{\mu},\gamma^{\nu}\right]F_{\mu\nu}\right)^2\right]$$
$$= \frac{-1}{32\pi^2}\begin{pmatrix} g^{\alpha\mu}g^{\beta\rho}g^{\nu\sigma} - g^{\alpha\mu}g^{\beta\sigma}g^{\nu\rho} + g^{\alpha\nu}g^{\beta\sigma}g^{\mu\rho} - g^{\alpha\nu}g^{\beta\rho}g^{\mu\sigma} \\ +g^{\beta\mu}g^{\alpha\sigma}g^{\nu\rho} - g^{\beta\mu}g^{\alpha\rho}g^{\nu\sigma} + g^{\beta\nu}g^{\alpha\rho}g^{\mu\sigma} - g^{\beta\nu}g^{\alpha\sigma}g^{\mu\rho}\end{pmatrix}Tr\left[F_{\mu\nu}F_{\rho\sigma}\right] \quad (A9)$$

In terms of the symmetry of metric and antisymmetry of 4-dimensional field strength tensor, we expand the anomaly function and find that it equals zero. So that, the Jacobian becomes

$$J^{[\alpha\beta]}=1 \tag{A10}$$

By the parallel procedure, for the case of the transformation Eq.(38), the axial vector anomaly function is given by

$$\begin{aligned}
A^{[\alpha\beta 5]}(x) &= \lim_{M\to\infty}\int\frac{d^4k}{(2\pi)^4}e^{-ikx}\sigma^{\alpha\beta}\gamma^5 f\left(\frac{\left(i\gamma^{\mu}D_{\mu}\right)^2}{M^2}\right)e^{ikx} \\
&= \lim_{M\to\infty}\int\frac{d^4k}{(2\pi)^4}e^{-ikx}\sigma^{\alpha\beta}\gamma^5\sum_n\frac{1}{n!}f^{(n)}\left(\frac{D^2}{M^2}\right)\left(-i\frac{1}{4M^2}\left[\gamma^{\mu},\gamma^{\nu}\right]F_{\mu\nu}\right)^n e^{ikx} \\
&= \lim_{M\to\infty}\int\frac{d^4k}{(2\pi)^4}e^{-ikx}Tr\left[\sigma^{\alpha\beta}\gamma^5 f^{(2)}\left(\frac{-D^2}{M^2}\right)\left(\frac{i}{4M^2}\left[\gamma^{\mu},\gamma^{\nu}\right]F_{\mu\nu}\right)^2\right] \\
&= \frac{-i}{32\pi^2}\begin{pmatrix}-g^{\alpha\mu}\varepsilon^{\beta\nu\rho\sigma}+g^{\alpha\nu}\varepsilon^{\beta\mu\rho\sigma}+g^{\beta\mu}\varepsilon^{\alpha\nu\rho\sigma}-g^{\beta\nu}\varepsilon^{\alpha\mu\rho\sigma} \\ +g^{\nu\rho}\varepsilon^{\alpha\beta\mu\sigma}-g^{\mu\rho}\varepsilon^{\alpha\beta\nu\sigma}+g^{\mu\sigma}\varepsilon^{\alpha\beta\nu\rho}-g^{\nu\sigma}\varepsilon^{\alpha\beta\mu\rho}\end{pmatrix}Tr\left[F_{\mu\nu}F_{\rho\sigma}\right] \\
&= -\bar{A}^{[\alpha\beta 5]}(x)
\end{aligned} \tag{A11}$$

The corresponding Jacobian is

$$J^{[\alpha\beta 5]}=1 \tag{A12}$$

In the above calculation, we have employed the following operator identities

$$\begin{aligned}
Tr\left(\prod_{j=1}^{6}\gamma_{\mu_j}\right) &= 4g_{\mu_1\mu_2}\left\{g_{\mu_3\mu_4}g_{\mu_5\mu_6}+g_{\mu_3\mu_6}g_{\mu_4\mu_5}-g_{\mu_3\mu_5}g_{\mu_4\mu_6}\right\} \\
&\quad -4g_{\mu_1\mu_3}\left\{g_{\mu_2\mu_4}g_{\mu_5\mu_6}+g_{\mu_2\mu_6}g_{\mu_4\mu_5}-g_{\mu_2\mu_5}g_{\mu_4\mu_6}\right\} \\
&\quad +4g_{\mu_1\mu_4}\left\{g_{\mu_2\mu_3}g_{\mu_5\mu_6}+g_{\mu_2\mu_6}g_{\mu_2\mu_6}-g_{\mu_2\mu_5}g_{\mu_3\mu_6}\right\} \\
&\quad -4g_{\mu_1\mu_5}\left\{g_{\mu_2\mu_3}g_{\mu_4\mu_6}+g_{\mu_2\mu_6}g_{\mu_3\mu_4}-g_{\mu_2\mu_4}g_{\mu_3\mu_6}\right\} \\
&\quad -4g_{\mu_1\mu_6}\left\{g_{\mu_2\mu_3}g_{\mu_4\mu_5}+g_{\mu_2\mu_5}g_{\mu_3\mu_4}-g_{\mu_2\mu_4}g_{\mu_3\mu_5}\right\}
\end{aligned} \tag{13}$$

$$Tr\left(\gamma^5\prod_{j=1}^{6}\gamma_{\mu_j}\right)=4i\left\{\begin{matrix}g_{\mu_1\mu_2}\varepsilon_{\mu_3\mu_4\mu_5\mu_6}-g_{\mu_1\mu_3}\varepsilon_{\mu_2\mu_4\mu_5\mu_6}+g_{\mu_2\mu_3}\varepsilon_{\mu_1\mu_4\mu_5\mu_6} \\ +g_{\mu_4\mu_5}\varepsilon_{\mu_1\mu_2\mu_3\mu_6}-g_{\mu_4\mu_6}\varepsilon_{\mu_1\mu_2\mu_3\mu_5}+g_{\mu_5\mu_6}\varepsilon_{\mu_2\mu_2\mu_3\mu_4}\end{matrix}\right\} \tag{A14}$$

Obviously the results in Eq.(A10) and Eq.(A12) is perfectly consistent with result of derivation of transverse vector and axial vector anomalies in four-dimensional $U(1)$ gauge theory using perturbative methods[21,22].

Now we have completed our computation in Euclidean space, it is necessary to transform the conclusions back into Minkowski space. According to the Wick substitutions,then, there is no change in the form of the final formula.

7. Acknowledgment

We have benefited from several conversations with Prof. Hai-Jun Wang and Prof. Shi-Hao Chen. We would particularly like to thank Prof. Sergey Ketov for heuristic suggestions.The work is supported in part by National Natural Science Foundation of China under Grant No.10775059.

8. References

[1] Weinberg, S.(2001). *The Quantum Theory of Fields(2)*, Cambridge University Press, ISBN 7-5062-6638-5/o.486, 111-160, 359-408.

[2] Grammer, G. & Yennie, Jr. and D.R.(1973). Improved treatment for the infrared-divergence problem in quantum electrodynamics, *Phys.Rev.D*, Vol.8(12), 4332-4344.

[3] Yong, B.L.(1987). *Introduction to Quantum Field Theory*, Science Press, ISBN7-03-000386-1, 341-420.

[4] Peskin, M.E. & Schroeder, D.V.(1995). *An Introduction to Quantum Field Theory*, Westview Press, ISBN7-5062-7294-6/o.556, 651-686.

[5] Bertlmann, R.A. (1996). *Anomalies in Quantum Field Theory*, Oxford University Press, New York, ISBN0-19-8507623, 246-297.

[6] Fujikawa, K.(1984). Evaluation of the chiral anomaly in gauge theories with γ^5 couplings, *Phys.Rev.D*, Vol.29, 285-292.

[7] Alvarez, L. & Witten, E.(1983). Gravitational anomalies, *Nucl.Phys.B*, Vol.234, 269-330

[8] Atiyah, M.F. & SingerI.M.(1984).*Dirac Operators Coupled to Vector Potentials*, ISBN978-7-5062-9213-9/o. 604, Proc.Nat.Acad.Sci.U.S.A., 81, 2597-2600.

[9] Zumino, B., Wu, Y.S. & Zee, A.(1984). Chiral anomalies, higher dimensions and differential geometry, *Nucl.Phys. B*, Vol.239, 477-507.

[10] Alvarez, O., Singer, I.M. & Zumino, B.(1984). Gravitational anomalies and the family's index theorem, *Commun.Math.Phys.*, Vol.96, 409-417.

[11] Nicole Verline, Ezra Getzler & Michele Vergne (2004).*Heat Kernels and Dirac operators*, Springer-Verlag Berlin Heidelberg ISBN 978-7-5062-9213-9/o.604, 99-298.

[12] Bao, A.D. & Wu, S.S.(2007). Various full Green functions in QED, *Z.Phys.C* , Vol.46(12), 3093-3108.

[13] Ticciati, R. (1999). *Quantum Field Theory for Mathematicians*, Cambridge University Press, ISBN 7-5062-5095-0/o.325, 323-437.

[14] Bao, A.D., Yao, H.B. & Wu, S.S.(2009). Topological approach to examine the singularity of the axial-vector current in an Abelian gauge field theory (QED), *Chinese phys.C* , Vol.33, 177-180.

[15] Bick, E. & Steffen, F.D.(2005). *Topology and geometry in physics*, Springer-Verlag Berlin Heidelberg, ISBN978-7-03-018786-4, 167-231.

[16] Palais, R.S., (1965). *Seminar on the Atiyah-Singer index theorem, Ann.of Maths. studies*, Princeton Univ. Press Princeton, N.J., Vol.57.

[17] Raoul Bott & Loring W. Tu, R. (1982). *Differential forms in algebraic topology*, Springer Verlag, New York, ISBN 0-387-90613-4, 273-291.

[18] Atiyah, M.F. & Singer, I.M.(1968). The index of elliptic operators, *Ann.Math.* , Vol.87, 484-530.

Atiyah, M.F. & Singer, I.M.(1968). The index of elliptic operators, *Ann.Math.* , Vol.87, 546-604.

Atiyah, M.F. & Singer, I.M.(1971). The index of elliptic operators, *Ann.Math.* , Vol.93, 119-138.
[19] Atiyah, M.F. Bott, R. & Patodi, V.K.(1973).On the heat equation and the index theorem, *Invent. Math.* , Vol.119, 279-330.
[20] Einhorn, M.B. & Jones, D.R.T.(1984). Comment on Fujikawa's path-integral derivation of the chiral anomaly, *Phys.Rev.D*, Vol.29, 331-333.
[21] He, H.X. (2001). Identical relations among transverse parts of variant Green's functions in gauge theories , *Phys.Rev.C* , Vol.63, 025207.
[22] Sun, W.M., Zong, H.S. & Chen, X.S. (2003). A note on transverse axial vector and vector anomalies in U(1) gauge theories, *Phys.Lett.B*, Vol. 569(2), 211-218,

9

Quantum Field Theory and Knot Invariants

Chun-Chung Hsieh
Institute of Mathematics, Academia Sinica, Nankang, Taipei, Taiwan

1. Introduction

Massey-Milnor linking theory was developed by J. Milnor [Mi1, Mi2] and W. Massey [Ma, Po] algebraically and homological-theoretically in 1960's, but still remains quite mysterious as the explicit formulae thereof is missing.

Chern-Simons-Witten configuration space integrals are the Feynman graphs in the aspect of perturbative quantum field theory, and are developed by E. Witten [ADW, AF1, AF2, At, Aw, Ba1, Ba2, HM, MV, RT, Tu, Wi] in 1990's. But, as in almost all quantum field theories to compute Feynman graphs explicitly is always beyond any rigorous mathematical attack for the time being [PS]. Nevertheless in this paper, in the aspect of the first nonvanishing Massey-Milnor linking [Ma, Mi1, Mi2, Po] we compute explicitly the related Chern-Simons-Witten configuration space integrals [HKT, Hs1, Hs2, Hs3, Hs4, Hs5, Hs6, Hs7, HY], from which we derive the combinatorial formulae of the Massey-Milnor linking when the link under study is represented as a link diagram on the plane $\mathbb{R}^2$.

2. Set-up

In this section for the forthcoming presentation of Massey-Milnor linking theory and Chern-Simons-Witten graphs in perturbative quantum field theory [AF1, AF2, Ba2, Wi], we define the related concept as follows.

For the set-up suppose that a given link $L = \{L_0, L_1, \ldots, L_n\}$ oriented with base points $\{x_j \in L_j | j = 0, 1, \ldots, n\}$, is in a general position with pairwise crossings specified in $\mathbb{R}^2$ and is represented schematically as

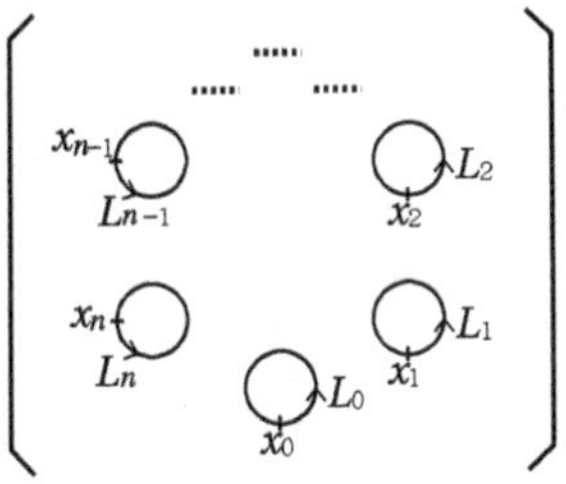

To be more precise, each component L_j is represented schematically by a trivial circle with the base point x_j placed outer the most on L_j of $L = \{L_0, L_1, \ldots, L_n\}$ which is arranged

counter-clockwise with L_0 as the root. Moreover each L_j is oriented counter-clockwise as shown schematically.

Also for the link $L = \{L_0, L_1, \ldots, L_n\}$ of $(n+1)$ components, to define the invariant $L_{n,n-1,\ldots,1,0}$ below we assume that all invariants of strictly lower degrees vanish namely that $L_{m,m-1,\ldots,1,0} = 0$, for any permutation of any subset $\{L_0^*, L_1^*, \ldots, L_m^*\} \subseteq L$ of $(m+1)$ components and for any $m \leq n-1$.

3. Chern-Simons-Witten graphs

In this section we present the key concept of Chern-Simons-Witten configuration space integrals in the framework of perturbative quantum field theory. Beyond that we define our first knot invariant $L_{n,n-1,\ldots,1}$ for a link $\{L_0, L_1, \ldots, L_n\}$, for which all invariants of strictly lower degrees vanish as in the setup.

Definition 1. *(1) Given an oriented link* $L = \{L_0, L_1, \ldots, L_{n-1}, L_n\}$ *as above, a Chern-Simons-Witten graph* Γ *supported on L is a uni-trivalent rooted tree with all univalent vertices supported on L.—Notice that our trees are "honest" trees in strict sense that all edges rooted at a vertex are all going upward therefrom.*

(2) Given a Chern-Simons-Witten graph on L we define its degree to be

degree $\Gamma = \#\{$ *edges of* $\Gamma\} - \#\{$ *trivalent vertices of* $\Gamma\}$.

(3) Given a Chern-Simons-Witten graph Γ *supported on L we define the associated Chern-Simons-Witten configuration space to be the space as follows.*

(3-1) For each trivalent vertex we assign a copy of $\mathbb{R}^3$.

(3-2) For a univalent vertex supported on the component L_j *of L we assign* L_j *to it.*

And if some univalent vertices $\{U_1, U_2, \ldots, U_k\}$ *ordered linearly with respect to the orientation and the base point* $x_j \in L_j$ *are supported on* L_j*, then we assign to* $\{U_1, U_2, \ldots, U_k\}$ *the subset of* $(L_j)^k$ *which respects the linear order of* L_j*, namely the subset* $\{(y_1, y_2, \ldots, y_k) | x_j \leq y_1 \leq y_2 \leq \cdots \leq y_k \leq x_j\} \subseteq (L_j)^k$ *with the induced orientation.*

As the configuration space of Γ *we take the abstract product of the spaces in (3-1) and (3-2), but with the orientation specified (or the ordering of the factors of the product) as follows:*

(3-2-1) Always start with the root L_0.

(3-2-2) Going up for each edge.

(3-2-3) From the right edge to the left edge at each trivalent vertex.

(3-2-4) Endow the connected subgraphs with the above orientations and then take the product of the orientation of all components. It does not matter how to get the product of the orientation of the components as all components are even dimensional spaces.

(4) For an edge joining vertices A and B in Γ*, we assign a differential 2-form to it by pulling-back the standard area form on the unit sphere in* $\mathbb{R}^3$ *by the map* $\dfrac{A-B}{|A-B|}$ *where vertex A sits below vertex B in* Γ*. And we define the differential form associated to* Γ *to be the product of the 2-forms indexed by all edges of* Γ*. Notice that it does not matter how we "arrange" the order of the product, as all these*

differential forms are 2-forms. Also notice that if univalent vertices $U_1, \ldots, U_k$ *are supported on a component* L_j*, then they are "positioned" on* L_j *with the same "height".*

(5) Finally for a Chern-Simons-Witten graph Γ *supported on a given link* $L = \{L_0, L_1, \ldots, L_{n-1}, L_n\}$*, we define the associted Chern-Simons-Witten configuration space integral to be the integral of the differential form constructed in (4) over the configuration space constructed in (3).*

Next in Definition 2 we define the first non-vanishing invariant $L_{n,n-1,\ldots,1,0}$ for the link $\{L_0, L_1, \ldots, L_n\}$ represented diagrammatically as in the setup. We coin the construction as HIST-transform where HI comes from the IHX-relation and ST comes from the STU-relation in the perturbative Chern-Simons-Witten quantum field theory [Oh, Wi, Ye].

Definition 2. *(1) We define the HI-transform as:*

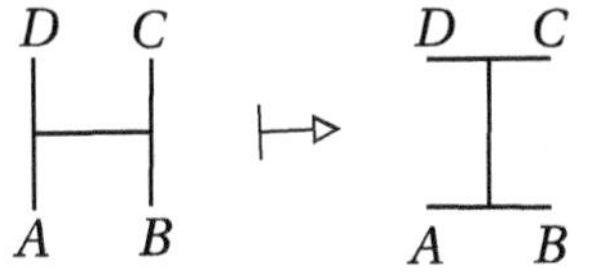

where vertices A, B, C and D are generic vertices of a Chern-Simons-Witten graph which are not necessarily uni-valent ones.

(2) We define the ST-transform as:

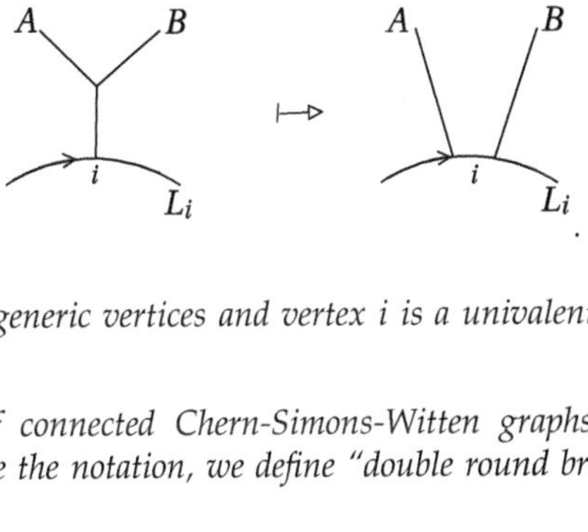

where vertices A and B are generic vertices and vertex i is a univalent vertex sitting on some knot component.

(3) For the construction of connected Chern-Simons-Witten graphs of degree n supported on $\{L_0, L_1, \ldots, L_n\}$*, and to ease the notation, we define "double round brackets"* $((n, n-1, \ldots, 2, 1))$ *as follows.*

(3-1) $((2,1)) = 2 \wedge 1$

(3-2) $((3,2,1)) = ((3,2)) \wedge 1 + 3 \wedge ((2,1))$

(3-3) $((4,3,2,1)) = ((4,3,2)) \wedge 1 + ((4,3)) \wedge ((2,1)) + 4 \wedge ((3,2,1))$

$\ldots \quad \ldots$

(3-4) $((n, n-1, \ldots, 2, 1)) = ((n, n-1, \ldots, 2)) \wedge 1 + ((n, n-1, \ldots, 3)) \wedge ((2,1)) + \cdots + ((n, n-1)) \wedge ((n-2, n-3, \ldots, 2, 1)) + n \wedge ((n-1, n-2, \ldots, 2, 1))$

(4) The construction of connected Chern-Simons-Witten graphs of degree n on $\{L_0, L_1, \ldots, L_n\}$ *is as follows.*

(4-1) For generic vertices A and B, we assign 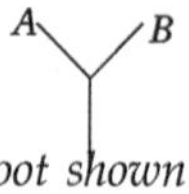

to $(A \wedge B)$*, this is we connect vertices A and B arranged as above, to a new vertex—the root shown here—by a "Y". And by definition the new vertex—the root—is denoted as* $(A \wedge B)$*, if we need to continue doing this construction therefrom.*

(4-2) For two connected Chern-Simons-Witten graphs A and B, to $(A + B)$ *we assign the disjoint copies of A and B.*

(5) On $\{L_0, L_1, \ldots, L_n\}$ *for the construction of the first non-vanishing Chern-Simons-Witten graph* $L_{n,n-1,\ldots,1,0}$*: Start with the connected ones in (4) and apply ST-transform exactly once to get the set of all Chern-Simons-Witten graphs of two components; repeat ST-transforms till we get finally the Chern-Simons-Witten graphs of exactly n components. And* $L_{n,n-1,\ldots,1,0}$ *is the "sum" of the Chern-Simons-Witten graphs constructed above.*

Before giving some examples we make the following remarks to make more sense of Definition 2.

Note 1. *(1) It is easy to see that the connected Chern-Simons-Witten graphs constructed in (4) of Definition 2 is closed under the HI-transforms.*

(2) If we coin the connected Chern-Simons-Witten graphs as 0-connected, those ones of two components as (-1)*-connected, the ones of three components as* (-2)*-connected and so forth and so on, then it is easy to see that for any* l *the set of* l*-connected Chern-Simons-Witten graphs of degree* n *supported on* $L = \{L_1, \ldots, L_n\}$ *are closed under HI-transform; moreover, the set of* l*-connected Chern-Simons-Witten graphs constructed as above will produce exactly the set of* $(l-1)$*-connected ones after doing ST-transform exactly once.*

(3) Our Chern-Simons-Witten graphs are always not edge-overlapping, that is when edges of the graphs are represented as line segment in $\mathbb{R}^2$ *they never intersect with one another except obviously at the vertices proper.*

Here are some examples to show the idea.

Example 1. *For* $n = 2$, $L = \{L_0, L_1, L_2\}$*, the connected Chern-Simons-Witten graph of degree 2 is:*

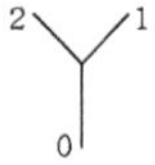

. As usual to ease the notation we use numerials $i, j, k, \ldots$ *for the knot component* L_i, L_j, L_k *etc. The set of* (-1)*-connected ones are*

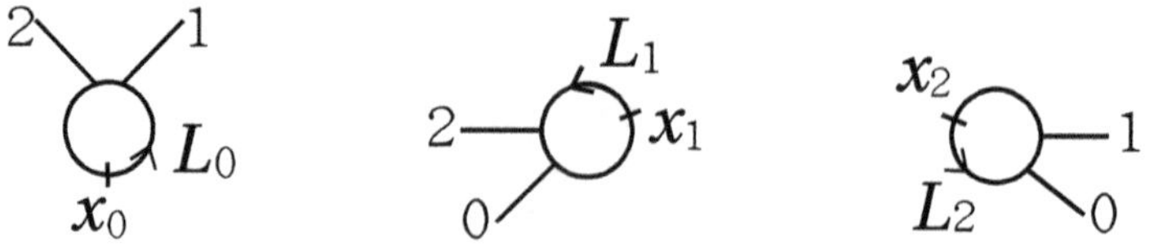

It is easy to see that starting with the connected Chern-Simons-Witten graphs we get the set of (-1)-connected ones after doing ST-transform exactly once.

Example 2. *For* $n = 3$, $L = \{L_0, L_1, L_2, L_3\}$*, from the construction in (3), (4) of Definition 2 it is to see that the connected Chern-Simons-Witten graphs of degree 3 are:* 

and 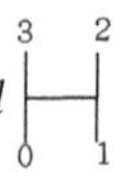

.

For the (-1)-connected Chern-Simons-Witten graphs,we apply ST-transform exactly once to the set of connected ones to get:

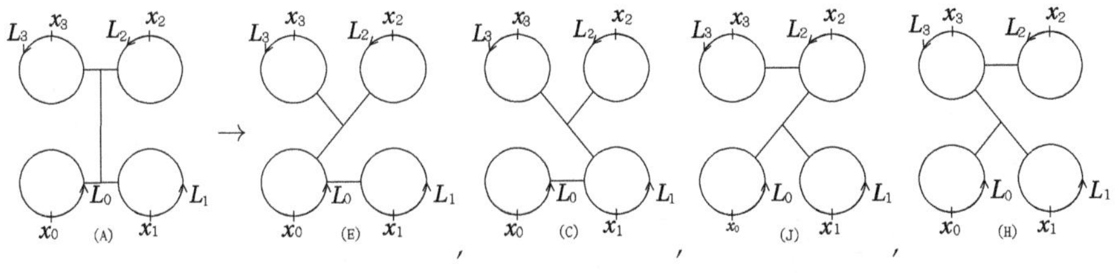

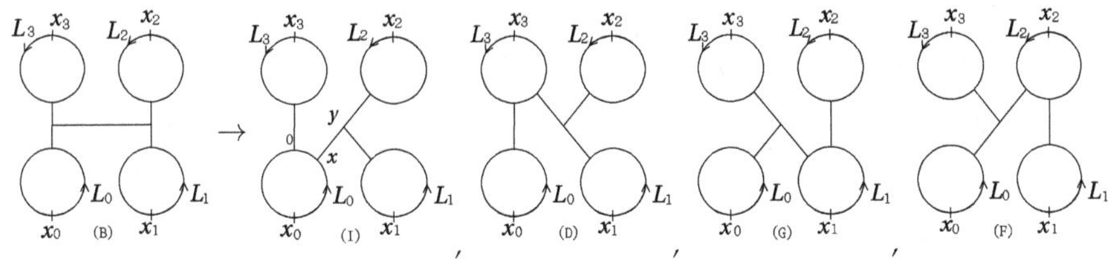

For the set of (-2)-connected Chern-Simons-Witten graphs, we apply ST-transform exactly once to the set of (-1)-connected ones to get the graphs of three components in $L_{3,2,1,0}$ listed below.

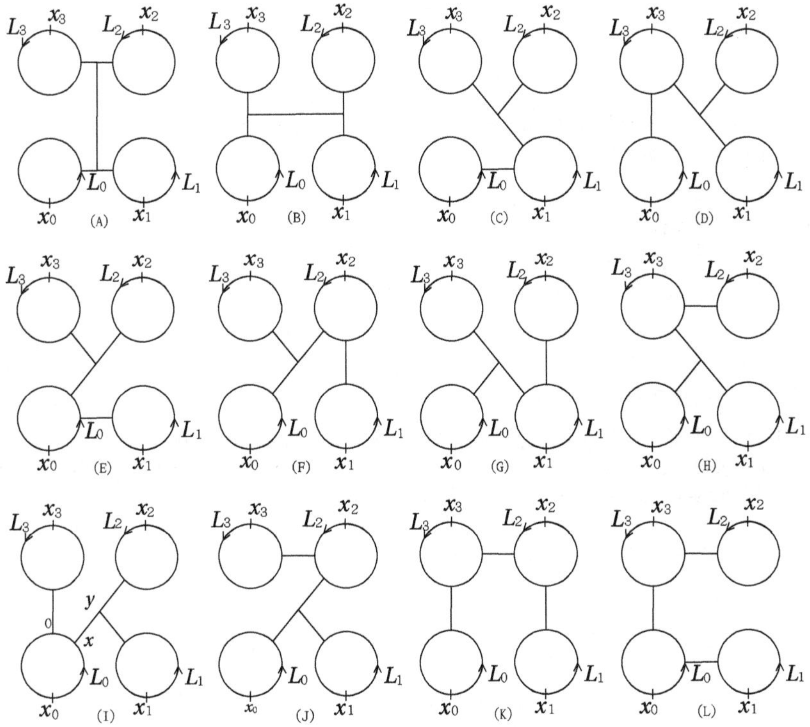

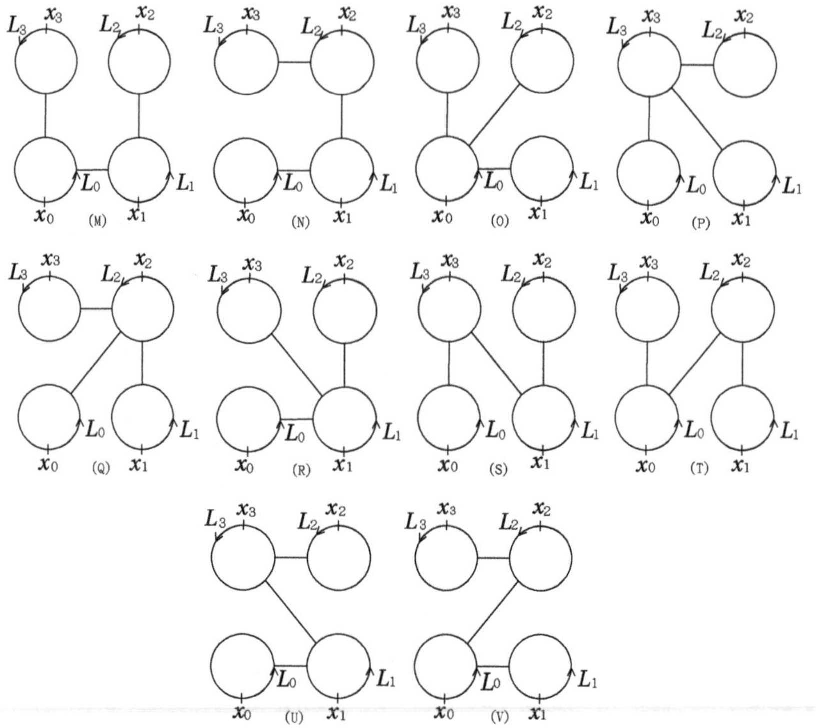

Also by Definition 1, the relevant Chern-Simons-Witten configuration space integrals of the above example are shown in example 5.

Example 3. *For $n = 1$, $L = \{L_0, L_1\}$, the invariant $L_{1,0}$ is nothing but the integration of the 2-form corresponding to the edge, over the configuration space $L_0 \times L_1$:*

$$L_{1,0} = \frac{1}{4\pi} \int_{L_0} \int_{L_1} \det \begin{pmatrix} y_0 - y_1 \\ dy_0 \\ dy_1 \end{pmatrix} \frac{1}{|y_0 - y_1|^3},$$

which is exactly the classic Gauss linking.

Example 4. *$n = 2$, $L = \{L_0, L_1, L_2\}$ for which the invariants of strictly lower degrees $L_{i,j} = 0$, $\forall i \neq j$, then $L_{2,1,0}$ is the sum of the Chern-Simons-Witten configuration space integrals corresponding to the Chern-Simons-Witten graphs listed in Example 1.*

$$\int_{L_0} \int_{\mathbb{R}^3} \int_{L_1} \int_{L_2} (0 - x) \wedge (x - 1) \wedge (x - 2),$$

$$\int_{L_0} \int_{x_0}^{0} \int_{L_1} \int_{L_2} (x - 1) \wedge (x - 2),$$

$$\int_{L_0} \int_{L_1} \int_{x_1}^{1} \int_{L_2} (0 - 1) \wedge (x - 2),$$

$$\int_{L_0}\int_{L_1}\int_{L_2}\int_{x_2}^{2}(0-2)\wedge(x-1).$$

Example 5. *$n = 3$, $L = \{L_0, L_1, L_2, L_3\}$ for which the invariants of strictly lower degrees vanish: $L_{i,j} = 0$, $L_{i,j,k} = 0$ for all distinct i, j, k, then the invariant $L_{3,2,1,0}$ is the sum of the following 22 Chern-Simons-Witten configuration space integrals which are the relevant integrals of the Chern-Simons-Witten graphs listed in Example 2.*

$(A) = \int_{L_0}\int_{\mathbb{R}^3}\int_{L_1}\int_{\mathbb{R}^3}\int_{L_2}\int_{L_3}(0-x)\wedge(x-1)\wedge(x-y)\wedge(y-2)\wedge(y-3),$

$(B) = -\int_{L_0}\int_{\mathbb{R}^3}\int_{L_3}\int_{\mathbb{R}^3}\int_{L_1}\int_{L_2}(0-x)\wedge(x-3)\wedge(x-y)\wedge(y-1)\wedge(y-2),$

$(C) = \int_{L_0}\int_{L_1}\int_{x_1}^{1}\int_{\mathbb{R}^3}\int_{L_2}\int_{L_3}(0-1)\wedge(x-y)\wedge(y-2)\wedge(y-3),$

$(D) = \int_{L_0}\int_{L_3}\int_{x_1}^{3}\int_{\mathbb{R}^3}\int_{L_1}\int_{L_2}(0-x)\wedge(3-y)\wedge(y-1)\wedge(y-2),$

$(E) = \int_{L_0}\int_{x_0}^{0}\int_{L_1}\int_{\mathbb{R}^3}\int_{L_2}\int_{L_3}(x-1)\wedge(0-y)\wedge(y-2)\wedge(y-3),$

$(F) = (+)\int_{L_0}\int_{\mathbb{R}^3}\int_{L_2}\int_{x_2}^{2}\int_{L_1}\int_{L_3}(0-x)\wedge(x-y)\wedge(y-1)\wedge(x-2),$

$(G) = (+)\int_{L_0}\int_{\mathbb{R}^3}\int_{L_1}\int_{x_1}^{1}\int_{L_2}\int_{L_3}(0-x)\wedge(x-1)\wedge(y-2)\wedge(x-3),$

$(H) = \int_{L_0}\int_{\mathbb{R}^3}\int_{L_1}\int_{L_3}\int_{x_3}^{3}\int_{L_2}(0-x)\wedge(x-1)\wedge(x-y)\wedge(y-2),$

$(I) = \int_{L_0}\int_{x_0}^{0}\int_{\mathbb{R}^3}\int_{L_1}\int_{L_2}\int_{L_3}(x-y)\wedge(y-1)\wedge(y-2)\wedge(0-3),$

$(J) = \int_{L_0}\int_{\mathbb{R}^3}\int_{L_1}\int_{L_2}\int_{x_2}^{2}\int_{L_3}(0-x)\wedge(x-1)\wedge(x-2)\wedge(y-3),$

$(K) = \int_{L_0}\int_{L_3}\int_{x_3}^{3}\int_{L_2}\int_{x_2}^{2}\int_{L_1}(0-x)\wedge(3-y)\wedge(2-1),$

$(L) = \int_{L_0}\int_{x_0}^{0}\int_{L_1}\int_{Lx_3}\int_{x_3}^{3}\int_{L_2}(x-1)\wedge(0-y)\wedge(3-2),$

$(M) = \int_{L_0} int_{x_0}^{0}\int_{L_1}\int_{x_1}^{1}\int_{L_2}\int L3(x-1)\wedge(y-2)\wedge(1-3),$

$(N) = \int_{L_0}\int_{L_1}\int_{x_1}^{1}\int_{L_2}\int_{x_2}^{2}\int_{L_3}(0-1)\wedge(x-2)\wedge(y-3),$

$(O) = \int_{L_0}\int_{x_0}^{0}\int_{x_0}^{y}\int_{L_1}\int_{L_2}\int_{L_3}(x-1)\wedge(y-2)\wedge(0-3),$

$(P) = \int_{L_0}\int_{L_3}\int_{x_3}^{3}\int_{x_3}^{y}\int_{L_1}\int_{L_2}(0-x)\wedge(y-1)\wedge(3-2),$

$(Q) = \int_{L_0}\int_{L_2}\int_{x_2}^{2}\int_{x_2}^{y}\int_{L_1}\int_{L_3}(0-y)\wedge(x-3)\wedge(2-1),$

$(R) = \int_{L_0}\int_{L_1}\int_{x_1}^{1}\int_{x_1}^{y}\int_{L_2}\int_{L_3}(0-1)\wedge(x-2)\wedge(y-3),$

$(S) = \int_{L_0}\int_{L_3}\int_{x_3}^{3}\int_{L_1}\int_{x_1}^{1}\int_{L_2}(0-x)\wedge(3-1)\wedge(y-2),$

$(T) = \int_{L_0}\int_{x_0}^{0}\int_{L_3}\int_{L_2}^{y}\int_{x_2}^{2}\int_{L_1}(0-3)\wedge(x-y)\wedge(2-1),$

$(U) = \int_{L_0}\int_{L_1}\int_{x_1}^{1}\int_{L_3}\int_{x_3}^{3}\int_{L_2}(0-1)\wedge(x-y)\wedge(3-2),$

$(V) = \int_{L_0}\int_{x_0}^{0}\int_{L_1}\int_{L_2}\int_{x_2}^{2}\int_{L_3}(x-1)\wedge(0-2)\wedge(y-3),$

4. Massey-Milnor linking theory

Although Massey-Milnor linking theory was developed in 1960's [Fe, Ma, Mi1, Mi2, Po], the explicit and combinatorial formulae thereof is still missing. In this section we develop the "absolute version" of homological theory of Massey-Milnor linking [HKT, Hs1, Hs2, Hs3, Hs4, Hs5, Hs6, Hs7, HY] in contrast to the relative homological theory of Massey's [Fe, Ma, Po] for the purpose of the first non-vanishing linking $L^*_{n,n-1,\ldots,1,0}$ and its combinatorial formulae for link $L = \{L_0, L_1, \ldots, L_n\}$ as in the set-up. First we need the following definition.

Definition 3. *(1) For each component $L_j \in L = \{L_0, L_1, \ldots, L_n\}$, we define a closed 1-form denoted as $j(x)$ associated to it as:*

$$j(x) = \frac{1}{4\pi} \int\limits_{L_j} \det \begin{pmatrix} x-y \\ dx \\ dy \end{pmatrix} \frac{1}{|x-y|^3} \overset{\triangle}{=} \int\limits_{L_j} (x-y),$$

which is a smooth 1-form as long as $x \notin L_j$. Notice also that by convention and by notation we set

$$(x-y) \overset{def}{=} \frac{1}{4\pi} \det \begin{pmatrix} x-y \\ dx \\ dy \end{pmatrix} \frac{1}{|x-y|^3} \overset{def}{=} \frac{1}{4\pi} \det \begin{pmatrix} x_1-y_1 & x_2-y_2 & x_3-y_3 \\ dx_1 & dx_2 & dx_3 \\ dy_1 & dy_2 & dy_3 \end{pmatrix} \frac{1}{|x-y|^3}$$

$$\begin{aligned} =&\frac{1}{4\pi}\frac{1}{|x-y|^3}((x_1-y_1)(dx_2 \wedge dy_3 - dx_3 \wedge dy_2) \\ &+ (x_2-y_2)(dx_3 \wedge dy_1 - dx_1 \wedge dy_3) \\ &+ (x_3-y_3)(dx_1 \wedge dy_x - dx_2 \wedge dy_1)). \end{aligned}$$

This is nothing but the pull-back of the standard area form of the unit sphere $S^2 \subseteq \mathbb{R}^3$ by the map $\frac{x-y}{|x-y|}$, x, y in $\mathbb{R}^3$.

(2) Whenever the Gauss linking (here for convenience, we coin it as the Massey-Milnor linking of degree one)

$$L^*_{i,j} = \int\limits_{L_i} \int\limits_{L_j} (x-y) = \frac{1}{4\pi} \int\limits_{L_i} \int\limits_{L_j} \det \begin{pmatrix} x-y \\ dx \\ dy \end{pmatrix} \frac{1}{|x-y|^3}$$

vanishes, we define a well-defined function as: $d^{-1}i(x) = \int\limits_{x_j}^{x} \int\limits_{i} (t-y)$, where $t \in L_i$, $x \in L_j$ and $x_j \in L_j$ is the base point of L_j.

Similarly, whenever a given 1-form ψ defined on L_j with $\int_{L_j} \psi = 0$, we define the relevant function $\overline{\psi}(x)$, $x \in L_j$ as: $\overline{\psi}(x) = d^{-1}(\psi) = \int_{x_j}^{x} \psi$, where $x_j \in L_j$ is the base point. And we also coin a function of this sort as linking function.

(3) Given a closed 2-form ϕ in $\mathbb{R}^3$, for simplicity we set 1-form $\overline{\phi} \overset{def}{=} d^{-1}(\phi)(x,y,z) = \int_{\infty}^{x} \phi(t,y,z)$, where the path of integration is taken along the $\frac{\partial}{\partial x}$-direction starting from $(-\infty, y, z)$ up to (x,y,z), the point of interests.

(4) Using $\bar{i}(x) \overset{def}{=} d^{-1}(i)(x)$, $\overline{\phi}(x) = d^{-1}(\phi)(x)$, whenever they make sense we define inductively closed 2-form in $\mathbb{R}^3$: the "round brackets" $(m, m-1, \dots, 1)$, and closed 1-forms of interests the "sharp brackets" $< m, m-1, \dots, 1 >$ as follows.

(4-1) We define the round bracket $(j,i) = (j) \wedge (i) + \bar{j}(di) - \bar{i}(dj)$ is a closed 2-form in $\mathbb{R}^3$. Notice that di is a Dirac-like singular 2-form supported on the knot component L_i, and that $\bar{j}(x)$ in $\bar{j}(x)(di)$ refers to $x \in L_i$ and is a well-defined function thereon (a so-called linking function) as we assume that $L^*_{i,j} = 0$. Similar remarks hold for $\bar{i}(x)(dj)$, with $x \in L_j$.

Next, we define the sharp bracket $< j,i >= \overline{(j,i)} + j(x)\bar{i}(x)$ which is a closed 1-form in a tubular neighborhood of L_k with $k \neq i$, $k \neq j$.

(4-2) We define the round bracket

$$(k,j,i) = \overline{(k,j)} \wedge i + k \wedge \overline{(j,i)} + \overline{< k,j >}(di) - \overline{ki}(dj) + \overline{< i,j >}(dk)$$

which is a closed 2-form in $\mathbb{R}^3$. We notice that $(di)(x)$ is a Dirac-like singular 2-form supported on the knot component L_i, and that $\overline{< k,j >}(x)$ refers to $x \in L_i$ and is a well-defined function on L_i(a linking function), as we assume that the Massey-Milnor linkings of degree 2 vanish there. Similar remarks hold for $\overline{ki}(dj)$ and $\overline{< i,j >}(dk)$.

Next, we define the sharp bracket which is a closed 1-form $< k,j,i > (x)$ for $x \in L_l$, $l \neq i,j,k$ as:

$$< k,j,i > (x) = \overline{(k,j,i)}(x) + \overline{(k,j)}(x)i(x) + k(x)\overline{< j,i >}(x)$$

which is well-defined as $L^*_{i.l} = 0$ and $L^*_{j,i,l} = 0$ by the assumption of vanishing of Massey-Milnor linkings of lower degrees.

(4-3) Inductively, we define the closed 2-form $(j, j-1, \dots, 2, 1)$ in $\mathbb{R}^3$ as

$$\begin{aligned}
&(j, j-1, \dots, 2, 1) \\
=&\overline{(j, j-1, \dots, 2)} \wedge 1 + \overline{(j, j-1, \dots, 3)} \wedge \overline{(2,1)} \\
&+ \dots + \overline{(j, j-1)} \wedge \overline{(j-2, \dots, 1)} + j \wedge \overline{(j-1, j-2, \dots, \dots, 1)} \\
&+ \overline{< j, j-1, \dots, 2 >}(d1) - \overline{< j, j-1, \dots, 3 >}i(d2) \\
&+ \overline{< j, j-1, \dots, 4} \cdot \overline{< 1,2 >}(d3) + \dots + (01)^j \overline{< 1, 2, \dots, j-1 >}(dj),
\end{aligned}$$

which is well-defined by the assumption of vanishing of Massey-Milnor linkings of strictly lower degrees.

Next, we define the closed 1-form $< j, j-1, \dots, 1 > (x)$, $x \in L_l$ with $l \neq 1, 2, \dots, j$ as:

$$\begin{aligned}
&< j, j-1, \dots, 1 > (x) \\
=&\overline{(j, j-1, \dots, 1)}(x) + \overline{(j, j-1, \dots, 2}(x)\bar{1}(x) \\
&+ \overline{(j, j-1, \dots, 3)}(x)\overline{< 2,1 >}(x) + \dots \\
&+ \overline{(j, j-1)}(x)\overline{(j-2, j-3, \dots, 1)}(x) \\
&+ j(x)\overline{< j-1, j-2, \dots, 1 >}(x),
\end{aligned}$$

which is a well-defined function on L_l, as we assume that Massey-Milnor linkings of strictly lower degrees vanish.

(5) Finally for a link $L = \{L_0, L_1, \ldots, L_n\}$ of $(n+1)$ components for which all Massey-Milnor linkings L^*'s of strictly lower degrees vanish, we define the first non-vanishing Massey-Milnor linking

$$L^*_{n,n-1,\ldots,1,0} = \int_{L_0} < n, n-1, \ldots, 2, 1 > .$$

To conclude the presentation of Massey-Milnor linkings $L^*_{n,n-1,\ldots,1}$, we notice that $L^*_{n,n-1,\ldots,1,0}$ is an ambient isotopy invariant with respect to L_0 as $< n, n-1, \ldots, 1 >$ is a closed 1-form in a tubular neighborhood of L_0 which is disjoint from $\{L_1, L_2, \ldots, L_n\}$. But a beautiful theorem of J. Milnor [Mi1, Mi2] asserting that $L^*_{n,n-1,\ldots,0} = L^*_{n-1,n-2,\ldots,0,n} = L^*_{n-2,n-3,\ldots,0,n,n-1} = L^*_{0,n,n-1,\ldots,1}$ implies that the first non-vanishing Massey-Milnor linking $L^*_{n,n-1,\ldots,1,0}$ is an ambient isotopy invariant with respect to L_i for $i = 0, 1, \ldots, n$. And notice that the "double round brackets" $((n, n-1, \ldots, 1))$ defined in (3) of Definition 2 are nothing but the "connected" part of $(n, n-1, \ldots, 1)$, and also notice that the difference between $((n, n-1, \ldots, 1))$ and $(n, n-1, \ldots, 1)$ is a Dirac-like singular 2-form supported on $\{L_1, L_2, \ldots, L_n\}$.

5. Some calculus lemma

Armed with the link $L = \{L_0, L_1, \ldots, L_n\}$ as in the setup—for example the cyclic arrangement of L as shown there and the assumption of the vanishing of both invariants L's (Chern-Simons-Witten invariants defined in Section 3) and L^*'s (Massey-Milnor linkings defined in Section 4) of strictly lower degrees—and for the purpose of the equality $L = L^*$ and the combinatorial formulae thereof, we prepare some calculus lemmas which mostly are localization computation as our link L is represented as a link diagram lying entirely on the plane $\mathbb{R}^2$, except possibly the infinitesimally small neighborhood of crossings of L.

Lemma 1.

$$\begin{aligned}&d_A(A-B) - d_B(\underline{A}-B)\\ =&(-)\delta(A-B)(dA_1 \wedge dA_2 \wedge dB_3 + dA_2 \wedge dA_3 \wedge dB_1 + dA_3 \wedge dA_1 \wedge dB_2),\end{aligned}$$

where A, $B \in \mathbb{R}^3$ are dummy variables,

$$(A-B) \stackrel{def}{=} \left(\frac{1}{4\pi}\right) \frac{1}{|A-B|^3} \det \begin{pmatrix} A-B \\ dA \\ dB \end{pmatrix}$$

and

$$(\underline{A}-B) \stackrel{def}{=} \left(\frac{1}{8\pi}\right) \frac{1}{|A-B|^3} \det \begin{pmatrix} A-B \\ dA \\ dA \end{pmatrix};$$

and $\delta(A-B)$ is the Dirac function in $\mathbb{R}^3$.

Proof.

$$d_A(A-B)$$

$$= d_A\left(\frac{1}{4\pi}\right)\frac{1}{|A-B|^3} \begin{pmatrix} A_1-B_1 & A_2-B_2 & A_3-B_3 \\ dA_1 & dA_2 & dA_3 \\ dB_1 & dB_2 & dB_3 \end{pmatrix} = d_A \begin{pmatrix} \Gamma_1(A-B) & \Gamma_2(A-B) & \Gamma_3(A-B) \\ dA_1 & dA_2 & dA_3 \\ dB_1 & dB_2 & dB_3 \end{pmatrix},$$

if we set $\Gamma(x) = \frac{1}{4\pi|x|}$

$$
\begin{aligned}
=& d_A[\Gamma_1(dA_2 \wedge dB_3 - dA_3 \wedge dB_2) + \Gamma_2(dA_3 \wedge dB_1 - dA_1 \wedge dB_3) \\
& + \Gamma_3(dA_1 \wedge dB_2 - dA_2 \wedge dB_1)] \\
=& (\Gamma_{11}dA_1 + \Gamma_{12}dA_2 + \Gamma_{13}dA_3) \wedge (dA_2 \wedge dB_3 - dA_3 \wedge dB_2) \\
& + (\Gamma_{21}dA_1 + \Gamma_{22}dA_2 + \Gamma_{23}dA_3) \wedge (dA_3 \wedge dB_2 - dA_1 \wedge dB_3) \\
& + (\Gamma_{31}dA_1 + \Gamma_{32}dA_2 + \Gamma_{33}dA_3) \wedge (dA_1 \wedge dB_2 - dA_2 \wedge dB_1) \\
=& (\Gamma_{11} + \Gamma_{22} + \Gamma_{33}(dA_1 \wedge dA_2 \wedge dB_3 + dA_2 \wedge dA_3 \wedge dB_1) \\
& + dA_3 \wedge dA_1 \wedge dB_2) \\
& - (\Gamma_{11}dA_2 \wedge dA_3 \wedge dB_1 + \Gamma_{12}dA_2 \wedge dA_3 \wedge dB_2 + \Gamma_{13}dA_2 \wedge dA_3 \wedge dB_3) \\
& - \Gamma_{21}dA_3 \wedge dA_1 \wedge dB_1 + \Gamma_{22}dA_3 \wedge dA_1 \wedge dB_2 + \Gamma_{23}dA_3 \wedge dA_1 \wedge dB_3) \\
& - (\Gamma 31 dA_1 \wedge dA_2 \wedge dB_1 + \Gamma_{32}dA_1 \wedge dA_2 \wedge dB_2 + \Gamma_{33}dA_1 \wedge dA_2 \wedge dB_3) \\
=& -\delta(A - B) + d_B(\Gamma_1 dA_2 \wedge dA_3 + \Gamma_2 dA_3 \wedge dA_1 + \Gamma_3 dA_1 \wedge dA_2) \\
=& -\delta(A - B) = \frac{1}{8\pi} d_B \frac{1}{|A - B|^3} \det \begin{pmatrix} A - B \\ dA \\ dA \end{pmatrix}
\end{aligned}
$$

□

Lemma 2. *For the variation with respect to A, we have*

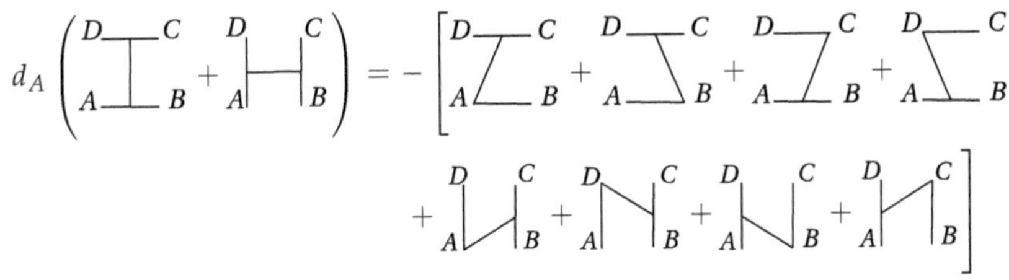

where A, B, C, and D are distinct indices (or dummy variables) in $\{0, 1, 2, \ldots, n\}$.

Proof. By definition, the configuration space integrals corresponding to 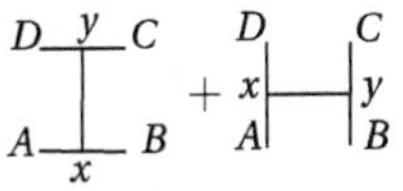 are:

$$\int_{A_x B_y CD} (A - x) \wedge (x - B) \wedge (x - y) \wedge (y - C) \wedge (y - D)$$

$$+ \int_{D_x A_y BC} (D - x_{\wedge}(x - A) \wedge (x - y) \wedge (y - B) \wedge (y - C).$$

Taking d_A, the exterior derivative of the LHS with respect to A, by our convention, is to take d_A of the associative configuration space integrals omitting the integration over A-chain, thereby we get a 2-form in dA. To be more explicit,

$$\begin{aligned}
& d_A \int_{xByCD} (A-x)\wedge(x-B)\wedge(x-y)\wedge(y-C)\wedge(y-C) \\
& \quad - d_A \int_{xDyBC} (A-x)\wedge(x-D)\wedge(x-y)\wedge(y-B)\wedge(y-C) \\
& = (-) \int_{ByCD} (A-B)\wedge(A-y)\wedge(y-C)\wedge(y-D) \\
& \quad + \int_{DyBC} (A-D)\wedge(A-y)\wedge(y-B)\wedge(y-C) \\
& \quad + \int_{xByCD} d_x(\underline{A}-x)\wedge(x-B)\wedge(x-y)\wedge(c-C)\wedge(y-D) \\
& \quad - \int_{xDyBC} d_x(\underline{A}-x)\wedge(x-D)\wedge(x-y)\wedge(y-B)\wedge(y-C)
\end{aligned}$$

$$\begin{aligned}
= & - \int_{ByCD} (A-B)\wedge(A-y)\wedge(y-C)\wedge(y-D) \int_{DyBC} (A-D)\wedge(A-y)\wedge(y-B)\wedge(y-C) \\
& - \int_{xByCD} (\underline{A}-x)\wedge(d_x(x-B))\wedge(x-y)\wedge(y-C)\wedge(y-D) \\
& + \int_{xDyBC} (\underline{A}-x)\wedge(d_x(x-D))\wedge(x-y)\wedge(y-D)\wedge(y-C) \\
& - \int_{xByCD} (\underline{A}-x)\wedge(x-B)\wedge(d_x(x-y))\wedge(y-C)\wedge(y-D) \\
& + \int_{xDyBC} (\underline{A}-x)\wedge(x-D)\wedge(d_x(x-y))\wedge(y-B)\wedge(y-C) \\
= & - \int_{ByCD} (A-B)\wedge(A-y)\wedge(y-C)\wedge(y-D) \\
& + \int_{DyBC} (A-D)\wedge(A-y)\wedge(y-B)\wedge(y-C) \\
& + \int_{ByCD} (\underline{A}-B)\wedge(B-y)\wedge(y-C)\wedge(y-D) \\
& - \int_{DyBC} (\underline{A}-D)\wedge(D-y)\wedge(y-B)\wedge(y-C) \\
& + \int_{ByCD} (\underline{A}-y)\wedge(y-B)\wedge(y-C)\wedge(y-D) \\
& - \int_{DyBC} (\underline{A}-y)\wedge(y-D)\wedge(y-B)\wedge(y-C) \\
& - \int_{xByCD} (\underline{A}-y)\wedge(y-D)\wedge(y-B)\wedge(y-C) \\
& - \int_{xByCD} (\underline{A}-x)\wedge(x-B)\wedge(d_y(\underline{x}-y))\wedge(y-C)\wedge(y-D) \\
& + \int_{xDyBC} (\underline{A}-x)\wedge(x-D)\wedge(D_y(\underline{x}-y))\wedge(y-B)\wedge(y-C)
\end{aligned}$$

$$= -\int_{ByCD}(A-B)\wedge(A-y)\wedge(y-C)\wedge(y-D)$$

$$+\int_{DyBC}(A-D)\wedge(A-y)\wedge(y-B)\wedge(y-C)$$

$$+\int_{ByCD}(\underline{A}-B)\wedge(B-y)\wedge(y-C)\wedge(y-D)$$

$$-\int_{DyBC}(\underline{A}-D)\wedge(D-y)\wedge(y-B)\wedge(y-C)$$

$$+\int_{xByCD}(\underline{A}-x)\wedge(x-B)\wedge(x-y)\wedge(D_y(y-C))\wedge(y-D)$$

$$-\int_{xDyBC}(\underline{A}-x)\wedge(x-D)\wedge(\underline{x}-y)\wedge(D_y(y-B)\wedge(y-C)$$

$$+\int_{xByCD}(\underline{A}-x)\wedge(x-B)\wedge(\underline{x}-y)\wedge(y-C)\wedge(D_y(y-D))$$

$$-\int_{xDyBC}(\underline{A}-x)\wedge(x-D)\wedge(\underline{x}-y)\wedge(y-B)\wedge(D_y(y-D))$$

$$= -\int_{ByCD}(A-B)\wedge(A-y)\wedge(y-C)\wedge(y-D)$$

$$+\int_{DyBC}(A-D)\wedge(A-y)\wedge(y-B)\wedge(y-C)$$

$$+\int_{ByCD}(\underline{A}-B)\wedge(B-y)\wedge(y-C)\wedge(y-D)$$

$$-\int_{yBCD}(\underline{A}-D)\wedge(D-y)\wedge(y-B)\wedge(y-C)$$

$$-\int_{xDBC}(\underline{A}-x)\wedge(x-B)\wedge(\underline{x}-C)\wedge(C-D)$$

$$+\int_{xDBC}(\underline{A}-x)\wedge(x-D)\wedge(\underline{x}-B)\wedge(B-C)$$

$$-\int_{xBCD}(\underline{A}-x)\wedge(x-B)\wedge(\underline{x}-D)\wedge(D-C)$$

$$+\int_{xDBC}(\underline{A}-x)\wedge(x-D)\wedge(\underline{x}-C)\wedge(C-B).$$

The last eight integrals are difficult to read in general. So for the purpose of book-keeping we use the graphic representation for them as in the above Lemma 2. Obviously the correspondence and the association of Chern-Simons-Witten configuration space integrals to edge-contracted Chern-Simons-Witten graphs, and vice versa are well-defined.

By definition the eight terms after the last equality sign in the above are the right hand side of Lemma 2. □

Lemma 3. $d_0($ 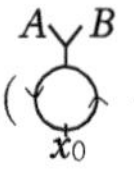

$+$ 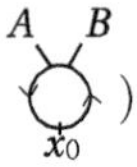

$) = 0$, *if* $A \neq B$ *in* $\{0, 1, 2, \ldots, n\}$.

Proof. By the definition of the associated Chern-Simons-Witten configuration space integrals both sides of the above equality are 2-forms in dL_0. That is

$$\begin{aligned}
LHS =& d_0 \int_{\mathbb{R}^3} \int_{BA} (0 - x) \wedge (x - B) \wedge (x - A) + d_0 \int_{x_0}^{0} \int_{BA} (x - B) \wedge (0 - A) \\
=& (-) \int_{BA} (0 - B) \wedge (0 - A) - \int_{BA} (\underline{0} - A) \wedge (A - B) \\
& - \int_{BA} (\underline{0} - B) \wedge (B - A) + \int_{BA} (0 - B) \wedge (0 - A) \\
=& - \int_{BA} (\underline{0} - A) \wedge (A - B) - \int_{BA} (\underline{0} - B) \wedge (B - A).
\end{aligned}$$

Observe that $(\underline{0} - A)$ and $(\underline{0} - B)$ are both functions in A and B which do not involve any differential forms in (dA) or (dB) respectively; so to carry out the above two integrations we may regard $(\underline{0} - A)$ and $(\underline{0} - B)$ as weight functions by computing first $\int_{BA}(A - B)$ and $\int_{BA}(B - A)$ to get Gauss signs of crossings of $L_A \cap L_B$ when $L = \{L_0, L_1, \ldots, L_n\}$ is in a generic position in $\mathbb{R}^2$ with pairwise crossings specified. Also observe that

$$- \int_{BA} (\underline{0} - A)(A - B) = - \sum_{L_A \cap L_B} (\underline{0} - A)(A, B)$$

and

$$\int_{BA} (\underline{0} - B)(B - A) = + \sum_{L_A \cap L_B} (\underline{0} - B)(A, B)$$

where (A, B), for the moment, stands for the Gaussian signs of crossings $L_A \cap L_B$ which is a finite sum supported on $L_A \cap L_B$; and this concludes the proof. □

Lemma 4. $d_0($ 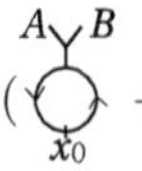

$+$ 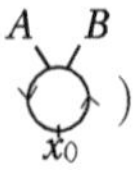

$) = 0$, *if* A, B *are two sharp brackets.*

Proof. The proof is exactly the same as that of Lemma 3. By integration-by-parts and by Lemma 1, we could regard both A and B as d-closed since both of them are sharp brackets. □

From now on we assume that $L = \{L_0, L_1, \ldots, L_n\}$ is in a generic position in $\mathbb{R}^2$ with pairwise crossings specified.

First we introduce some more notations: $\overline{(2,1)^*} = 2 \wedge 1$, $\overline{(3,2,1)^*} = \overline{\overline{(3,2)^*} \wedge 1 + 3 \wedge \overline{(2,1)^*}}$, and in general, $\overline{(n, n-1, \ldots, 1)^*} = \overline{(n, \ldots, 2)^*} \wedge 1 + \overline{(n, \ldots, 3)^*} \wedge \overline{(2,1)^*} + \cdots + n \wedge \overline{(n-1, \ldots, 1)^*}$.

Lemma 5. *The outer edge contractions of* 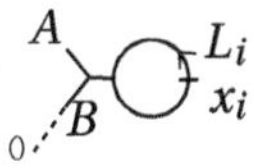

$+$ 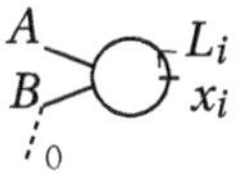

at L_i of the variation cancel each other, where the graph components containing vertex B connect L_i to L_0.

Proof. Starting with the exterior differentiation at L_0, d_0 the integration by parts as in Lemma 2 up to x in the first graph, which is a dummy variable on L_i to get the same resulting integrals for the above graph of two components. Then do the integrations by part again at x and at y, to get the same outer edge-contraction at L_i except for a difference of $(-)$ sign, so these two outer edge contractions at L_i cancel each other. This concludes the poof. □

Lemmas 1 to 5 are essentially preparatory computation for the calculus relevant to $L_{n,n-1,\ldots,1,0}$—the Chern-Simons-Witten invariants; next we do some preparatory computation for $L^*_{n,n-1,\ldots,1,0}$—the Massey-Milnor linking theory.

Lemma 6. *Denote(* 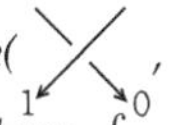

, 

respectively) as $(+)$ ($(-)$, respectively) for crossings of $L = \{L_0, L_1\}$, then we have $\int_{(+)}\overset{\frown}{(0-1)} = \frac{1}{2}$ and $\int_{(-)}(0-1) = -\frac{1}{2}$ where to ease the notation we keep the convention and notation in (1) of Definition 3.

Proof. Without loss of generality, we need only to prove

$$\frac{1}{4\pi}\int_{(+)} \begin{pmatrix} x-y \\ dx \\ dy \end{pmatrix} \frac{1}{|x-y|^3} = \frac{1}{2} \times \lim_{\substack{\epsilon\to 0 \\ \delta\to 0-\epsilon \\ \delta \ll \epsilon}} \int^{\epsilon} \frac{\delta dx dy}{(x^2+y^2+\delta^2)} = \frac{1}{2}$$

□

Note 2. *As our link $L = \{L_0, L_1, \ldots, L_n\}$ is represented by a link diagram in the plane $\mathbb{R}^2$ with arbitrarily small "germs" of pairwise crossings specified, it is easy to see that the only contribution of Gauss linking—which is equal to both $L^*_{i,j}$ and $L_{i,j}$—comes from the crossing part of L_i and L_j, and is denoted as (i,j) or (j,i).*

Lemma 7. *For two knot components L_i and L_j in $\{L_i, L_j, L_k\}$ with the associated 1-form $i(x)$, $j(x)$ as above the 1-form*

$$\overline{i \wedge j}(x) = \int_{-\infty}^{x} i(t) \wedge j(t)$$

$x \in \mathbb{R}^2$ is a Dirac-like 1-form supported on the horizontal positive $\frac{\partial}{\partial x}$-ray through the crossing of L_i and L_j in the link diagram of $\{L_i, L_j, L_k\}$. Moreover,

$$\int_{L_k} \overline{i \wedge j}(x) \stackrel{def}{=} (i,j)_k$$

$$= \begin{cases} \frac{1}{4} - \text{if} \\ -\frac{1}{4} - \text{if} \end{cases}$$

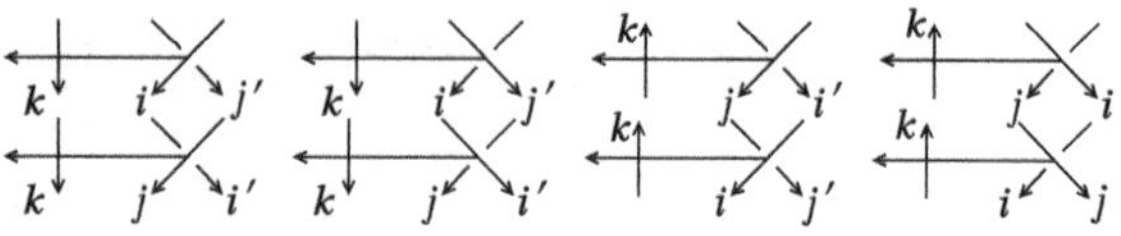

Proof. By the very defintion of integration along positive $\frac{\partial}{\partial x}$-direction in the plane $\mathbb{R}^2$, we have $\int_{-\infty}^{x} i(t) \wedge j(t) \equiv (\int_{-\infty}^{x} i(t)) \wedge j(x) - i(x) \wedge \int_{-\infty}^{x} j(t))$. And also notice that $\int_{-\infty}^{x} i(t) \wedge j(t)$ as a 1-form in $x \in \mathbb{R}^2$, is supported on the horizontal $\frac{\partial}{\partial x}$-ray passing through the crossing of the link diagram of L_i and L_j by a simple localization computation. Another simple localized estimate shows that $\int_{L_k} \overline{i \wedge j} = \int_{L_k} (\int_{-\infty}^{x} i(t)) \wedge j(x) - \int_{L_k} i(x) \wedge (\int_{-\infty}^{x} j(t)) = (i,j)_k$ as claimed. □

6. Main theorem

In this section we will state the main theorem of this paper and some proofs for the case of low degrees: $L_{1,0} = L^*_{1,0}$, $L_{2,1,0} = L^*_{2,1,0}$ and $L_{3,2,1,0} = L^*_{3,2,1,0}$. We will come back to the proof of the main theorem in Section 7 in full generality.

Here is the main theorem and as usual we use numerals $i, j, k, \ldots$, etc. for the corresponding knot components $L_i, L_j, L_k, \ldots$, etc. to ease the notation.

Theorem 1. *Given a link $L = \{L_0, L_1, \ldots, L_n\}$ of $(n+1)$ components arranged diagrammatically as in the setup for which all Chern-Simons-Witten graphs $L_{m,m-1,\ldots,1,0} = 0$, and all Massey-Milnor linkings $L^*_{m,m-1,\ldots,1,0} = 0$, where the sublink $\{L^*_0, L^*_1, \ldots, L^*_m\} \subseteq \{L_0, L_1, \ldots, L_n\}$ is any ordered subset of $(m+1)$ components, $m \leq n-1$, then for the first non-vanishing invariants $L_{n,n-1,\ldots,1,0}$ and ${}^*_{n,n-1,\ldots,1,0}$ we have*

(1) $L_{n,n-1,\ldots,1,0} = L^*_{n,n-1,\ldots,1,0}$, and

(2) Both $L_{n,n-1,\ldots,1,0}$ and $L^*_{n,n-1,\ldots,1,0}$ are independent of the base point $x_j \in L_j$, for $j = 0, 1, 2, \ldots, n$.

We will prove this theorem in full generality in Section 7. Here to make the presentation smoother and to show the idea, we do the "detailed" proof of: $L_{1,0} = L^*_{1,0}$, $L_{2,1,0} = L^*_{2,1,0}$ and $L_{3,2,1,0} = L^*_{3,2,1,0}$ as follows.

Example 6. *For $n = 1$ and $L = \{L_0, L_1\}$, by the very definition of $L_{1,0}$ and $L^*_{1,0}$ it is obvious that both $L_{1,0}$ and $L^*_{1,0}$ are exactly the Gauss linking and are formulated combinatorically as $L_{1,0} = L^*_{1,0} = \sum_{L_1 \cap L_0} (1,0)$. Here we follow the convention and notation of the* **Note 2** *after Lemma 6: $(1,0) = (\pm\frac{1}{2})$ according to the arrangement of L_1 and L_0 around the crossing "$L_1 \cap L_0$" in the link diagram of $\{L_0, L_1\}$.*

Example 7. *For $n = 2$ and $L = \{L_0, L_1, L_2\}$, we assume that all pairwise invariants $L_{i,j} = 0 = L^*_{i,j}$ then for the first non-vanishing invariants: $L_{2,1,0} = L^*_{2,1,0}$ and they can be computed combinatorically as follows. First recall that invariants $L_{2,1,0}$—the Chern-Simons-Witten graph defined in Section 4—is the sum of 4 configuration space integrals defined in Example 1; and $L^*_{2,1,0}$—the Massey-Milnor linking defined in Definition 4—is*

$$L^*_{2,1,0} = \int_0 \overline{2 \wedge 1} + \int_0 \overline{\overline{2} \wedge d1} - \int_0 \overline{\overline{1}d2} + \int_0 2\overline{1}$$

So to prove $L_{2,1,0} = L^*_{2,1,0}$, we need only to prove that the configuration space integral of "Y-graph"—which is nothing but the first integral in Example 4—in Example 1 is $\int_0 \overline{2 \wedge 1}$ in $L^*_{2,1,0}$.

Claim 1. *If we set the 1-form*

$$\phi(0) \stackrel{def}{=} \int_{\mathbb{R}^3} \int_{L_1} \int_{L_2} (0 - x) \wedge (x - 1) \wedge (x - 2)$$

which is a part of the configuration space integrals of the Chern-Simons-Witten graph "Y", then $d\phi(0) = (2 \wedge 1)(0)$.

Proof. With the link $L = \{L_0, L_1, L_2\}$ represented by a link diagram in the plane $\mathbb{R}^2$ and in the spirit of localization computation of Lemma 6 and Lemma 7, a direct computation concludes the proof by using the integration by part in Lemma 1.

And so,

$$\int_0 \overline{2 \wedge 1} = \int_0 \overline{d\phi} = \int_0 \overline{d\phi} + \int_0 d\overline{\phi} = \int_0 \phi = \text{ the configuration space integral of "Y"}.$$

And this concludes the proof that $L_{2,1,0} = L^*_{2,1,0}$. And by using the combinatoric formulae of $L^*_{2,1,0}$—that in Lemma 6 and Lemma 7—we derive the explicit formulae of $L_{2,1,0} = L^*_{2,1,0}$

$$L_{2,1,0} = (2,1)_0 + \sum_{\substack{1<2 \\ 0}} (1,0)(2,0) + \sum_{\substack{2<0 \\ 1}} (2,1)(0,1) + \sum_{\substack{0<1 \\ 2}} (0,2)(1,2)$$

□

Example 8. *For $n = 3$, $L = \{L_0, L_1, L_2, L_3\}$, for which all pairwise $L_{i,j} = L^*_{i,j} = 0$ and all triple-wise $L_{i,j,k} = L^*_{i,j,k} = 0$, then for the first non-vanishing invariants: we have $L_{3,2,1,0} = L^*_{3,2,1,0}$ and also they can be computed as listed below correspondingly, both graphically and combinatorically as follows:*

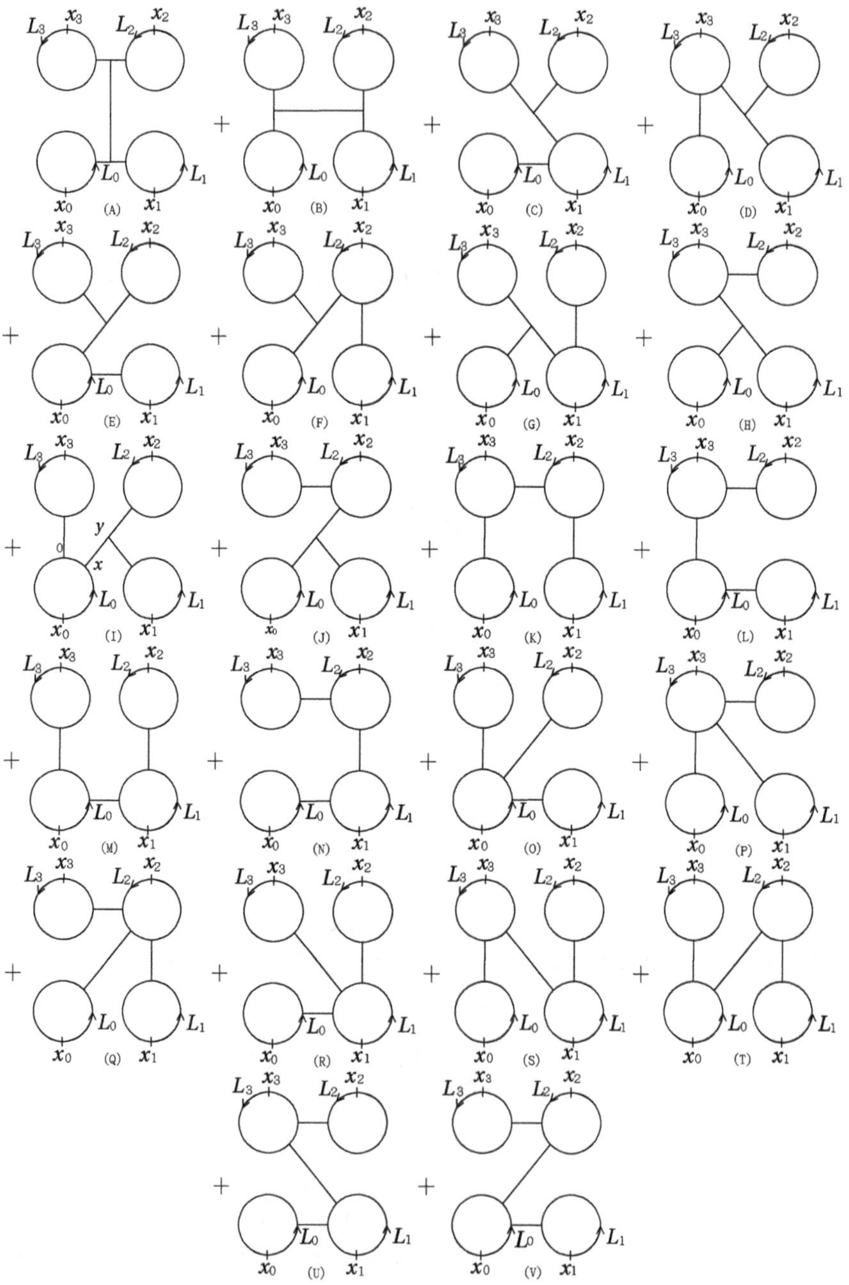
(A)
(B)
(C)
(D)
(E)
(F)
(G)
(H)
(I)
(J)
(K)
(L)
(M)
(N)
(O)
(P)
(Q)
(R)
(S)
(T)
(U)
(V)

and

$$\sum_{(2,1)\underset{0}{<}3}(2,1)_0(3,0)+\sum_{1\underset{0}{<}(3,2)}(3,2)_0(1,0)+\sum_{(3,2)\underset{1}{<}0}(3,2)_1(0,1)$$

$$+\sum_{2\underset{1}{<}(0,3)}(0,3)_1(2,1)+\sum_{(0,3)\underset{2}{<}1}(0,3)_2(1,2)+\sum_{3\underset{2}{<}(1,0)}(1,0)_2(3,2)$$

$$+\sum_{(1,0)\underset{3}{<}2}(1,0)_3(2,3)+\sum_{0\underset{3}{<}(2,1)}(2,1)_3(0,3)+\sum_{1\underset{0}{<}2\underset{0}{<}3}(0,1)(0,2)(0,3)$$

$$+\sum_{2\underset{1}{<}3\underset{1}{<}0}(1,2)(1,3)(1,0)+\sum_{3\underset{2}{<}0\underset{2}{<}1}(2,3)(2,0)(2,1)+\sum_{0\underset{3}{<}1\underset{3}{<}2}(3,0)(3,1)(3,2)$$

$$+\sum_{\substack{1\underset{0}{<}3\\ 2\underset{1}{<}0}}(1,0)(3,0)(2,1)+\sum_{\substack{2\underset{1}{<}0\\ 3\underset{2}{<}1}}(2,1)(0,1)(3,2)+\sum_{\substack{3\underset{2}{<}1\\ 0\underset{3}{<}2}}(3,2)(1,2)(0,3)$$

$$+\sum_{\substack{0\underset{3}{<}2\\ 1\underset{0}{<}3}}(0,3)(2,3)(1,0)+\sum_{\substack{2\underset{0}{<}3\\ 0\underset{2}{<}1}}(2,0)(3,0)(1,2)+\sum_{\substack{2\underset{1}{<}3\\ 0\underset{3}{<}1}}(2,1)(3,1)(0,3)$$

$$+\sum_{\substack{3\underset{1}{<}0\\ 1\underset{3}{<}2}}(3,1)(0,1)(2,3)+\sum_{\substack{1\underset{0}{<}2\\ 3\underset{2}{<}0}}(1,0)(2,0)(3,2)\ .$$

For this example we try to be as explicit as possible to show the general scheme of the proof of the main theorem.

First we derive the related combinatorical formulae of the Massey-Milnor linkings $L^*_{n,n-1,\ldots,1,0}$ as defined in Section 4 from which we do use the vanishing of Massey-Milnor linkings of strictly lower degrees. Also to ease the notation we use: $\bar{j}(x)=\int_{x_i}^{x} j(t)$, where $x_i, x \in L_i$ and x_i is the base point of L_i, whenever $L_{i,j}=0$. And $\overline{<j,k>}(x)=\int_{x_i}^{x} <j,k>(t)$ where $x_i, x\in L_i$ and x_i is the base point of L_i, whenever $L_{j,k,i}=0$. And obviously by induction they could be represented graphically as follows:

$\bar{j}(x) =$ 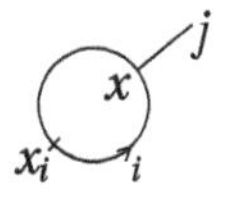

and

$\overline{<j,k>}(x) =$ 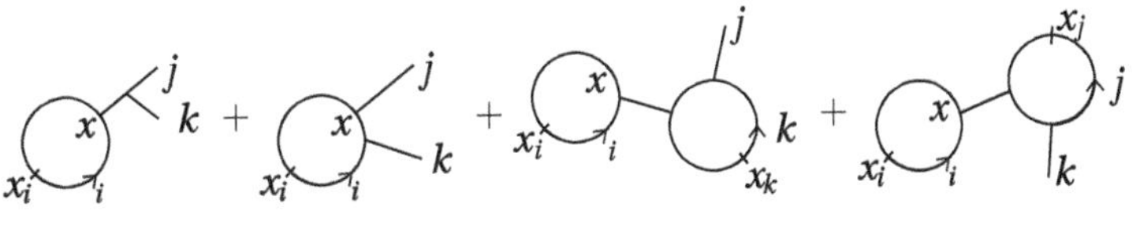

Recalling that:

$$\begin{aligned}
L^*_{3,2,1,0} &= \int_0 <3,2,1> = \int_0 \overline{(3,2,1)} + \int_0 \overline{(3,2)}\bar{1} + \int_0 3\overline{<2,1>} \\
&= \int_0 \overline{\overline{(3,2)} \wedge 1} + \int_0 \overline{3 \wedge \overline{(2,1)}} + \int_0 \overline{\overline{<3,2>}d1} \\
&\quad - \int_0 \overline{\bar{3}\bar{1}d2} + \int_0 \overline{\overline{<1,2>}d3} + \int_0 \overline{(3 \wedge 2)}\bar{1} \\
&\quad + \int_0 \overline{\bar{3}d2}\bar{1} - \int_0 \overline{\bar{2}d3}\bar{1} + \int_0 3\overline{<2,1>} \\
&= \int_0 \overline{\overline{(3 \wedge 2)} \wedge 1} + \int_0 \overline{\overline{(\bar{3}d2)} \wedge 1} - \int_0 \overline{\overline{(\bar{3}d2)} \wedge 1} \\
&\quad + \int_0 \overline{3 \wedge \overline{(2 \wedge 1)}} + \int_0 \overline{3 \wedge \overline{(\bar{2}d1)}} - \int_0 \overline{3 \wedge \overline{(\bar{1}d2)}} \\
&\quad + \int_0 \overline{\overline{<3,2>}d1} - \int_0 \overline{\bar{3}\bar{1}d2} + \int_0 \overline{\overline{<1,2>}d3} + \int_0 \overline{(3 \wedge 2)}\bar{1} \\
&\quad + \int_0 \overline{\bar{3}d2}\bar{1} - \int_0 \overline{\bar{2}d3}\bar{1} + \int_0 3\overline{<2,1>};
\end{aligned}$$

and using the above inductive scheme and by induction we have combinatorically,

$$\begin{aligned}
&L^*_{3,2,1,0} \\
&= \int_0 \overline{\overline{(3 \wedge 2} \wedge 1} + \int_0 \overline{3 \wedge \overline{(2 \wedge 1)}} + \sum_{3 \underset{2}{<} (1,0)} (1,0)_2(3,2) \\
&\quad + \sum_{(1,0) \underset{3}{<} 2} (1,0)_3(2,3) + \sum_{2 \underset{1}{<} (0,3)} (0,3)_1(2,1) + \sum_{(0,3) \underset{2}{<} 1} (0,3)_2(1,2) \\
&\quad + \sum_{(3,2) \underset{1}{<} 0} (3,2)_1(0,1) + \sum_{2 \underset{1}{<} 3 \underset{1}{<} 0} (2,1)(3,1)(0,1) + \sum_{\substack{2 \underset{1}{<} 0 \\ 3 \underset{2}{<} 1}} (2,1)(0,1)(3,2) \\
&\quad + \sum_{\substack{3 \underset{1}{<} 0 \\ 1 \underset{3}{<} 2}} (3,1)(0,1)(2,3) + \sum_{3 \underset{2}{<} 0 \underset{2}{<} 1} (3,2)(0,2)(1,2) + \sum_{1 \underset{3}{<} (2,1)} (2,1)_3(0,3) \\
&\quad + \sum_{0 \underset{3}{<} 1 \underset{3}{<} 2} (0,3)(1,3)(2,3) + \sum_{\substack{3 \underset{2}{<} 1 \\ 0 \underset{3}{<} 2}} (3,2)(1,2)(0,3) + \sum_{\substack{2 \underset{1}{<} 3 \\ 0 \underset{3}{<} 1}} (2,1)(3,1)(0,3) \\
&\quad + \sum_{1 \underset{0}{<} (3,2)} (3,2)_0(1,0) + \sum_{\substack{1 \underset{0}{<} 2 \\ 3 \underset{2}{<} 0}} (1,0)(2,0)(3,2) + \sum_{\substack{1 \underset{0}{<} 3 \\ 0 \underset{3}{<} 2}} (1,0)(3,0)(2,3) \\
&\quad + \sum_{(2,1) \underset{0}{<} 3} (2,1)_0(3,0) + \sum_{1 \underset{0}{<} 2 \underset{0}{<} 3} (1,0)(2,0)(3,0) + \sum_{\substack{2 \underset{1}{<} 0 \\ 1 \underset{0}{<} 3}} (2,1)(0,1)(3,0) \\
&\quad + \sum_{\substack{2 \underset{0}{<} 3 \\ 0 \underset{2}{<} 1}} (2,0)(3,0)(1,2)
\end{aligned}$$

For the reamining two integrals above we have

Claim 2. $\int_0 \overline{\overline{(3\wedge 2)}\wedge 1} = 0$ *and* $\int_0 \overline{3\wedge\overline{(2\wedge 1)}} = 0.$

Proof. Firstly we do the Massey-Milnor linking: $L^*_{3,2,1,0}$. By notation for a given over-barred 2-form $\overline{\phi}$ which is 1-form after doing the horizontal $\frac{\partial}{\partial x}$-integration which is the inverse of exterior differentiation, (after gauge-fixing) evaluated on the plane $\mathbb{R}^2$—as all integrations defined in the Massey-Milnor linkings $L^*_{n,n-1,\ldots,1,0}$ are carried out in the plane $\mathbb{R}^2$. We need to define the suitable d^{-1} as the average of horizontal $\frac{\partial}{\partial x}$-integrations, both infinitesimally above $\mathbb{R}^2$ and infinitesimally below $\mathbb{R}^2$. And so for each wedge product 2-form like $\overline{(3\wedge 2)}\wedge 1$, once we do the horizontal $\frac{\partial}{\partial x}$-integration again (the average of two, one above $\mathbb{R}^2$ infinitesimally and one below $\mathbb{R}^2$ infinitesimally) we always get zero by simply observing that $\overline{\overline{(3\wedge 2)}\wedge 1} = (-)\overline{(3\wedge 2)}\overline{1} = 0$ $\square$

Next we do the Chern-Simons-Witten graph: $L_{3,2,1,0}$. By induction the new and nontrivial graphs appearing in the Chern-Simons-Witten graphs $L_{3,2,1,0}$ are: 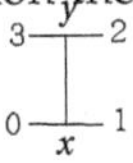 and 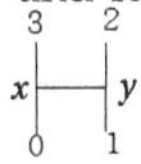 of which the Chern-Simons-Witten configuration space integrals are defined in Example 5. Now by shifting gear we regard these two configuration space integrals as two suitable 1-forms in $d0$ before carrying out the "final" integration over L_0; namely the two relevant 1-forms in $d0$ are:

$$\phi(0) = \int_{\mathbb{R}^3}\int_1\int_{\mathbb{R}^3}\int_2\int_3 (0-x)\wedge(x-1)\wedge(x-y)\wedge(y-2)\wedge(y-3),$$

and

$$\psi(0) = \int_{\mathbb{R}^3}\int_{\mathbb{R}^3}\int_1\int_2\int_3 (0-x)\wedge(x-y)\wedge(y-1)\wedge(y-2)\wedge(y-3).$$

By stoke's theorem:

$$\int_0 \phi(0)+\psi(0) = \int_0 \overline{d\phi(0)} + \overline{d\psi(0)} = \text{ sum of following eight integrals}$$

(which will be made explicit in lemma 10 in Section 7):

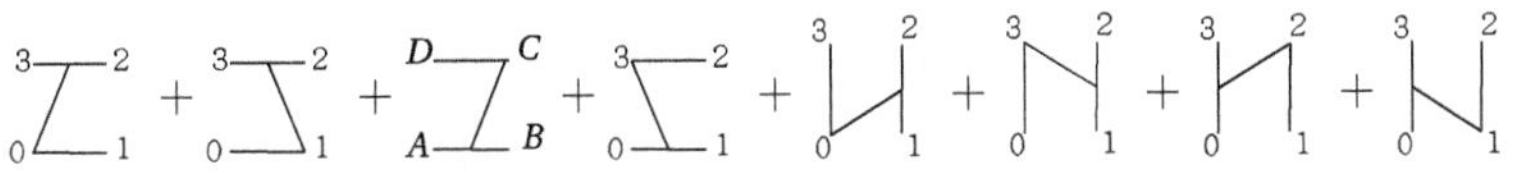

In $L_{3,2,1,0}$, corresponding to the above eight contracted graphs we have eight lower Chern-Simons-Witten linking together with an extra 1-chord:

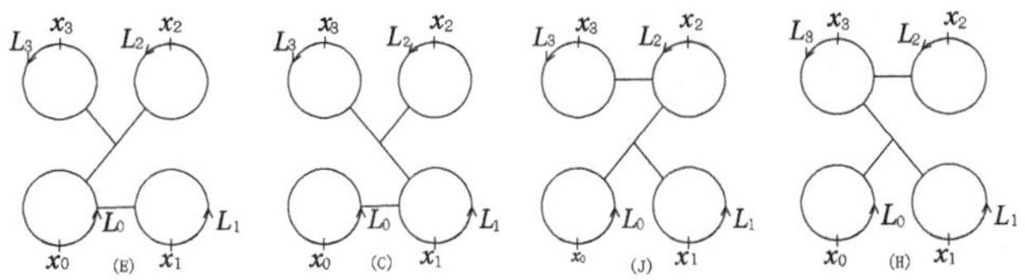

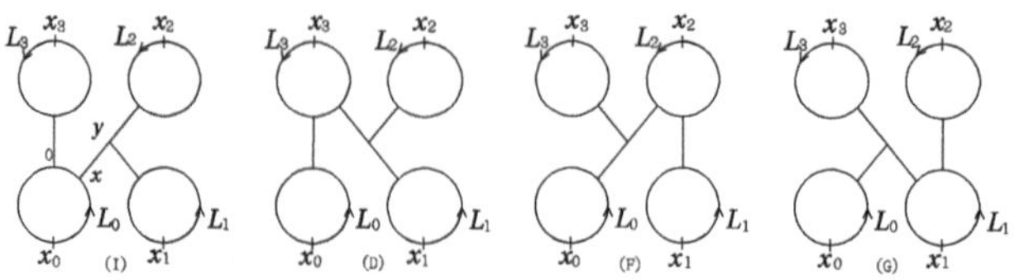

But to compute the latter 8 Chern-Simons-Witten configuration space integrals aiming at the combinatorial formulae we need to take into account the boundary terms for each related Chern-Simons-Witten graphs of degree 2—say the first graph along L_0. For example by induction $L_{2,1,0} = L^*_{2,1,0}$, but not now; as we don't have full cycles as before (as L_0 is cut into finitely many pieces of arcs by L_3). And the beauty of computing the extra correction due to the end points of the set of arcs along L_0 is that: the correction is compensated by those 8 integrals in $\int_0 \phi(0) + \psi(0)$ which is just a discrete sum on the corresponding end points of the pieces of arcs—cut by the attached chord in the contracted Chern-Simons-Witten graphs—of the configuration space integrals of Y-graphs thereof.

In greater details: take 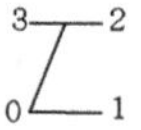 and 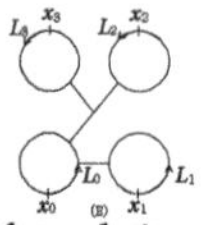 as an example to show the details proper. For simplicity we assume that a piece of arc lying on L_0 is cut out by L_1 as shown.

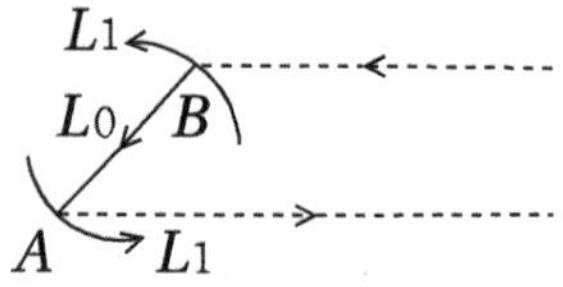

By induction: applying linking of degree 2 to this piece of arc as shown we need to artificially attach two horizontal rays to both A and B to make it a cycle as shown above. Then on this full cycle, we could compute both integrals of 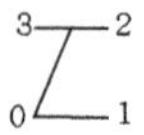 and 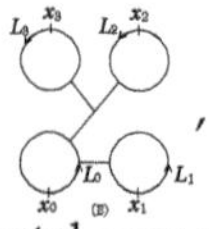, by replacing the original sub-arc $\overline{AB}$ with the full "cycle" made up of two horizontal rays and $\overline{AB}$ to get two extra corrections of opposite signs along the two added horizontal rays at A and B in the above two integrals. More precisely these two corrections—up to signs—are nothing but the configuration space integrals of Y-graph: 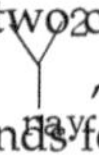, where ray at the root denotes the added ray and numeral 2 (respectively numeral 3) stands for L_2 (respectively L_3).

With the same computation for the other 7 pairs of Chern-Simons-Witten graphs listed above, we are done with the proof of $L^*_{3,2,1,0} = L_{3,2,1,0}$ if we have proved the following.

Claim 3. *Consider* 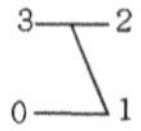 *as either a 2-form in d0:* $\eta(0) = \int(\underline{0}-1)(1-x)(x-2)(x-3)$ *or a 2-form in d1:* $\eta(1) = \int(0-1)(1-x)(x-2)(x-3)$ *where we carry out all related integrations except that over* L_0 *in* $\eta(0)$ *and except that over* L_1 *in* $\eta(1)$*, then we have* $\int_{L_0}\overline{\eta(0)} = \int_{L_1}\overline{\eta(1)}$.

Note 3. *Before giving the proof we notice that this scheme of proof works fine for the other contracted Chern-Simons-Witten graphs in* $L_{3,2,1,0}$. *And this proves that* $L_{3,2,1,0} = L^*_{3,2,1,0}$, *and also derives the combinatorial formulae of* $L^*_{3,2,1,0}$.

Note 4. *We will repeat the same trick for connected Chern-Simons-Witten graphs in* $L_{n,n-1,\ldots,1,0}$ *in Lemma 10, and will be more explicit in detail in this aspect. More precisely there are three aspects in this trick: firstly, we do contraction on the set of connected Chern-Simons-Witten graphs to get a bunch of 2-forms in* $d0$ *(namely those ones corresponding to the connected Chern-Simons-Witten graphs contracted at uni-valent vertices); secondly, for a connected Chern-Simons-Witten graph contracted at a vertex* L_j *with* $j \neq 0$, *we repeat the proof of the above claim to show:* $\int_{L_0} \overline{\eta(0)} = \int_{L_j} \overline{\eta(j)}$; *thirdly, we compute explicitly* $\int_{L_j} \overline{\eta(j)}$ *and prove that it compensates the corresponding correction coming from the disconnected Chern-Simons-Witten graph which has the same contracted graph after being contracted at vertex* L_j.

Now we come to the proof of the claim.

Proof. Recall that the over-bars on 2-forms, by definition, are horizontal $\frac{\partial}{\partial x}$-integrations from $x = -\infty$ up to points of interests; and that if the point of interests is a planar point, the horizontal $\frac{\partial}{\partial x}$-integration should be defined as the average of the one infinitesimally above $\mathbb{R}^2$ and the one infinitesimally below $\mathbb{R}^2$.

(A) First we define and state contracted Chern-Simons-Witten graphs and the corresponding 2-forms—either in $d0$ or $d1$ explicitly. But this is nothing but the content of Lemma 1 and Lemma 2 in Section 5.

(B) Next we compute: $\int_{L_1} \overline{\eta(1)} = \Sigma_{L_0 \cap L_1} (0,1) \int_{\text{hori}} \int_2 \int_3 \int_{\mathbb{R}^3} (h-x)(x-2)(x-3)$, where hori stands for horizontal $\frac{\partial}{\partial x}$-ray starting/ending at the crossings of L_1 and L_0, $(0,1)$ is the Gauss linking of L_1 and L_0, and h denotes the dummy variable on the horizontal $\frac{\partial}{\partial x}$-rays stated above.

(C) Next we compute $\int_{L_0} \overline{\eta(0)}$. To apply the calculus Lemma 1 and Lemma 2 to an sub-arc on L_1 cut-off by crossing with L_0, we need to artificially add two horizontal $\frac{\partial}{\partial x}$-rays at the ends of this arc to make it a full cycle. Now carry out the fundamental theorem of calculus as stated in Lemma 1 and Lemma 2 for this sub-arc of L_1 to get exactly the extra boundary evaluation at the ends of the configuration space integrals of Y-graph 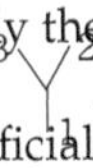

, and then identify Y-graph with symbols $(3,2)_1 = (2,1)_3 = (1,3)_2$ for L_2, L_3 and the artificial cycle L_1 (which is the union of the above sub-arc and two horizontal $\frac{\partial}{\partial x}$-rays) to conclude the proof of the claim. □

Note 5. *The proof of the claim applies also to any double round brackets in Definition 2, the so-called connected Chern-Simons-Witten graphs. In short in the Massey-Milnor linkings* $L^*_{n,n-1,\ldots,1,0}$, *connected Chern-Simons-Witten graphs always contribute nothing except those connected* Y*-graphs of degree 2.*

Note 6. *We notice that we could only do the trick of contraction for connected graphs as above, but not for non-connected ones.*

Note 7. *In next section we will repeat the same trick of contracting connected Chern-Simons-Witten graphs to compensate the corrections on the "cut-points"—due to lower linking functions (by induction, a discrete sum) along a specific knot component of interests—of suitable configuration space integrals involving artificial horizontal rays starting/ending at the cut-points, and other knot components proper.*

Note 8. *In essence the proof and the computation for* $L^*_{3,2,1,0} = L_{3,2,1,0}$ *and the combinatorial formulae thereof prevail in general case, because we assume the vanishing of linkings; and so we could compute the related lower linking functions along some specific knot component—which by induction are just some discrete sums along that knot component.*

7. Proof of the main theorem—general case

In this section we prove the main theorem of this paper. Here are some preparatory lemmas.

Lemma 8. *In the Massey-Milnor linking* $L^*_{n,n-1,\ldots,1,0}$ *or the associated 1-form* $< n, n-1, \ldots, 2, 1 >$ *on* L_0*, the Dirac-like singular one-form when evaluated at* $x \in \mathbb{R}^2$*,* $\overline{dj}(x)$ *is nothing but the one-form* $j(x) = \int_{L_j} (x-t)$ *where* $(x-t) = \frac{1}{4\pi}\frac{1}{|x-t|^3} det \begin{pmatrix} x-t \\ dt \\ dx \end{pmatrix}$*, and the over-bar denotes the horizontal* $\frac{\partial}{\partial x}$*-integration as defined in Definition 3.*

Proof. By de Rham theory for any 1-form $j(x)$ in $\mathbb{R}^3$ we have $j(x) = \overline{dj} + d\bar{j}$, where as usual the over-bar stands for the inverse of exterior differentiation—after gauge-fixing, which is the horizontal $\frac{\partial}{\partial x}$-integration from $x = -\infty$ up to the point of interests. But in Massey-Milnor linking the integrations of either $j(x)$ or $\overline{dj}(x)$ are carried out in $\mathbb{R}^2$, so the suitable horizontal $\frac{\partial}{\partial x}$-integration (namely the over-bar) in $\overline{dj}$ should be defined as the average of the two natural ones—either infinitesimally above $\mathbb{R}^2$ or infinitesimally below $\mathbb{R}^2$. Hence $\bar{j}(x) = 0$ when restricted to the plane $\mathbb{R}^2$, and this concludes the proof. □

Note 9. *From Lemma 8 when restricted to the plane* $\mathbb{R}^2$ $\overline{dj} \wedge i$ *is nothing but* $j(x) \wedge i(x)$*, so in particular* $\int_0 \overline{\overline{dj} \wedge j} = \int_0 \overline{j \wedge i} = (j,i)_0$*, in the notation of Lemma 7.*

Lemma 9. *In Massey-Milnor linking* $L^*_{n,n-1,\ldots,1,0}$*, all connected Chern-Simons-Witten graphs of degree larger than 2 contribute nothing.*

Proof. For connected Chern-Simons-Witten graphs of degree larger than 2 we have double round brackets of length larger than 2. Take a double round bracket of length 3 such as $\int_0 \overline{\overline{(3 \wedge 2 \wedge 1}} = +\int_0 \overline{\overline{(3 \wedge 2)}} \wedge 1 - \int_0 \overline{(3 \wedge 2)} \wedge \bar{1} = 0$, we get nothing simply because the term $\overline{\overline{(3 \wedge 2)}}$ has repeated $\frac{\partial}{\partial x}$-integrations and $\bar{1}(x)$ vanishes by the proof of Lemma 8. So any double round bracket of length larger than 2 also contributes nothing as it contains double round brackets of length 3. □

For the Chern-Simons-Witten perspective $L_{n,n-1,\ldots,1,0}$, we have the following.

Lemma 10. *(A) Regard the set of connected Chern-Simons-Witten graphs in $L_{n,n-1,\ldots,1,0}$—namely those corresponding to double round brackets in Definition 2—as 1-forms in $d0$ after doing all related integrations except the one over the component L_0. And if we take the exterior differentiation once with respect to $d0$ of the sum of these 1-forms, then we are left with only those Chern-Simons-Witten graphs contracted at the uni-vertices.*

(B) For any contracted Chern-Simons-Witten graph in (A), if the contracted vertex—only one such for each contracted Chern-Simons-Witten graph—is L_j, $j \neq 0$, then it is the same as the contracted Chern-Simons-Witten graph when considered as taking the exterior differential with respect to the coordinate j(the dummy variable for L_j). More precisely, for example if the contracted connected Chern-Simon-Witten graph Γ looks like

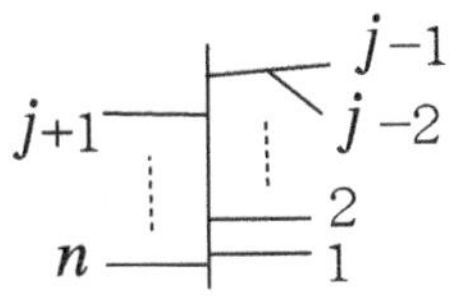

Then $\int\int_{D_0} \Gamma = \int\int_{D_j} \Gamma$, where the Γ inside the integral sign on the left hand side is considered as a 2-form in $d0$, say $\eta(0)$—as we take an exterior differentiation with respect to $d0$ once; and where the Γ in the integral sign on the right hand side is considered as a 2-form in dj, say $\eta(j)$—as, this time we take an exterior differentiation with respect to djonce. In short, $\int\int_{D_0} \eta(0) = \int_{L_0} \overline{\eta(0)} = \int\int_{D_j} \eta(j) =$ $int_{L_j} \overline{\eta(j)}$.

Proof. (A) The content (A) is nothing but a corollary of Lemma 1 and Lemma 2 in Section 5.

(B) We repeat the trick and the proof of the claims in Example 8. That is to each connected Chern-Simons-Witten graph contracted at a vertex L_j, with $j \neq 0$, we associated two 2-forms (one in $d0 \wedge d0$, and one in $dj \wedge dj$), $\eta(0)$ and $\eta(j)$. We will proceed by induction: $\int_{L_0} \overline{\eta(0)} =$ $\int_{L_j} \overline{\eta(j)}$, where the over-bars are the average of horizontal $\dfrac{\partial}{\partial x}$-integration: one infinitesimally above $\mathbb{R}^2$ and one infinitesimally below $\mathbb{R}^2$ from $x = -\infty$ up to the points of interests in $\mathbb{R}^2$.

(B-1) *We treat $\int_{L_j} \overline{\eta(j)}$ first.* The connected Chern-Simons-Witten graph contracted at the vertex L_j is considered now in another perspective, as the contraction of two distinct connected graphs of strictly lower degree at the vertex L_j. By induction a connected component of the above two distinct graphs could be regarded as a part of linking function of lower degree at L_j, and hence would cut off L_j into finitely many sub-arcs. Repeat the trick of the claims in Example 8 for the set of knot components attached to the other graph compoment and a specific subarc of L_j to get the horizontal $\dfrac{\partial}{\partial x}$-rays start/end at the ends of the sub-arc of L_j cut-off by the combinatorial formulae (a part of the lower linking function associated to that graph component).

Repeat the same computaton with the roles of these two graph components switched. That is obvious as $\overline{\eta(j)} = \overline{\alpha(j) \wedge \beta(j)} = \overline{\alpha(j)}\beta(j) - \alpha(j)\overline{\beta(j)}$ which is a nontrivial and key formulae in our discrete computation of horizontal $\frac{\partial}{\partial x}$-integration, but it is easy to see in the following set-up: cut the knot component L_j by the discrete contributions due to these two connected Chern-Simons-Witten graphs (which give rise to two 1-forms $\alpha(j)$ and $\beta(j)$ above); and as above we add artificially horizontal $\frac{\partial}{\partial x}$-rays strating/ending at these cut-point so that we could apply induction scheme for these two subgraphs of lower degree to suitable knot components attached thereon and the artificial cycles, each of which consists of a subarc of L_j and the two horizontal $\frac{\partial}{\partial x}$-rays added onto the boundary points of the subarc. Finally do the obvious discrete sum along L_j to get the net contribution as claimed.

(B-2) *Next we treat* $\int_{L_0} \overline{\eta(0)}$.

As the graph component containing L_0 is of strictly lower degree than n, by induction and after carrying out all the integrations of the configuration space integral of that graph component we are left with an honest function (not a differential form) on L_j. This also can be regarded as the boundary evaluation of the associated configuration space integral when adding the artificial horizontal $\frac{\partial}{\partial x}$-ray to the sub-arc of L_j to make a full cycle so that we could apply Lemma 1 and Lemma 2. And this boundary evaluation is exactly the horizontal $\frac{\partial}{\partial x}$-integration of the configuration space integral by the trick of Example 8.

Similarly switch the roles of these two graph components to get the other horizontal $\frac{\partial}{\partial x}$-integration of the configuration space integral of one graph component; while the other graph component plays the role of lower linking function and so is just a discrete evaluation there; and hence cuts off L_j into sub-arcs whose boundary points support the artificially horizontal $\frac{\partial}{\partial x}$-rays as above. In short the boundary evaluation at the ends of a specific subarc of L_j of the configuration space integral of the subgraph containing L_0, by Lemma 1 and Lemma 2 is exactly the horizontal $\frac{\partial}{\partial x}$-integration of the artificial rays as above simply by computing the Chern-Simons-Witten configuration space integrals in two ways—one, we integrate over just that specific sub-arc to get the extra boundary evaluation; and the other, we integrate over the full cycle consisting of the above sub-arc and two added horizontal $\frac{\partial}{\partial x}$-rays. □

With all the preparatory lemmas we come to the proof of Theorem 1 in Section 6.

Step 1. *Massey-Milnor linking* $L^*_{n,n-1,\ldots,1,0}$.

By assumption all Massey-Milnor linkings of lower degree vanish and by Lemma 9 all connected Chern-Simons-Witten graphs of degree larger than 2 contribute nothing to $L^*_{n,n-1,\ldots,1,0}$. The combinatorial formulae of $L^*_{n,n-1,\ldots,1,0}$ could be read out directly from the expansion of the sharp bracket $< n, n-1, \ldots, 2, 1 >$. More precisely, regard lower linkings as linking functions supported on some knot component L_j which corresponds to some dj in $< n, n-1, \ldots, 1 >$. By induction only those Chern-Simons-Witten connected graphs of degree one and two contribute to the combinatorial formulae of $L^*_{n,n-1,\ldots,1,0}$. Also for a connected Chern-Simons-Witten graph of degree 2  which is not symmetric with respect to its vertices $\{i, j, k\}$, the symbol $(k, j)_i$ is naturally and canonically assigned to this Y-graph without ambiguity simply by inspecting the related linking function therewith.

Step 2. *Chern-Simons-Witten graph* $L_n, n-1, \ldots, 1, 0$.

(A) Consider the set of connected Chern-Simons-Witten graphs first. Mimicking the trick of contraction of connected Chern-Simons-Witten graphs in $L_{3,2,1,0}$ in Section 6, we regard this set of contracted connected Chern-Simons-Witten graphs a sum of 2-forms with only one contracted vertex for each, along which we will construct artificially horizontal $\frac{\partial}{\partial x}$-rays as before. And as in the Section 6 this kind of Chern-Simons-Witten configuration space integrals that involves the artificial horizontal rays will compensate those boundary evaluation of the associated lower linking functions at the contracted vertex (equivalently the knot component corresponding to the vertex numeral). In short even in Chern-Simons-Witten graphs, the connected graphs of degree larger than two do not contribute anything to the combinatorial formulae thereof. This is obvious by the above argument and by induction.

(B) For Chern-Simons-Witten graphs of degree 3 $L_{3,2,1,0}$, we have explicit combinatorial formulae as given in Example 8. And so by induction only connected Chern-Simons-Witten graphs of degree one and two contribute to the combinatorial formulae of $L_{n,n-1,\ldots,1,0}$. And for Y-graph 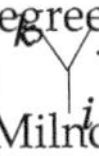, joining 3 knot components $\{L_i, L_j, L_k\}$ we need to apply the identification of Massey-Milnor linkings and Chern-Simons-Witten graphs of strictly lower degrees to determine the suitable symble $(k, j)_i$ for this Y-graph.

Step 3. *Massey-Milnor linkings = Chern-Simons-Witten graphs; that is* $L^*_{n,n-1,\ldots,1,0} = L_{n,n-1,\ldots,1,0}$.

(A) We have done this for $n = 1, 2$ in Section 6.

(B) By induction.

Assume that we are done for all such of degrees less than n, that is we have identified all those non-connected Chern-Simons-Witten graphs once they are grouped together in the manner of lower linking function which correspond to those over-barred sharp brackets with some dj attached besides, in $L^*_{n,n-1,\ldots,1,0}$. And obviously, we are left with only those connected Chern-Simons-Witten graphs of degree n in both Massey-Milnor linking and Chern-Simons-Witten graph set-ups. In the former, connected Chern-Simons-Witten graphs correspond to double round brackets, and so contribute nothing by Lemma 10 to $L^*_{n,n-1,\ldots,1,0}$; in the latter, the connected Chern-Simons-Witten graphs will compensate the artificial horizontal $\frac{\partial}{\partial x}$-rays derived from the lower linkings functions and the suitable sub-graphs contracted there on.

Step 4. *Both* $L_{n,n-1,\ldots,1,0}$ *and* $L^*_{n,n-1,\ldots,1,0}$ *are independent of the base points* $x_j \in L_j$, $j = 0, 1, 2, \ldots, n$.

Due to cyclic symmetry of $L^*_{n,n-1,\ldots,1,0}$—as is obvious from the Chern-Simons-Witten perspective—we need only to prove that $L^*_{n,n-1,\ldots,1,0}$ is independent of the choice of $x_0 \in L_0$. And we denote the change or the variation due to the change of base point $x_0 \in L_0$ by the notation δ. For example, δ may stand for changing the old base point $x_0 \in L_0$ to the new base point $x_0^* \in L_0$; and within $[x_0, x_0^*] \subseteq L_0$, we may have

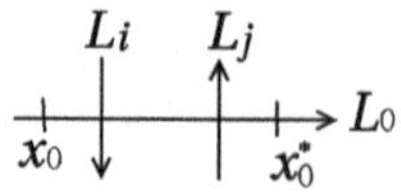

or

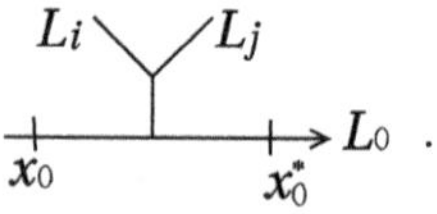

By definition, $L^*_{n,n-1,\dots,1,0} = \int_0 \overline{(n,n-1,\dots,1} + \int_0 \overline{(n,\dots,2)1} + \int_0 \overline{(n,\dots,3)} \cdot \overline{<2,1>} + \cdots + \int_0 n\overline{<n-1,\dots,1>}$, and if for some k such that $\delta\overline{<k,k-1,\dots,1>} = c \neq 0$ and $\delta\overline{<k-1,\dots,1>} = 0 = \delta\overline{<k-2,\dots,1>} = \cdots = \delta\overline{<3,2,1>} = \delta\overline{<2,1>} = \delta\overline{1} = 0$, then we have

Claim 4.

$$\delta\overline{<k+1,k,\dots,1>} = C\overline{(k+1)},$$
$$\delta\overline{<k+2,k+1,\dots,1>} = C\overline{<k+2,k+1>},$$
$$\delta\overline{<k+3,k+2,\dots,1>} = C\overline{<k+3,k+2,k+1>},$$
$$\cdots \quad \cdots \quad \cdots$$
$$\delta\overline{<n-1,n-2,\dots,1>} = C\overline{<n-1,n-2,\dots,k+1>}.$$

Proof. By the very definition of $\overline{<k+1,\dots,1>}, \overline{<k+2,\dots,1>}, \cdots, \overline{<n-1,n-2,\dots,1>}$, for $x \in L_0$ we have (being the dumming variable of L_0), $\delta\overline{<k+1,\dots,1>}(x) = \delta\int_{x_0}^{x}\overline{(k+1,\dots,1)} + \delta\int_{x_0}^{x}\overline{(k+1,\dots,2)1} + \delta\int_{x_0}^{x}\overline{(k+1,\dots,3)<21>} + \cdots + \delta\int_{x_0}^{x}(k+1)\overline{<k,\dots,1>} = C\int_{x_0}^{x}(k+1) = C\overline{(k+1)}(x)$, where as usual, for $j \in \{1,2,\dots,n\}$, $j(x)$ is regarded as 1-form.

Similarly, $\delta\overline{<k+2,k+1,\dots,1>} = \delta\int_{x_0}^{x}\overline{(k+2,\dots,1)} + \delta\int_{x_0}^{x}\overline{(k+2,\dots,2)1} + \cdots + \delta\int_{x_0}^{x}(k+2)\overline{<k+1,\dots,1>} = C\int_{x_0}^{x}\overline{(k+2,k+1)} + C\int_{x_0}^{x}(k+2)\overline{<k+1>} = C\overline{<k+2,k+1>}(x)$.

In general, $\delta < l,l-1,\dots,1 >= \delta\int_{x_0}^{x}\overline{(l,l-1,\dots,1)} + \delta\int_{x_0}^{x}\overline{(l,l-1,\dots,2)1} + \cdots + \delta\int_{x_0}^{x} l\overline{<l-1,l-2,\dots,1>} = C\int_{x_0}^{x}\overline{(l,l-1,\dots,k+1)} + C\int_{x_0}^{x}\overline{(l,l-1,\dots,k+2)(k+1)} + \cdots + C\int_{x_0}^{x} l\overline{<l-1,\dots,k+1>} = C\overline{<l,l-1,\dots,k+1>}(x)$. And this concludes the proof of the claim.

Now, we come back to the proof of independence of the choice of base point $x_0 \in L_0$.

By the very definition of $L^*_{n,n-1,\dots,1,0}$, we have

$$
\begin{aligned}
&\delta L^*_{n,n-1,\dots,1,0}\\
=&\delta \int_0 \overline{(n,n-1,\dots,1)} + \delta \int_0 \overline{(n,n-1,\dots,2)1}\\
&\quad + \delta \int_0 \overline{(n,n-1,\dots,3)<2,1>} + \cdots\\
&\quad + \delta \int_0 n\overline{<n-1,n-2,\dots,1>}\\
=&C \int_0 \overline{(n,n-1,\dots,k+1)}\\
&\quad + C \int_0 \overline{(n,n-1,\dots,k+2)<k+1>} + \cdots\\
&\quad + C \int_0 \overline{(n,n-1)<n-2,\dots,k+1>}\\
&\quad + C \int_0 n\overline{<n-1,n-2,\dots,k+1>}\\
=&L^*_{n,n-1,\dots,k+2,k+1,0}\\
=&0,
\end{aligned}
$$

by the assumption that all Massey-Milnor linkings of lower degrees vanish.

And this concludes the proof of Theorem 1. □

8. References

Y. Akutsu, T. Deguchi, M. Wadati, *Exactly solvable models and new link polynomials I-V*, J. Phys. Soc. Japan 56 (1987), 3039–3051; 3464–3497; 57 (1988), 757–776; 1173–1185; 1905–1923.

D. Altschuler, L. Freidel, *On universal Vassiliev invariant*, Comm. Math. Phys. 170 (1995), 41–62.

D. Altschuler, L. Freidel, *Vassiliev knot invariants and Chern-Simons perturbative theory to all orders*, Comm. Math. Phys. 187 (1997), 261–287.

M.F. Atiyah, *The geometry and physics of knots*, Cambridge University Press, 1990.

Y. Akutsu, M. Wadati, *Knots, links, braids and exactly solvable models in statistical mechanics*, Comm. Math. Phys. 117 (1988), 243–259.

D. Bar-Natan, *On the Vassiliev link invariants*, Topology 34 (1995), 423–472.

D. Bar-Natan, *Perturbative Chern-Simons theory*, J. of knot theory and its ramification 4 (1995), 503–547.

R. Fenn, *Techniques of geometric topology*, Cambridge University Press, 1983.

C.C. Hsieh, L.H. Kauffman, M.C. Tsau, *Massey-Milnor linking in knot theory*, Preprint.

N. Habegger, G. Masbaum, *The Kontsevich integral and Milnor's invariants*, Topology 39 (2000), 1253–1289.

C.C. Hsieh, *Linking in knot theory*, Journal of knot theory and its ramification, 15 (2006), 957–962.

C.C. Hsieh, *Combinatoric and diagrammatic study in knot theory*, Journal of knot theory and its ramification, 16 (2007), 1235–1253.

C.C. Hsieh, *Massey-Milnor linking=Chern-Simons-Witten graphs*, Journal of knot theory and its ramification, 17 (2008), 877–903.

C.C. Hsieh, *Borromean rings and linkings*, Journal of geometry and physics, 60 (2010), 823–831.

C.C. Hsieh, *First non-vanishing self-linkings of knots(III), Massey-Milnor theory*, Journal of knot theory and its ramification, 20 (2011), 677–687.

C.C. Hsieh, *Combinatoric Massey-Milnor invariant theory*, Journal of knot theory and its ramification, 20 (2011), 927–938.

C.C. Hsieh, *First non-vanishing self-linkings of knots(I), Combinatoric and diagrammatic study*, to appear in Journal of knot theory and its ramification.

C.C. Hsieh, S.W. Yang, *Chern-Simons-Witten configuration space integrals in knot theory*, Journal of knot theory and its ramifications, 14 (2005), 689–711.

W.S. Massey, *Higher order linking numbers*, Conf. on algebraic topology (University of Illinois, at Chicago Circle, 1969), University of Illinois at Chicago Circle, 174–205.

J. Milnor, *Link groups*, Ann. of Math. 59 (1954), 177–195.

J. Milnor, *Isotopy of links*, Algebraic geometry and topology, A symposium in honor of S. Lefschetz, 280–306, Princeton University Press, 1957.

G. Masbaum, A. Vaintrob, *Milnor numbers, spanning trees and Alexander-Conway polynomial*, Preprint.

T. Ohtsuki, *Quantum invariants—A study of knots, 3-manifolds and their sets*, World Scientific, 2002.

R. Porter, *Milnor u-invariants and Massey products*, Trans. Amer. Math. Soc. 257 (1980), 39–71.

M.E. Peskin, D.V. Schroeder, *An introduction to quantum field theory*, Addison-Wesley, 1995.

N.Y. Reshetikhin, V.G. Turaev, *Invariants of 3-manifolds via link polynomials and quantum groups*, Invent. Math. 103 (1991), 547–597.

V.G. Turaev, *Quantum invariants for knots and 3-manifolds*, de Gruyter, Berlin, 1994.

E. Witten, *Quantum field theory and Jones polynomial*, Comm. Math. Phys. 121 (1989), 351–399.

D. Yetter, *Functorial knot theory*, World Scientific, 2001.

Permissions

The contributors of this book come from diverse backgrounds, making this book a truly international effort. This book will bring forth new frontiers with its revolutionizing research information and detailed analysis of the nascent developments around the world.

We would like to thank Prof. Dr. Sergey Ketov, for lending his expertise to make the book truly unique. He has played a crucial role in the development of this book. Without his invaluable contribution this book wouldn't have been possible. He has made vital efforts to compile up to date information on the varied aspects of this subject to make this book a valuable addition to the collection of many professionals and students.

This book was conceptualized with the vision of imparting up-to-date information and advanced data in this field. To ensure the same, a matchless editorial board was set up. Every individual on the board went through rigorous rounds of assessment to prove their worth. After which they invested a large part of their time researching and compiling the most relevant data for our readers. Conferences and sessions were held from time to time between the editorial board and the contributing authors to present the data in the most comprehensible form. The editorial team has worked tirelessly to provide valuable and valid information to help people across the globe.

Every chapter published in this book has been scrutinized by our experts. Their significance has been extensively debated. The topics covered herein carry significant findings which will fuel the growth of the discipline. They may even be implemented as practical applications or may be referred to as a beginning point for another development. Chapters in this book were first published by InTech; hereby published with permission under the Creative Commons Attribution License or equivalent.

The editorial board has been involved in producing this book since its inception. They have spent rigorous hours researching and exploring the diverse topics which have resulted in the successful publishing of this book. They have passed on their knowledge of decades through this book. To expedite this challenging task, the publisher supported the team at every step. A small team of assistant editors was also appointed to further simplify the editing procedure and attain best results for the readers.

Our editorial team has been hand-picked from every corner of the world. Their multi-ethnicity adds dynamic inputs to the discussions which result in innovative outcomes. These outcomes are then further discussed with the researchers and contributors who give their valuable feedback and opinion regarding the same. The feedback is then collaborated with the researches and they are edited in a comprehensive manner to aid the understanding of the subject.

Apart from the editorial board, the designing team has also invested a significant amount of their time in understanding the subject and creating the most relevant covers. They scrutinized every image to scout for the most suitable representation of the subject and create an appropriate cover for the book.

The publishing team has been involved in this book since its early stages. They were actively engaged in every process, be it collecting the data, connecting with the contributors or procuring relevant information. The team has been an ardent support to the editorial, designing and production team. Their endless efforts to recruit the best for this project, has resulted in the accomplishment of this book. They are a veteran in the field of academics and their pool of knowledge is as vast as their experience in printing. Their expertise and guidance has proved useful at every step. Their uncompromising quality standards have made this book an exceptional effort. Their encouragement from time to time has been an inspiration for everyone.

The publisher and the editorial board hope that this book will prove to be a valuable piece of knowledge for researchers, students, practitioners and scholars across the globe.

List of Contributors

Sergei V. Ketov
Department of Physics, Graduate School of Science, Tokyo Metropolitan University, Hachioji-shi, Tokyo, Japan
Institute for the Physics and Mathematics of the Universe, The University of Tokyo, Kashiwa-shi, Chiba, Japan

Genta Michiaki and Tsukasa Yumibayashi
Department of Physics, Graduate School of Science, Tokyo Metropolitan University, Hachioji-shi, Tokyo Japan

A. V. Shebeko
Institute for Theoretical Physics, National Research Center "Kharkov Institute of Physics & Technology", Ukraine

James P. Vary
Department of Physics and Astronomy, Iowa State University, Ames, Iowa, USA

M. Erementchouk, V. Turkowski and Michael N. Leuenberger
NanoScience Technology Center and Department of Physics, University of Central Florida, USA

Brian D. Serot and Xilin Zhang
Department of Physics and Center for Exploration of Energy and Matter, Indiana University-Bloomington, USA

C.A. Linhares
Instituto de Física, Universidade do Estado do Rio de Janeiro, Rio de Janeiro, Brazil

A.P.C. Malbouisson and I. Roditi
Centro Brasileiro de Pesquisas Físicas, Rio de Janeiro, Brazil

E. Brüning
University of KwaZulu-Natal, South Africa

S. Nagamachi
The University of Tokushima, Japan

Ai-Dong Bao and Shi-Shu Wu
Department of Physics Northeast Normal University Chang Chun, People's Republic of China
Center for Theoretical Physics Department of Physics Jilin University Chang Chun, People's Republic of China

Chun-Chung Hsieh
Institute of Mathematics, Academia Sinica, Nankang, Taipei, Taiwan